small gas engines

fundamentals, service, troubleshooting, repairs

by

ALFRED C. ROTH

Assistant Professor, Industrial Technology
Eastern Michigan University
Ypsilanti, Michigan

RONALD J. BAIRD

Professor, Industrial Technology
Eastern Michigan University
Ypsilanti, Michigan

South Holland, Illinois
THE GOODHEART-WILLCOX COMPANY, INC.
Publishers

Cutaway view of one cylinder, four cycle small gas engine
used in rotary lawn mower applications. (Deere & Co.)

Library of Congress Cataloging in Publication Data
Roth, Alfred C.
  Small gas engines.
  Includes index.
  Summary: Basic information on small engine construc-
tion, maintenance, service, and repair.
  1. Internal combustion engines, Spark ignition—
Juvenile literature. [1. Internal combustion engines]
I. Baird, Ronald J.   II. Title.
TJ790.R72 1985   621.43'4   85-807
ISBN 0—87006—498—3

INTRODUCTION

SMALL GAS ENGINES provides students, do-it-yourselfers and apprentice mechanics with basic information on small engine construction, how the systems operate, lubrication requirements, preventive maintenance practices, servicing techniques and rebuilding procedures.

Today, the small gas engine serves as a power source for almost every conceivable type of labor-saving device. Around the home, we have power mowers, lawn edgers, riding tractors, garden tillers and snow blowers. On the farm, small gas engines power chain saws, portable pumps, post hole drillers, brush cutters, portable conveyer systems and cement mixers.

In industry, small engine applications include portable electric generating systems, hydraulic pumps, fork lifts, portable winches, cement finishers, portable tampers, in-field welding rigs, horizontal boring units and trenchers for laying underground pipe. Other usages are portable paint sprayers for applying highway traffic stripes, flame blowers (to dry the paint), stripe erasers and a variety of cleaners and sweepers.

Recreational vehicle applications represent the strongest area of growth of the small engine market. Snowmobiles now number in the millions. Outboard engines are offered in a broad range of horsepower ratings for use in all water sports. Other recreational applications include motorcycles, mini-bikes, go-carts, gyrocopters, ATVs (all-terrain-vehicles) and ultra-light aircraft.

Regardless of application, however, small engines require maintenance, service and repair. With this in mind, SMALL GAS ENGINES is designed to provide detailed, technical information on one and two-cylinder, two and four cycle gasoline engines.

Rotary engines are covered in some detail in Chapter 14. Diesel and LP-Gas engines are described in this text. Review Questions and Suggested Activities are featured at the end of every chapter.

Alfred C. Roth

IMPORTANT SAFETY NOTICE

Proper service and repair is important to the safe, reliable operation of small gas engines and related equipment. Procedures recommended and described in this book are effective methods of performing service operations. This book also contains various safety procedures and cautions that must be followed to minimize the risk of personal injury and part damage. These notices and cautions are not exhaustive. Those performing a given service procedure or using a particular tool must first satisfy themselves that safety is not being jeopardized.

THE MANY APPLICATIONS OF SMALL GAS ENGINES

Two and four cycle gasoline engines of low horsepower and small size play a familiar and important role in almost everyone's life. When power needs take us out of reach of the electric cord or beyond the application of an electric motor, we depend on small gas engines to ease our chores and power our recreational vehicles.

We use small gas engine-powered equipment to cut grass, remove snow, cultivate gardens, cut wood, drive generators, pump water, and sweep factory floors. We also use them for snowmobiling, go-carting, motorcycling and boating. Pictured here and on pages 6, 7 and 8 are representative applications. Chapter 15 covers small gas engine applications in detail.

Snowmobiles are a relatively new application of small gas engines. In colder regions of the country, they are very common and popular.

CONTENTS

A small four cycle engine drives this garden cultivator . . . and another provides both sweeping and suction power for this cleaning machine used mainly in industrial applications.

A small gas engine-propelled riding mower can be used to pull a sprayer whose pressure is supplied by an engine-driven compressor.

Left. Rotary lawn mowers account for a large share of small engines in use. Center. Gas engine-powered mudjack lifts sunken sections of sidewalk. Right. Small gas engine powers rotor and driving wheels of this snow blower.

A powerful small gas engine is used to drive this amphibious all-terrain-vehicle.

Riding mowers are used where lawns are spacious. Property owners often require the services and talents of small gas engine mechanics to provide maintenance and repair services for engines and vehicles.

Chapter 1
ENGINE CONSTRUCTION AND PRINCIPLES OF OPERATION

A gasoline-fueled engine is a mechanism designed to transform the chemical energy of burning fuel into mechanical energy. In operation, it controls and applies this energy to mow lawns, cut trees, propel tractors and perform many other laborsaving jobs.

A gasoline engine is an internal combustion engine. The gasoline is combined with air and burned inside the engine. In its simplest form, this engine consists of a ported cylinder, piston, connecting rod and crankshaft, Fig. 1-1.

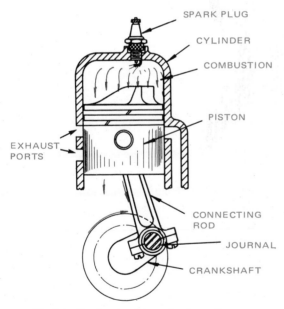

Fig. 1-1. Combustion forces the piston down to rotate crankshaft.

The piston is a close "fit" inside the cylinder, yet it is free to slide on the lubricated walls. One end of the connecting rod is attached to the piston; the other end is fastened to an offset crankpin, or journal, of the crankshaft. As the crankshaft turns, the journal and the lower end of the connecting rod follow a circular path around the center line of the shaft, pushing and pulling the piston up and down in the cylinder.

SIMPLE ENGINE IN OPERATION

When the engine is cranked, the gasoline is atomized (reduced to minute particles) and mixed with air. This mixture is forced through an intake port into the cylinder, where it is compressed by the piston on the upstroke and ignited by an electrical spark.

Then, burning rapidly, the heated gases trapped within the cylinder (combustion chamber) expand and apply pressure to the walls of the cylinder and to the top of the piston. This pressure drives the piston downward on the power stroke, Fig. 1-1, causing the crankshaft to turn.

As the piston and connecting rod push the crankshaft journal to the bottom of the stroke, the pressure of the burned gases is released through an exhaust port. Meanwhile, a fresh air-fuel charge enters the cylinder and the momentum of the power stroke turns the crankshaft journal through bottom dead center (BDC) and into the upstroke on another power cycle.

GASOLINE

Gasoline is a hydrocarbon fuel (mixture of hydrogen and carbon), refined from petroleum. Petroleum is a dark, thick liquid that comes from the earth by way of oil wells. Luckily for the internal combustion engine, petroleum is the second most plentiful liquid in the world; only water is available in greater quantity. But gasoline, when used, cannot be recycled and used over again as water can. Therefore, it is imperative that we conserve gasoline and use it wisely.

Gasoline contains a great amount of energy. For engine use, it should:
1. Ignite readily, burn cleanly and resist detonation

(violent explosion).

2. Vaporize easily, without being subject to vapor lock (vaporizing in fuel lines, impeding flow of liquid fuel to carburetor).
3. Be free of dirt, water and abrasives.

Gasoline is assigned an octane number corresponding to its ability to resist detonation. Premium grade gasoline ("high test" or "ethyl") burns slower than regular gasoline. It has a high octane number and is used in engines with high compression. Regular grade gasoline has a lower octane number and burns relatively fast. Generally, regular gasoline is used in small one and two cylinder gasoline engines with low compression.

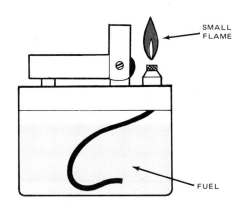

Fig. 1-2. A small flame is produced, due to small area of exposed fuel.

GASOLINE MUST BURN QUICKLY

Gasoline placed in a container and ignited will produce a hot flame, yet it will not burn fast enough to produce the rapid release of heat necessary to run an engine. Even though a considerable quantity of fuel may be involved, Fig. 1-2, a large flame will not necessarily result.

NOTE: Under no circumstances should experiments illustrated in this chapter be performed. Gasoline can be a very dangerous fuel and must be handled with caution. Illustrations and examples discussed here are meant to demonstrate how gasoline is prepared and used in an engine.

In Fig. 1-2, the surface area of the wick in the lighter is small. Vapor from the surface of the liquid, combined with oxygen, is what burns readily. If the surface of the liquid is small, relatively little vapor will be given off to provide combustion. Since the liquid must change to vapor before it is burned, it would take considerable time to use up the fuel at this rate.

By placing the same amount of fuel in a shallow but wide container, much more surface area is in contact with the air. Now the fuel will burn rapidly, Fig. 1-3.

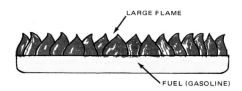

Fig. 1-3. A large flame is produced by a large area of exposed fuel.

FUEL IS ATOMIZED

Understandably then, the more surface area of gasoline exposed to the air, the faster a given amount will burn. To produce the rapid burning required in an engine, the gasoline must be broken up into tiny droplets and mixed with air. This is called "atomizing."

Once the entire surface of each droplet of the air-fuel mixture is exposed to the surrounding air, a huge burning area becomes available. Given a spark, the entire amount of gasoline will flash into flame almost instantly. *In effect, atomization causes a sudden, explosive release of heat energy, Fig. 1-4.*

EXPLOSION MUST BE CONTAINED

To perform useful work, the explosive force caused by the burning gas must be contained and controlled. To illustrate, imagine that a metal lid is suspended on a string, supported several inches from the ground. If a mixture of gasoline and air (atomized) were sprayed under it and ignited, the lid would be raised a short distance by the force of the explosion. See Fig. 1-5.

The reason the lid hardly moved was because the explosion was not confined and directed toward the lid. Instead, the explosion exerted force in all directions, and much of the force was lost. If the gasoline and air mixture is sprayed inside a metal container with a lid, the full force of the explosion will be directed against the lid when the mixture is ignited. This will blow the lid high into the air, as in C, Fig. 1-6.

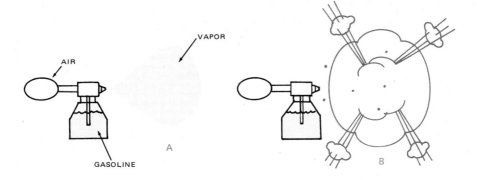

AIR

VAPOR

GASOLINE

A

B

Fig. 1-4. Atomized fuel exposes a large area of fuel which, when ignited, releases heat energy with an explosive force.

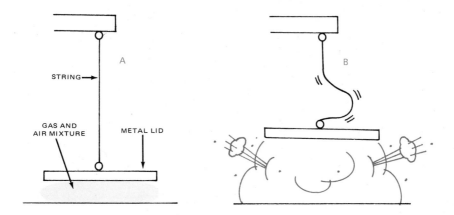

STRING

GAS AND
AIR MIXTURE

METAL LID

A

B

Fig. 1-5. A mixture of air and fuel ignited under a lid lifts lid a short distance.

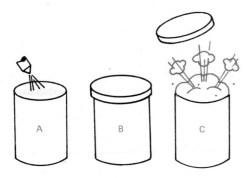

A

B

C

Fig. 1-6. A—Mixture of fuel and air is sprayed into a container. B—Lid is placed on top. C—Full force of explosion is directed toward base of lid when mixture is ignited, and lid is driven high into air.

FURTHER IMPROVEMENT

Even though the burning air-fuel mixture is confined by the container, once the lid starts to lift, a large amount of the force escapes to the sides. To eliminate

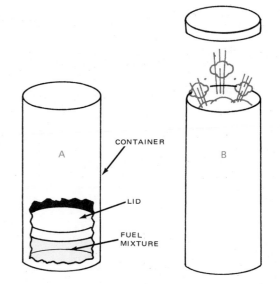

A

CONTAINER

LID

FUEL
MIXTURE

B

Fig. 1-7. A—Lid is placed in a long container. B—Most of energy of burning fuel is absorbed by lid, imparting greater speed to lid when explosion occurs.

this loss, a long cylindrical container may be used with the lid having a close sliding fit, Fig. 1-7. With the fuel mixture slightly compressed in the bottom of the container by the weight of the lid, the fuel will burn and direct most of the pressure against the lid as it travels up through the container. When the lid reaches the top, it will be traveling at a high rate of speed. The expansion of the gas will be nearly complete and little force will be lost, even after the lid clears the container.

BASIS FOR AN ENGINE

In using the setup shown in Fig. 1-7 to form an elementary engine, a crankshaft and connecting rod are needed. The bearings which support the crankshaft are called main bearings. A pulley is attached to one end. The lid (piston) is attached to the journal of the crankshaft by means of a connecting rod. When the

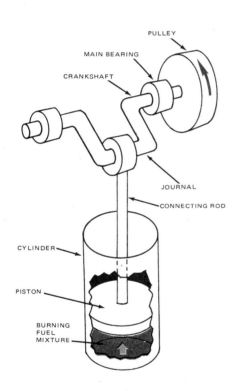

Fig. 1-8. Principles of operation illustrated here are the same as used in a modern gasoline engine. Note how burning fuel mixture forces lid (piston) upward to turn crankshaft and pulley.

air-fuel mixture in the cylinder is ignited, Fig. 1-8, it will drive the piston upward causing it to turn the crankshaft and rotate the pulley.

This elementary engine, although crude, illustrates the operating principles of a modern gasoline engine.

Study the names of the various parts shown in Fig. 1-8 and become acquainted with their application to engine design.

There are many faults with the engine pictured in Fig. 1-8. How will a fresh air-fuel charge be admitted to the cylinder? How will the charge be ignited? What holds the various parts in alignment? How will the engine be cooled and lubricated? What will "time" the firing of the air-fuel mixture so that the piston will push on the crankshaft when the journal is in the correct position? How will the burned charge be removed (exhausted) from the cylinder? What will keep the crankshaft rotating after the charge is fired, until another charge can be admitted and fired?

Other factors are to be considered, but the one mentioned are included in five basic areas: MECHANICAL (engine design and construction), CARBURETION (mixing gasoline and air, and admitting it to cylinder), IGNITION (firing and fuel charge), COOLING (heat dissipation), and LUBRICATION (oiling of moving parts).

In this chapter, emphasis will be placed on the mechanical aspects of engine design and construction. This will provide you with an opportunity to develop a workable engine. It will be assumed that the gasoline and air are being mixed correctly, the fuel charge is being fired at the right time, and the engine is properly cooled and lubricated.

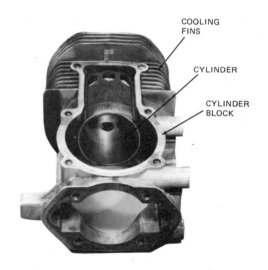

Fig. 1-9. Cylinder block is important because it keeps all moving parts in alignment. (Jacobsen Mfg. Co.)

PARTS ALIGNMENT — CYLINDER BLOCK

Because it keeps all parts of the engine in alignment, the cylinder block, Fig. 1-9, is a critical part of the engine. The block is usually a casting of iron, or an aluminum alloy. The cylinder formed in the block can be produced accurately by modern methods. It may be bored directly into the casting, or the block may be bored oversize and a steel cylinder sleeve inserted.

Aluminum cylinder blocks are cast around a steel sleeve. Aluminum, being a soft metal, would wear out quickly due to the friction of the piston. Advantages of aluminum are its light weight and ability to dissipate heat rapidly.

All air-cooled engines have cooling fins on the outside of the cylinder and cylinder head. *The size, thickness, spacing and direction of the cooling fins is carefully engineered for efficient air circulation and heat control.*

The cylinder block must be rigid and strong enough to contain the power developed by the expanding gases. In some cases, the block is a separate unit; in others, it is cast as part of the crankcase. Similarly, the cylinder head may be bolted to the block or it may be cast as one complete unit. The method employed depends on the intended application of the engine and the manufacturer's preference.

Fig. 1-10 shows a combined cylinder block and crankcase with a separate, bolted cylinder head. Note the gasket that seals the unit. A sleeved aluminum die cast cylinder is shown in Fig. 1-11.

Fig. 1-11. An aluminum cylinder block die cast around a steel sleeve. Note fins for air cooling.

CRANKSHAFT AND CRANKCASE

The crankshaft, Fig. 1-12, is the major rotating part of the engine. Generally, it is forged steel, with all bearing surfaces carefully machined and precision ground. Counterweights are used to balance the weight of the connecting rod, which is fastened to the journal. Since connecting rods are cast or forged of different weight materials, holes are often drilled in counterweights to balance the crankshaft and prevent vibration.

Fig. 1-13 shows a crankshaft being installed in a crankcase. Note the tapered roller bearings. The flywheel is keyed to the end of the shaft with a Woodruff key. This type of key cannot slip out during operation. A lock washer and nut hold the flywheel in place.

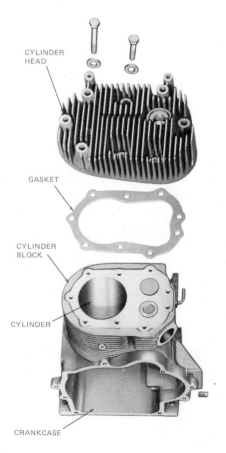

Fig. 1-10. A combined cylinder block and crankcase. Cylinder head and sealing gasket are bolted to cylinder block. (Wisconsin Motor Corp.)

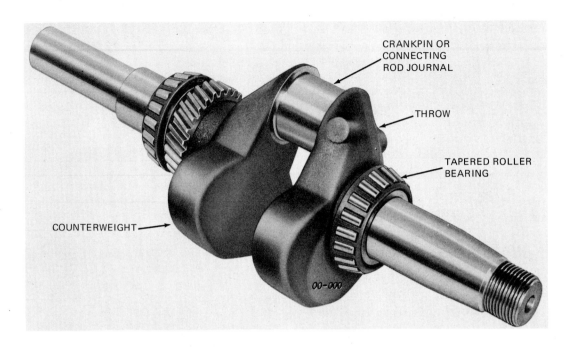

Fig. 1-12. Crankshaft for a single cylinder engine. Large counterweights opposite the crank journal balance rotational forces.

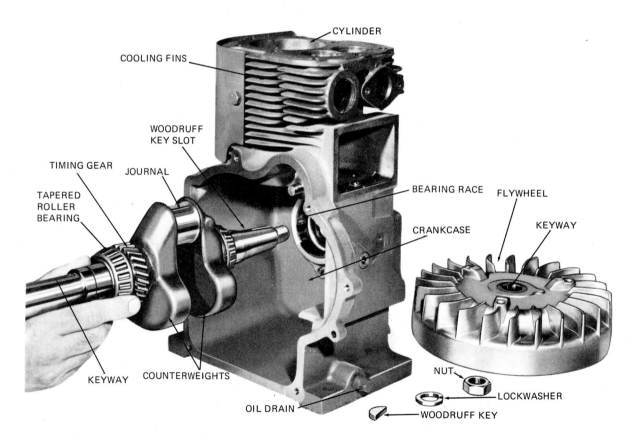

Fig. 1-13. Crankshaft must be clean and carefully installed in crankcase. Tapered end fits into flywheel, which is secured with a key, lockwasher and nut. (Wisconsin Motors Corp.)

The end of the crankshaft and the hole through the flywheel have matched tapers that provide good holding power. When roller bearings are used to support the crankshaft, highly polished, hardened alloy steel bearing races are pressed into the crankcase to reduce friction and provide good wearability.

The crankcase must be rigid and strong enough to withstand the rotational forces of the crankshaft, while keeping all parts in proper alignment. Oil for lubrication is contained in the crankcase on some engines. On others, a valve system is used which allows a fuel, air and oil mixture to enter. *The crankcase must be designed to protect the internal parts. Gaskets and oil seals are used to keep dirt out and clean oil in.*

The crankcase and the cylinder block may be cast (metal melted and poured into a form of the desired shape) as a unit or fastened together by bolts. Fig. 1-14 shows a two cylinder engine with the cylinder head

Fig. 1-14. Two cylinder crankcase and cylinder block being assembled. Stud bolts in crankcase will hold cylinder block in place when nuts are tightened.

being placed on the crankcase with the crankshaft already installed. Note the tapered end on the crankshaft that receives the flywheel.

Fig. 1-15. These engine components, when assembled, will become driving members of a lightweight chain saw.   (Beaird-Poulan, Inc.)

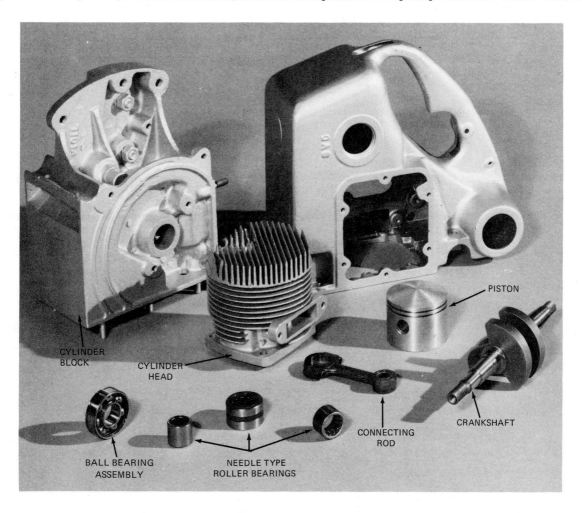

In certain engine applications, the crankcase is not only an important part of the engine but also an integral part of the apparatus being driven. Engine parts illustrated in Fig. 1-15 are for a chain saw. Fig. 1-16 shows a similar saw completely assembled.

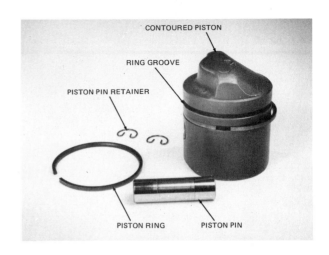

Fig. 1-17. Piston is largest, sliding, reciprocating part in engine. Piston rings seal combustion chamber from crankcase and must fit properly. (Jacobsen Mfg. Co.)

Fig. 1-16. A typical gasoline engine driven chain saw.

PISTONS

The piston is the straight line driving member of the engine. It is subjected to the direct heat of combustion and must have adequate clearance in the cylinder to allow for expansion. *The piston provides a seal between the combustion chamber and the crankcase.* This is accomplished by cutting grooves near the top of the piston and installing piston rings. The piston rings fit the grooves with slight side clearance and exert tension on the cylinder wall. Properly installed, piston rings prevent blow-by of exhaust gases into the crankcase and leakage of oil into the combustion chamber.

The number of piston rings per piston depends upon the type of engine and its design. Note the two piston rings in Fig. 1-17. The piston is hollow to reduce weight. The top may be flat, domed or contoured to provide efficient flow of gases entering and leaving the combustion chamber.

There is a hole in each side of the piston through which a piston pin, or wrist pin, is placed. This pin acts as a hinge between the connecting rod and piston, and serves to hold the two together. Generally, spring retainers hold the piston pin in place.

CONNECTING PISTON TO CRANKSHAFT

The sliding piston is connected to the rotating crankshaft with a metal link called a connecting rod. The big end that encircles the crankshaft journal contains a bearing to permit free movement. The upper, or small end, of the connecting rod also must be movable. Note how the metal piston pin is passed through the connecting rod and piston, Fig. 1-18.

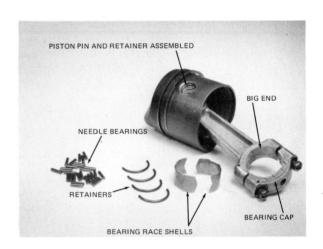

Fig. 1-18. Big end of connecting rod is "capped" to fit around crankshaft journal. Needle bearings are inserted to help reduce friction. Piston pin and retainers are shown assembled. (Jacobsen Mfg. Co.)

Also note the needle type roller bearings, bearing race shells and retainers which must be installed in the large end of the connecting rod when it is placed on the crank journal. The bearing cap holds the assembly together with connecting rod bolts or screws. Fig. 1-19 shows the relative position of the connecting rod and cap with the bearings in place.

Fig. 1-19. The relative positions of connecting rod parts are illustrated with roller bearings in place. (Kiekhaefer Mercury)

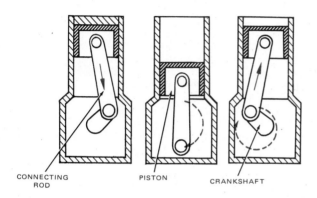

CONNECTING ROD   PISTON   CRANKSHAFT

Fig. 1-20. Connecting rod must be free to pivot on piston pin, while crank journal follows a rotary path. Connecting rod must withstand severe stress in operation.

Expanding gases push the piston toward the crankshaft, causing the connecting rod to turn the shaft. Fig. 1-20 shows how the reciprocating (up and down) movement of the piston is changed to rotary (revolving) motion by the crankshaft. Notice that the upper connecting rod bearing allows the connecting rod to swing back and forth while the lower bearing permits the crankshaft journal to rotate within the rod.

INTAKE AND EXHAUST

In developing an engine, we need to provide a way in which a fresh air-fuel mixture can be admitted to the engine and, once burned, the waste products exhausted. This can be done by using ports (openings) that are alternately covered and exposed by the piston (two-stroke cycle design), or by using poppet valves to open and close the port openings (four-stroke cycle design). Both two-stroke and the four-stroke designs are commonly used. Each has definite advantages and disadvantages. The engine being discussed here will be of the four-stroke cycle type.

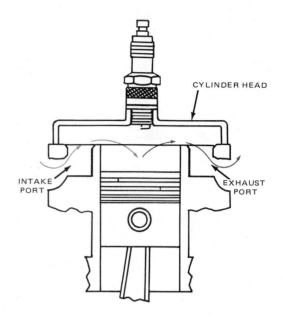

CYLINDER HEAD

INTAKE PORT   EXHAUST PORT

Fig. 1-21. Intake port permits air-fuel mixture to enter. Exhaust port allows burned gases to escape.

ADMITTING FUEL MIXTURE TO ENGINE

For a four-stroke cycle engine (see Chapter 2 for additional information on fundamentals), passages leading to and from cylinder area must be constructed. Fig. 1-21 shows two ports cast into cylinder block, one intake and one exhaust. The cylinder head is recessed to provide a passage from ports to the cylinder.

By installing a valve in each port, it is possible to control the flow of fresh fuel mixture into the cylinder, and provide a means of exhausting the burned gases. During the period of expansion of the burning gases that drive the piston downward, both valves are tightly closed, Fig. 1-22.

Angled face of each valve will close tightly against a smooth seat cut around each port opening. To align valve and assure accurate raising and lowering, in relation to seat, the valve stem passes through a machined hole in block. This hole is called a valve guide.

VALVE SPRING ASSEMBLY

A valve spring must be used on each valve to hold it firmly against the seat. Placed over the valve stem, the spring is compressed to provide tension, then it is connected to the valve stem by means of a washer and keeper or lock.

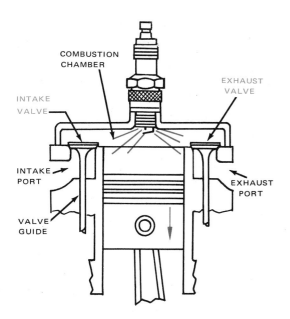

Fig. 1-22. Poppet valves seal intake and exhaust ports during power stroke. Valve guides keep valves aligned with valve seats.

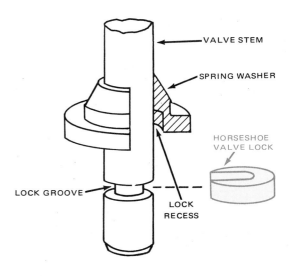

Fig. 1-24. Typical method of retaining valve spring on valve stem. Special tool generally is used to compress spring prior to removing horseshoe valve lock.

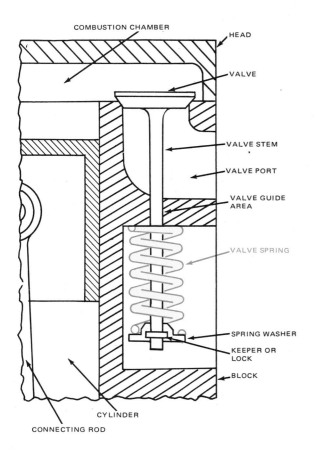

Fig. 1-23. Valve spring keeps tension on valve to insure proper seating. Valve spring keeper and washer hold spring in place and permit removal when necessary.

The spring allows the valve to be opened when necessary, and will close it when pressure is removed from the valve stem. Fig. 1-23 shows the location of the spring and keeper assembled on the valve. An enlarged view of the "horse shoe" valve lock system is shown in Fig. 1-24.

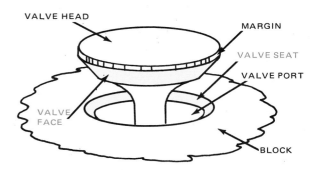

Fig. 1-25. Valve face and valve seat must be ground to correct angles and concentric to center line of guide to seal properly.

A valve in the open position is shown in Fig. 1-25. When pressure is removed from the end of the valve stem, the spring will draw the valve down against the seat and seal off the port from the combustion chamber. *For the engine to function properly, the valves must be opened the right amount at the right time. They must remain open for a specific period, then close at the correct instant.*

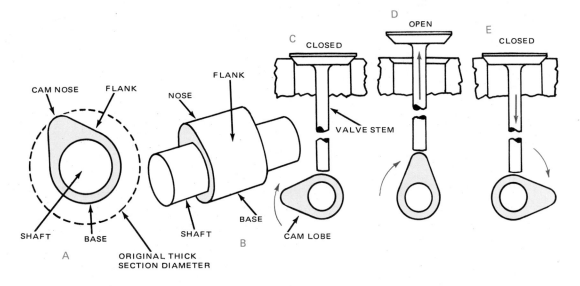

Fig. 1-26. A, B—By grinding a round shaft into a cam shape, a camshaft is formed. C, D, E—When camshaft is revolved, cam lobe will open valve.

By using a shaft with two thick sections spaced to align with the valve stems, a basic device for opening and closing the valves is provided. It is necessary to grind the thick sections into a cam shape. Then, when the shaft is revolved, the cam lobe will cause the valve to rise and fall, opening and closing the ports. Study Fig. 1-26.

VALVE LIFTER OR TAPPET

In actual practice, the cam lobe does not contact the valve stem directly. By locating the camshaft some distance below the valve stem end, it is possible to insert a valve lifter between the lobe and stem. The valve lifter may have an adjustment screw in the upper end to provide a means of adjusting valve stem-to-lifter clearance. Without this adjustment, proper clearance must be obtained by grinding the end of either the lifter or valve stem. Base of lifter may be made wider to provide a larger cam lobe-to-lifter contact area, Fig. 1-27.

By drilling a hole in the block above the camshaft, a guide is formed in which the lifter can operate, Fig. 1-28. The base of the lifter rides on the cam and the

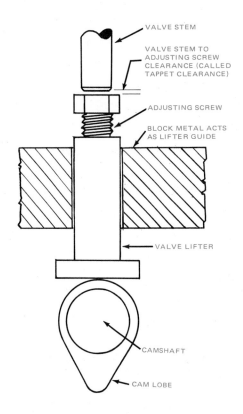

Fig. 1-28. As camshaft turns, cam lobe will operate valve lifter to open valve, then allow it to close.

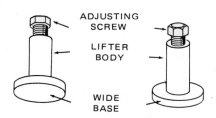

Fig. 1-27. Valve lifter may be called a tappet or cam follower. Adjustment screw allows setting of proper valve clearance. Wide base provides a larger contact area.

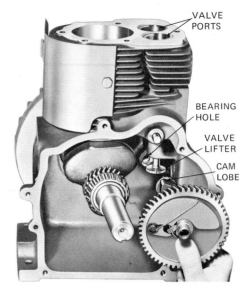

Fig. 1-29. Cam lobes are located directly under valve lifters. Camshaft turns in lubricated bearing holes. (Wisconsin Motors Corp.)

Fig. 1-30. Complete valve train. Study part names and their relationship to each other.

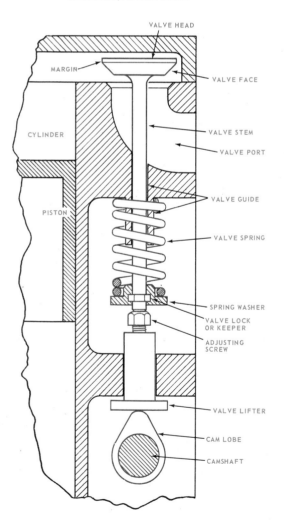

VALVE PORTS

BEARING HOLE

VALVE LIFTER

CAM LOBE

VALVE HEAD

MARGIN

VALVE FACE

CYLINDER

VALVE STEM

VALVE PORT

VALVE GUIDE

PISTON

VALVE SPRING

SPRING WASHER

VALVE LOCK OR KEEPER

ADJUSTING SCREW

VALVE LIFTER

CAM LOBE

CAMSHAFT

adjusting screw almost touches the end of the valve stem. As the camshaft revolves, the lifter will rise and fall, opening and closing the valve.

LOCATING THE CAMSHAFT

Generally, the camshaft is located in the crankcase, directly below the valve stems and valve lifters. The ends of the camshaft are supported in bearings in the block. The camshaft is shown being fitted into place in Fig. 1-29. The valve assembly is illustrated in Fig. 1-30. Study the relationship of one part to another. Fig. 1-31 shows the location of each valve part in relation to the rest of the engine block.

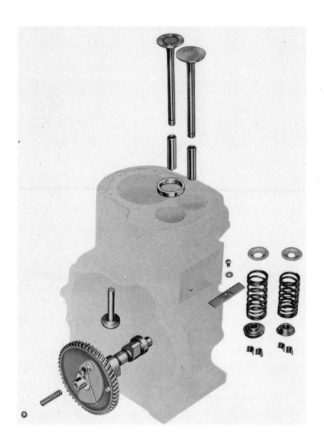

Fig. 1-31. Valve parts and their positions relative to cylinder block and crankcase.

The camshaft is driven by the crankshaft through gears. Fig. 1-32 shows the large camshaft gear meshed with the smaller crankshaft gear. *The camshaft gear is always twice as large as the crankshaft gear. This gear ratio will be explained in Chapter 2 under four-stroke cycle engine.*

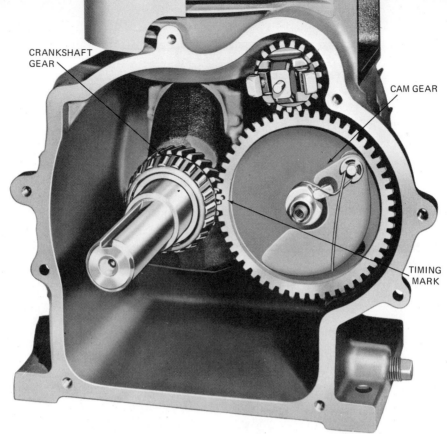

CRANKSHAFT GEAR

CAM GEAR

TIMING MARK

Fig. 1-32. Camshaft is gear driven from crankshaft. Camshaft gear is always twice as large as crankshaft gear for proper timing. During assembly, timing marks must be matched.

FLYWHEEL

CRANKSHAFT

Fig. 1-33. Flywheel is fastened to crankshaft. Its weight, when rotating, smooths engine operation. (Briggs and Stratton Corp.)

FLYWHEEL

Even though the crankshaft moves fast during the power stroke, it is relatively light and tends to slow down or stop before the next power stroke. This periodic application of power, followed by coasting, would cause the engine to speed up, slow down and run rough.

To improve the running quality of the engine, an additional weight in the form of a round "flywheel" is fastened to one end of the crankshaft, Fig. 1-33. During the non-power strokes, the inertia of the heavy flywheel keeps the crankshaft spinning and smooths engine operation. Metal fins on the flywheel act as a fan that forces air over the cylinder to cool the engine. Magnets cast into the flywheel produce electrical current for the ignition system.

REVIEW QUESTIONS – CHAPTER 1

1. Fuel must be atomized in the engine for the purpose of _____ .
2. Force on the piston is transmitted to the crankshaft by the _____ .
3. Name five desirable characteristics of gasoline for use in small engines.
4. Small gasoline engines generally use:
   a. High octane fuel.
   b. Low octane fuel.
5. The bearings which support the crankshaft are called _____ bearings.
6. The cylinder block is generally made of _____ or _____ .
7. When are sleeved cylinders used?
8. Why are sleeved cylinders used?
9. Why are metal fins made as a part of the cylinder?
10. _____ are designed into the crankshaft to provide for engine balancing?
11. Why do some pistons have a contoured face?
12. The _____ _____ will cause the valve to rise and fall, opening and closing the ports.
13. The angled face of each valve will close tightly against a smooth _____ cut around each port opening.
14. Which is larger, the camshaft gear or the crankshaft gear? How much larger?
15. What are three tasks performed by the flywheel?

SUGGESTED ACTIVITIES

1. Visit a local gasoline station and find out the following:
   a. Various fuel prices.
   b. Octane ratings of the fuels sold.
   c. Kind of containers fuel may be sold in.
   d. Quantities that legally may be stored at home.
2. Disassemble an engine and identify the parts discussed in this chapter. Carefully analyze the function of each part as they are related one to another.
3. If the engine used for disassembly is a used one, look for possible defects such as worn bearings, burned valves, a scored cylinder, broken or worn piston rings, a loose fitting piston pin.
4. Write to manufacturers of small gasoline engines requesting specifications for the models they produce. Write a report on the types of pistons, connecting rods and crankshafts they use.
5. Prepare a display of the major components of a small gasoline engine. Use actual parts, photos, drawings and cutaways to show the principal use of each part.

Chapter 2

TWO AND FOUR CYCLE ENGINES

A basic design feature that aids in small engine identification is how many piston strokes it takes to complete one operating (power) cycle. A four-stroke cycle engine, for example, requires four strokes per cycle; a two-stroke cycle engine requires two.

A stroke of the piston is its movement in the cylinder from one end of its travel to the other. Each stroke of the piston, then, is either toward the rotating crankshaft or away from it. And each stroke is identified by the job it performs (intake, exhaust, etc.).

FOUR-STROKE CYCLE ENGINE

In a four-stroke cycle engine (commonly called "four cycle"), four strokes are needed to complete the operating cycle. These strokes are termed: INTAKE, COMPRESSION, POWER, EXHAUST. Two strokes occur during each revolution of the crankshaft. Therefore, a four-stroke cycle requires two revolutions of the crankshaft.

Fig. 2-1 illustrates each of the four strokes taking place in proper sequence.

INTAKE STROKE

Drawing A in Fig. 2-1 shows the piston traveling downward in the cylinder on intake stroke. As piston moves down, the volume of space above it is increased. This creates a partial vacuum that sucks the air-fuel mixture through intake valve port into cylinder.

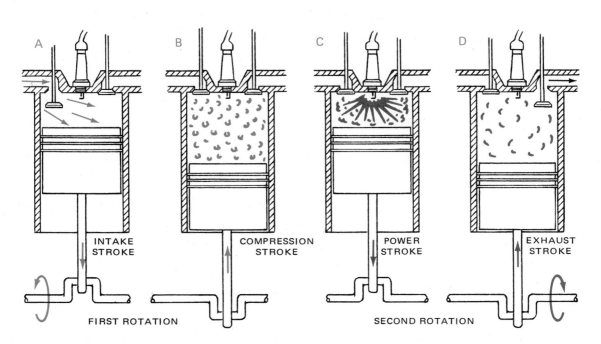

Fig. 2-1. Sequence of events in four-stroke cycle engine, requiring two revolutions of crankshaft and one power stroke out of four.

With the intake valve open during the intake stroke, atmospheric pressure outside the engine forces air through the carburetor. This gives a large boost to the air-fuel induction process. With nature balancing unequal pressures in this manner, it follows that the larger the diameter of the cylinder and the longer the stroke of the piston, the greater the volume of air entering the cylinder on the downstroke.

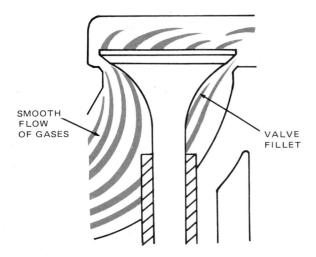

SMOOTH
FLOW
OF GASES

VALVE
FILLET

Fig. 2-2. Shape of valve smooths flow of gases around it. Note how flow follows fillet, speeding entry or expulsion. (Cedar Rapids Engineering Co.)

Bear in mind that the intake valve, Fig. 2-2, performs several key functions:

1. It must open at correct instant to permit intake of air-fuel mixture.
2. It must close at correct time and seal during compression.
3. Its shape must be streamlined, so flow of gases into combustion chamber will not be obstructed.

The intake valves are not subjected to as high temperatures as the exhaust valve. The incoming air-fuel mixture tends to cool the intake valve during operation.

COMPRESSION STROKE

The compression stroke, drawing B in Fig. 2-1, is created by the piston moving upward in the cylinder. Compression is a squeezing action while both valves are closed. On this stroke, the valves are tightly sealed and the piston rings prevent leakage past the piston.

As the piston moves upward, the air-fuel mixture is compressed into a smaller space. This increases the force of combustion for two reasons:

1. When atoms that make up tiny molecules of air and fuel are squeezed closer together, heat energy is created. Each molecule of fuel is heated very close to its flash point (point at which fuel will ignite spontaneously). When combustion does occur, it is practically instantaneous and complete for entire air-fuel mixture.
2. Force of combustion is increased because tightly packed molecules are highly activated and are striving to move apart. This energy combined with expanding energy of combustion provides tremendous force against piston. NOTE: It is possible to run an engine on uncompressed mixtures, but power loss produces a very inefficient engine.

POWER STROKE

During the power stroke, both valves remain in the closed position. See C in Fig. 2-1. As the piston compresses the charge and reaches the top of the cylinder, an electrical spark jumps the gap between the electrodes of the spark plug. This ignites the air-fuel mixture, and the force of the explosion (violent burning action) forces the piston downward.

Actually, the full charge does not burn at once. The flame progresses outward from the spark plug, spreading combustion and providing continued even pressure over the piston face throughout the power stroke.

The entire fuel charge must be ignited and expanded during an incredibly short time. Most engines have the spark timed to ignite the fuel slightly before the piston reaches TOP DEAD CENTER (TDC) of the compression stroke. This provides a little more time for the mixture to burn and accumulate its expanding force.

Basically, the amount of power produced by the power stroke depends on the volume of the air-fuel mixture in the cylinder and the compression ratio of the engine (proportionate difference in volume of cylinder and combustion chamber at bottom dead center and at top dead center). If the compression ratio is too high, the fuel may be heated to its flash point and ignite too early.

EXHAUST STROKE

After the piston has completed the power stroke, the burned gases must be removed from the cylinder before introducing a fresh charge. The exhaust valve opens and the rising piston pushes the exhaust gases from the cylinder. See D in Fig. 2-1.

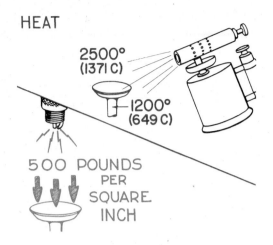

Fig. 2-3. Exhaust valve temperature may range from 1200 deg. F (649 C) to 2500 deg. F (1371 C) due to hot gases surrounding it. Pressure of combustion may be as high as 500 pounds per square inch. (Briggs and Stratton Corp.)

The exhaust valve has to function much like the intake valve. When closed, it must seal. When open, it must allow a streamlined flow of exhaust gases out through the port, Fig. 2-2. The removal of gases from the cylinder is called SCAVENGING.

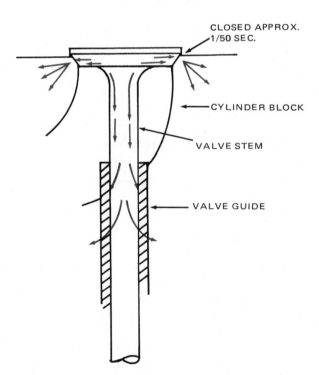

Fig. 2-4. Exhaust valve must cool during incredibly short period of 1/50 sec. at 3600 rpm. Heat is conducted from valve through seat to cylinder block. Some heat conducts down stem and to valve guide.

The passageway provided to carry away exhaust gases is referred to as the exhaust manifold or exhaust port. Like the intake manifold, the exhaust manifold must be designed for smooth flow of gases.

The heat absorbed by the exhaust valve must be controlled or the valve will deteriorate rapidly. Some valve heat is carried away by conduction through the valve stem to the guide. However, the hottest part of the valve, the valve head, transfers heat through the valve seat to the cylinder block, Fig. 2-4.

VALVE TIMING

The degree at which the valves open or close before or after the piston is at top dead center (TDC) or bottom dead center (BDC) varies with different engines. However, if the timing marks on the crankshaft and camshaft gears are aligned, the valve timing will take care of itself. NOTE: Engineers also specify the point at which the spark must occur (see Chapter 5).

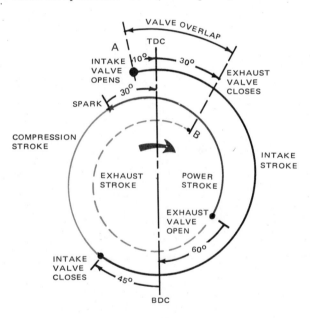

Fig. 2-5. Four-stroke cycle diagram shows exact number of degrees each valve is open or closed, and time spark ignition occurs. Note that both valves are open (overlap) through an arc of 40 deg., permitting exhausting gases to create a partial vacuum in cylinder and help initiate a mixture of fuel in cylinder.

Fig. 2-5 shows one complete operating cycle of a four cycle engine. Beginning at point A, the intake valve opens 10 deg. before TDC and stays open through 235 deg. Exhaust valve closes 30 deg. after TDC. VALVE OVERLAP is while both valves are open.

During the compression stroke, Fig. 2-5, the intake valve closes and ignition occurs 30 deg. before TDC. The power stroke continues through 120 deg. past TDC. The exhaust valve opens 60 deg. before BDC and stays open through 270 deg. During the last 40 deg., the intake valve is also open and the second cycle has begun.

LUBRICATION

Lubrication of the four cycle engine is provided by placing the correct quantity and grade of engine oil in the crankcase. Several methods are used to feed the oil to the correct locations. The two most common methods are the splash system and pump system. Some engines employ one or the other; others use a combination of both.

The multiple vee cylinder engine, Fig. 2-6, utilizes a combination of the splash and pressure lubrication system. The pump picks up the oil from the crankcase and circulates some oil through the filter and directly back to the crankcase. This keeps a clean supply available.

Oil is also pumped through a spray nozzle aimed at the crankshaft, Fig. 2-6. When the shaft rotates, it deflects the oil toward other moving parts. In addition, the splash finger on the bearing cap dips into the crankcase oil and splashes it on various internal surfaces.

Part of the engine oil is pumped through a tube to lubricate the governor assembly above the engine. Oil holes are provided in the connecting rod for lubricating the bearings and piston pin, Fig. 2-6.

Obviously, the oil in a four cycle engine must be drained periodically and replaced with clean oil. Also worth noting, four cycle engines must be operated in an

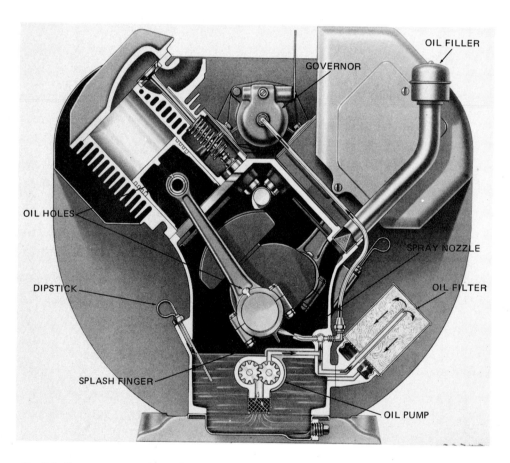

Fig. 2-6. Two common methods of supplying lubrication in four cycle engines are splash system and pressurized system. Engine shown employs both methods. Splash finger churns oil into a mist that makes its way into oil holes and other parts. Gear pump directs oil to remote parts and sprays some on critical parts. (Wisconsin Motors Corp.)

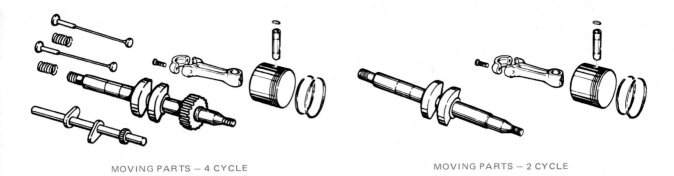

MOVING PARTS — 4 CYCLE   MOVING PARTS — 2 CYCLE

Fig. 2-7. Number of moving parts in a four cycle engine is greater than in a two cycle engine. Other differences are listed in chart at end of chapter.   (Lawn-Boy Power Equipment, Gale Products)

upright position or the oil will flow away from the pump or splash finger, preventing lubrication.

TWO-STROKE CYCLE ENGINE

The two-stroke cycle engine (commonly called "two cycle") performs the same cycle of events as the four cycle engine. The main difference is that intake, compression, power and exhaust functions take place during only two strokes of the piston. The two strokes occur during each revolution of the crankshaft. Therefore, it takes only one revolution of the shaft to complete one two-stroke cycle.

A two cycle engine has several advantages over a four cycle unit. It is much simpler in design than the four cycle engine because the conventional valves, tappets and camshaft are unnecessary. See Fig. 2-7.

In addition, a two cycle engine is smaller and lighter in weight than a four cycle engine of equivalent horsepower. Also in its favor, a two cycle engine will get adequate lubrication when operated at an extreme angle. It receives its lubrication as fuel mixed with oil is passed through the engine.

Installing the correct mixture of fuel and oil is a critical factor in maintaining a two cycle engine in good working condition. The prescribed type and grade of engine oil must be mixed with the fuel in proper proportion, then placed in the fuel tank.

In this way, there is a continuously new, clean supply of oil to all moving parts while the engine is running. The oil eventually burns in the combustion chamber and is exhausted with other gases.

Two cycle engines are popular in power lawn mowers, snowmobiles, dune buggies, chain saws, jet boats and other high rpm applications.

VARIATIONS IN DESIGN

Two basic types of two cycle engines are in general use. They are the cross-scavenged and loop-scavenged designs, Fig. 2-8.

The cross-scavenged engine has a special contour on the piston head which acts as a baffle to deflect the air-fuel charge upward in the cylinder. See A in Fig. 2-8. This prevents the charge from going straight out the exhaust port located directly across from the intake.

Cross-scavenged engines usually employ reed valves or a rotary valve attached to the flywheel. These valves hold the incoming charge in the crankcase so it can be compressed while the piston moves downward in the cylinder. With this design, the piston acts as a valve in opening and closing intake, exhaust and transfer ports. The transfer port permits passage of the fuel from the crankcase to the cylinder.

The loop-scavenged engine does not have to deflect the incoming gases, so it has a relatively flat or slightly domed piston, as shown in C in Fig. 2-8. The fuel transfer ports in loop-scavenged engines are shaped and located so that the incoming air-fuel mixture swirls. This controlled flow of gas helps force exhaust gases out and permits the new charge of air and fuel to enter.

PRINCIPLES OF OPERATION

The location of the ports in a two cycle engine is essential to correct timing of the intake, transfer and exhaust functions. The cutaway cylinder, A in Fig. 2-9, shows the exhaust port at the highest point, the transfer port next and the intake port lowest of the three. Some engines, particularly loop-scavanged engines, have more than one transfer port. See B in Fig. 2-9.

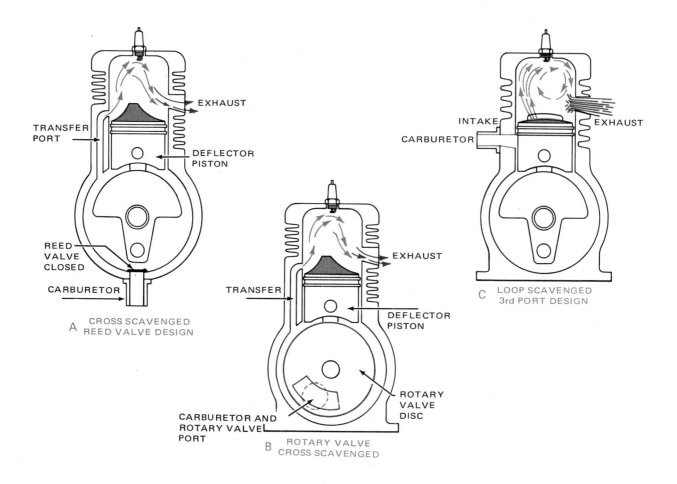

Fig. 2-8. Basically, two cycle engines are either cross scavenged or loop scavenged. Cross scavenged engines have a contoured baffle on top of piston to direct air-fuel mixture upward into cylinder while exhaust gases are being expelled. Loop scavenged engines have flat or domed pistons with more than one transfer port. Note three styles of crankcase intake valves. (Kohler Co.)

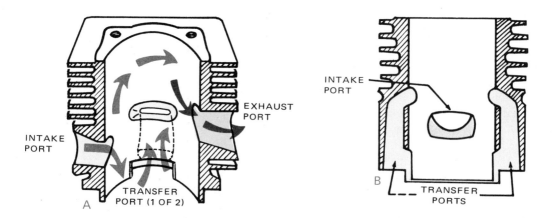

Fig. 2-9. A cutaway cylinder block shows location of intake, exhaust and transfer ports of a loop scavenged engine. A—Due to cutaway, only one of two transfer ports is shown. B—Section is revolved 90 deg. to show both ports.

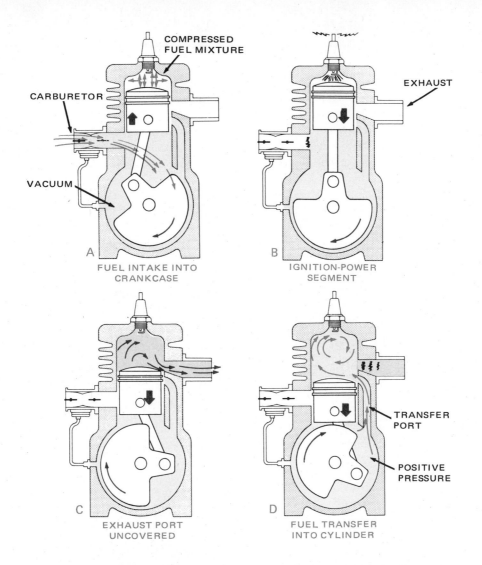

CARBURETOR

COMPRESSED
FUEL MIXTURE

VACUUM

A

FUEL INTAKE INTO
CRANKCASE

EXHAUST

B

IGNITION-POWER
SEGMENT

C

EXHAUST PORT
UNCOVERED

D

FUEL TRANSFER
INTO CYLINDER

TRANSFER
PORT

POSITIVE
PRESSURE

Fig. 2-10. Illustrations show sequence of events that take place in a two cycle engine in operation. Compression and intake occur simultaneously then ignition occurs. Exhaust precedes transfer of fuel during lower portion of power stroke. Piston functions as only valve.   (Rupp Industries, Inc.)

INTAKE INTO CRANKCASE

As the piston moves upward in the cylinder of a two cycle engine, crankcase pressure drops and the intake port is exposed. Atmospheric pressure is greater than the crankcase pressure, so air rushes through the carburetor and into the crankcase to equalize the pressures. See A in Fig. 2-10.

While passing through the carburetor, the intake air pulls a charge of fuel and oil along with it. This charge remains in the crankcase to lubricate ball and needle bearings until the piston opens the transfer ports on the downstroke.

FUEL TRANSFER

Drawings C and D in Fig. 2-10 show the piston moving downward, compressing the air-fuel charge in the crankcase. When the piston travels far enough on the downstroke, the transfer port is opened and the compressed air-fuel charge rushes through the transfer port into the cylinder. The new charge cools the combustion area and pushes or scavenges the exhaust gases out of the cylinder.

IGNITION-POWER

As the piston travels upward, A in Fig. 2-10, it compresses the air-fuel charge in the cylinder to about one tenth its original volume. The spark is timed to ignite the air-fuel mixture when the piston reaches TDC. See B in Fig. 2-10.

On some small engines, the spark occurs almost at TDC during starting, then automatically advances so the spark occurs earlier. This is done to get better efficiency from the force of combustion at higher speeds.

Peak combustion pressure is applied against the piston top immediately after TDC. Driving downward with maximum force, the piston transmits straight line motion through the connecting rod to create rotary motion of the crankshaft. See C in Fig. 2-10.

EXHAUST

Several things happen during the exhaust phase. See C in Fig. 2-10. As the piston moves to expose the exhaust port, most of the burned gases are expelled. Complete exhausting of gases from the cylinder and combustion chamber takes place when the transfer

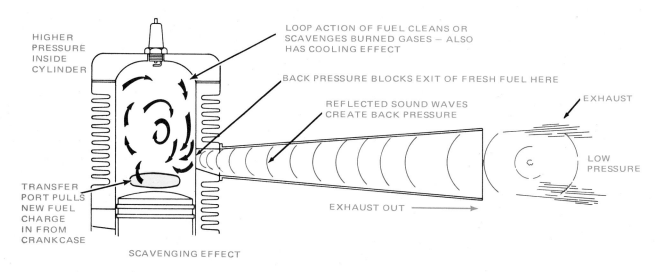

Fig. 2-11. Pressure pulse exhaust tuning is an effective way of increasing power and efficiency in two cycle engines. Exhaust sound waves reflected back into manifold creates a back pressure that stops fuel mixture from leaving cylinder before piston closes port. This system requires precise engineering.   (Kohler Co.)

Fig. 2-12. A straight pipe may sound louder and more powerful than tuned exhaust but actually is far less efficient. In this illustration, center of sound is too far away and lacks amplification to have any beneficial effect on engine.

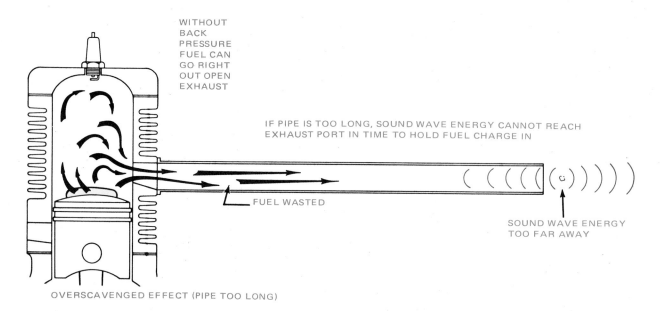

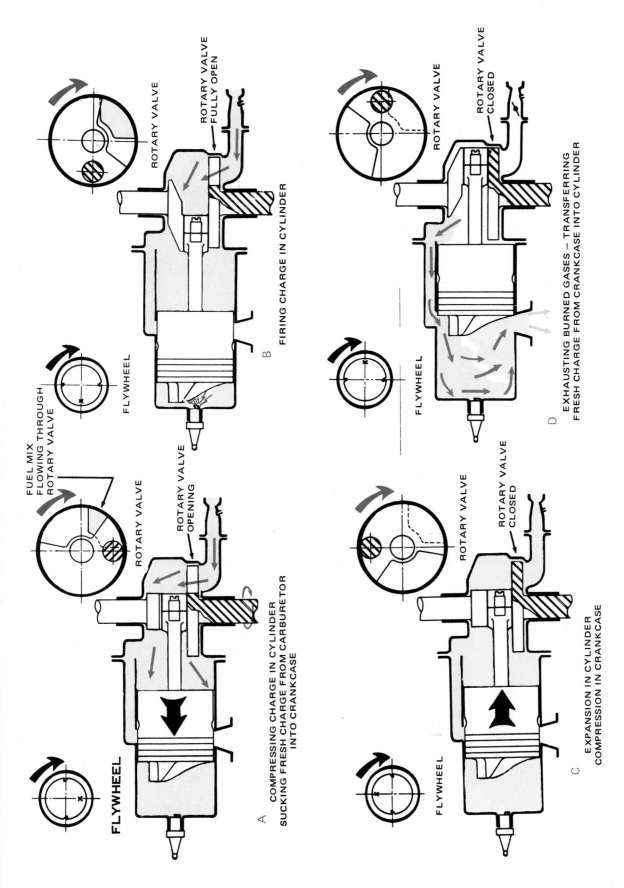

ROTARY VALVE

ROTARY VALVE FULLY OPEN

FLYWHEEL

FUEL MIX FLOWING THROUGH ROTARY VALVE

ROTARY VALVE

ROTARY VALVE OPENING

FLYWHEEL

A COMPRESSING CHARGE IN CYLINDER
SUCKING FRESH CHARGE FROM CARBURETOR
INTO CRANKCASE

B FIRING CHARGE IN CYLINDER

ROTARY VALVE

ROTARY VALVE CLOSED

FLYWHEEL

ROTARY VALVE

ROTARY VALVE CLOSED

FLYWHEEL

C EXPANSION IN CYLINDER
COMPRESSION IN CRANKCASE

D EXHAUSTING BURNED GASES — TRANSFERRING
FRESH CHARGE FROM CRANKCASE INTO CYLINDER

Fig. 2-13. Diagrams show how rotary valve operates in a two cycle engine. Since rotary disc is part of crankshaft, port is open only when hole in disc and crankcase port are in alignment. (Evinrude Motors)

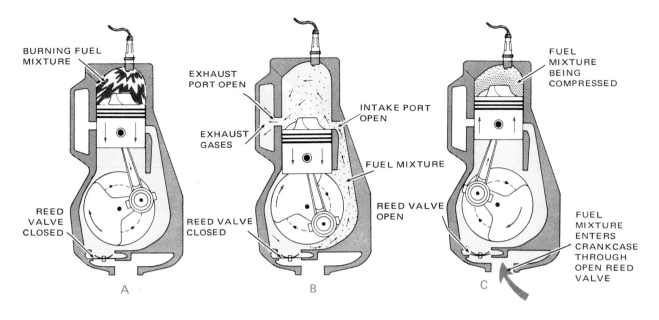

Fig. 2-14. A popular method of crankcase valving is reed valve designed to fit into crankcase wall. It relies upon difference between atmospheric pressure and crankcase pressure to be opened. At rest position is closed position.

ports are opened, D in Fig. 2-10, and the new air-fuel charge rushes in. This completes one cycle of operation.

SCAVENGING AND TUNING

When properly designed, the exhaust system scavenges all exhaust gases from the combustion chamber. It allows the new fuel charge to move in more rapidly for cleaner and more complete combustion.

For best efficiency, the fuel charge should be held in the cylinder momentarily while the exhaust port is open. This helps prevent fuel from being drawn out of the cylinder with exhaust gases.

Some well engineered exhaust systems use the energy of sound waves from the exhaust gases for proper tuning. Fig. 2-11 shows a megaphone-like device which amplifies the sound to speed up scavenging. The sound waves are reflected back into the megaphone to develop back pressure, which prevents the incoming air-fuel mixture from leaving with the exhaust gases. Compare this device with straight pipe operation shown in Fig. 2-12.

ROTARY DISC VALVE ENGINE

Fig. 2-13 illustrates a two cycle engine equipped with a rotary disc valve. The intake port is located directly in the crankcase, allowing room for additional transfer ports that promote better fuel transfer and scavenging.

REED VALVE ENGINE

The reed valve engine, Fig. 2-14, permits fuel intake directly into the crankcase. The reed is made of thin, flexible spring steel which is fastened at one end, Fig.

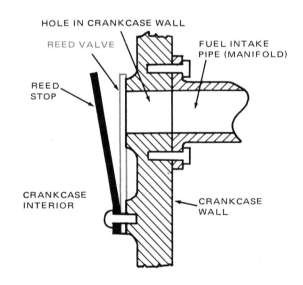

Fig. 2-15. Side view of a reed valve shows spring steel reed covering entry hole. Reed stop controls distance reed may open. This prevents permanent distortion and failure of reed to return snugly against port during time of crankcase compression.

2-15. The opposite end covers the intake port. The reed stop is thick and inflexible. It prevents the reed from opening too far and becoming permanently bent.

In operation, the reed is opened by atmospheric pressure during the intake stroke. It is closed by the springiness of the metal and the compression in the crankcase on the power stroke. Fig. 2-16 illustrates: A—Fuel mixture entering the crankcase. B—How the reed valve is closed by crankcase pressure.

There are many reed plate designs. Some are illustrated in Fig. 2-17.

FOUR CYCLE VS TWO CYCLE

The advantages and disadvantages of any engine are directly related to the purpose for which the engine is intended. It cannot be said that one type of engine is better than another without considering every aspect of its application.

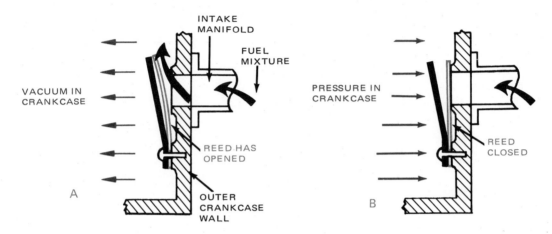

Fig. 2-16. Reed valve action. A—Vacuum in crankcase formed by upward moving piston causes atmospheric pressure to force air-fuel mixture through port opening. B—Downward piston movement compresses fuel mixture in the crankcase to a pressure greater than atmospheric pressure. Springy reed and crankcase pressure act together to close port.

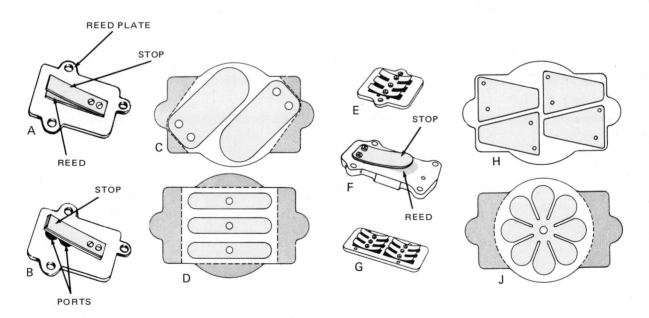

Fig. 2-17. Several forms of reed valves. A—Single reed, in closed postion. B—Single reed, open position. Note how the reed opening distance is controlled by the stop. C—Twin reed. D—Triple reed. E—Another form of triple reed. F—Single reed. G—Multiple reed. H—Four reed. J—Multiple reed.

The following chart lists the differences between two and four cycle engines.

CHARACTERISTICS	FOUR CYCLE ENGINE (equal hp) ONE CYLINDER	TWO CYCLE ENGINE (equal hp) ONE CYLINDER
1. Number of major moving parts	Nine	Three
2. Power strokes	One every two revolutions of crankshaft	One every revolution of crankshaft
3. Running temperature	Cooler running	Hotter running
4. Overall engine size	Larger	Smaller
5. Engine weight	Heavier construction	Lighter in weight
6. Bore size equal hp	Larger	Smaller
7. Fuel and oil	No mixture required	Must be pre-mixed
8. Fuel consumption	Fewer gallons per hour	More gallons per hour
9. Oil consumption	Oil recirculates and stays in engine	Oil is burned with fuel
10. Sound	Generally quiet	Louder in operation
11. Operation	Smoother	More erratic
12. Acceleration	Slower	Very quick
13. General Maintenance	Greater	Less
14. Initial cost	Greater	Less
15. Versatility of operation	Limited slope operation (Receives less lubrication when tilted)	Lubrication not affected at any angle of operation
16. General operating efficiency (hp wt. ratio)	Less efficient	More efficient
17. Pull starting	Two crankshaft rotations required to produce one ignition phase	One revolution produces an ignition phase
18. Flywheel	Requires heavier flywheel to carry engine through three non-power strokes	Lighter flywheel

REVIEW QUESTIONS — CHAPTER 2

1. Name the four strokes of a four cycle engine in proper order.
2. Name three important intake valve functions.
3. Explain why a four cycle engine runs cooler than a two cycle engine.
4. Why is there a difference in temperature between the intake and exhaust valve?
5. The exhaust valve is cooled mainly by:
   a. Radiation.
   b. Conduction.
   c. Convection.
   d. Air-fuel circulation.
6. How does compression increase the engine power?
7. The compression ratio must be limited in gasoline spark ignition engines because:
   a. There is no power advantage after compressing the fuel to a certain point.
   b. The engine becomes too difficult to start.
   c. Mechanically it is not possible to increase the compression ratio.
   d. The heat of compression will ignite the air-fuel mixture too soon.
8. What are the two methods employed for lubricating four cycle engines?
9. What are the two types of scavenging systems used in two cycle engines?
10. Why can two cycle engines be run in any position?
11. Name three valving systems employed in two cycle engines.
12. The baffle on the contoured piston is for the purpose of:
    a. Creating turbulent flow of gases.
    b. Slowing the air-fuel mixture entering the combustion chamber.
    c. Directing the flow of air-fuel mixture upward in the cylinder
    d. Directing oil evenly to the cylinder walls.
13. The _____ _____ type of two cycle engine requires a contoured piston.
14. In a properly tuned exhaust system, _____ _____ prevent the air-fuel mixture from leaving with the exhaust.
15. What advantage is there in having the intake port lead directly into the crankcase?
16. The time during the four-stroke cycle when both valves are open is called _____ _____.
17. A four cycle engine accelerates slower than a two cycle engine because:
    a. There is only one power stroke in four.
    b. The flywheel is heavier to carry the engine through three non-power strokes.
    c. There are more moving parts to be driven by the engine.
    d. All of the above are true.

SUGGESTED ACTIVITIES

1. Look up additional information about the development of internal combustion engines. Some names to look up are: Christian Huygens, Philip Lebon, Samuel Brown, William Barnett, Pierre Lenoir, Beau DeRochas, Dr. N. A. Otto, Atkinson, Gottlieb Daimler, Priestman and Hall, Herbert Akroyd Stuart, Rudolph Diesel.
2. Begin a collection of engine repair and service manuals by writing to various engine manufacturers.
3. Using a worn out engine, cut away portions that will make the working parts visible while still enabling them to move. Make a report on the operation and timing of each part.
4. After further study, electrify the spark plug of the cutaway engine with a small light bulb switched on and off by the breaker points to simulate ignition.
5. Make a bulletin board display that illustrates the main principles of four and two cycle engines.

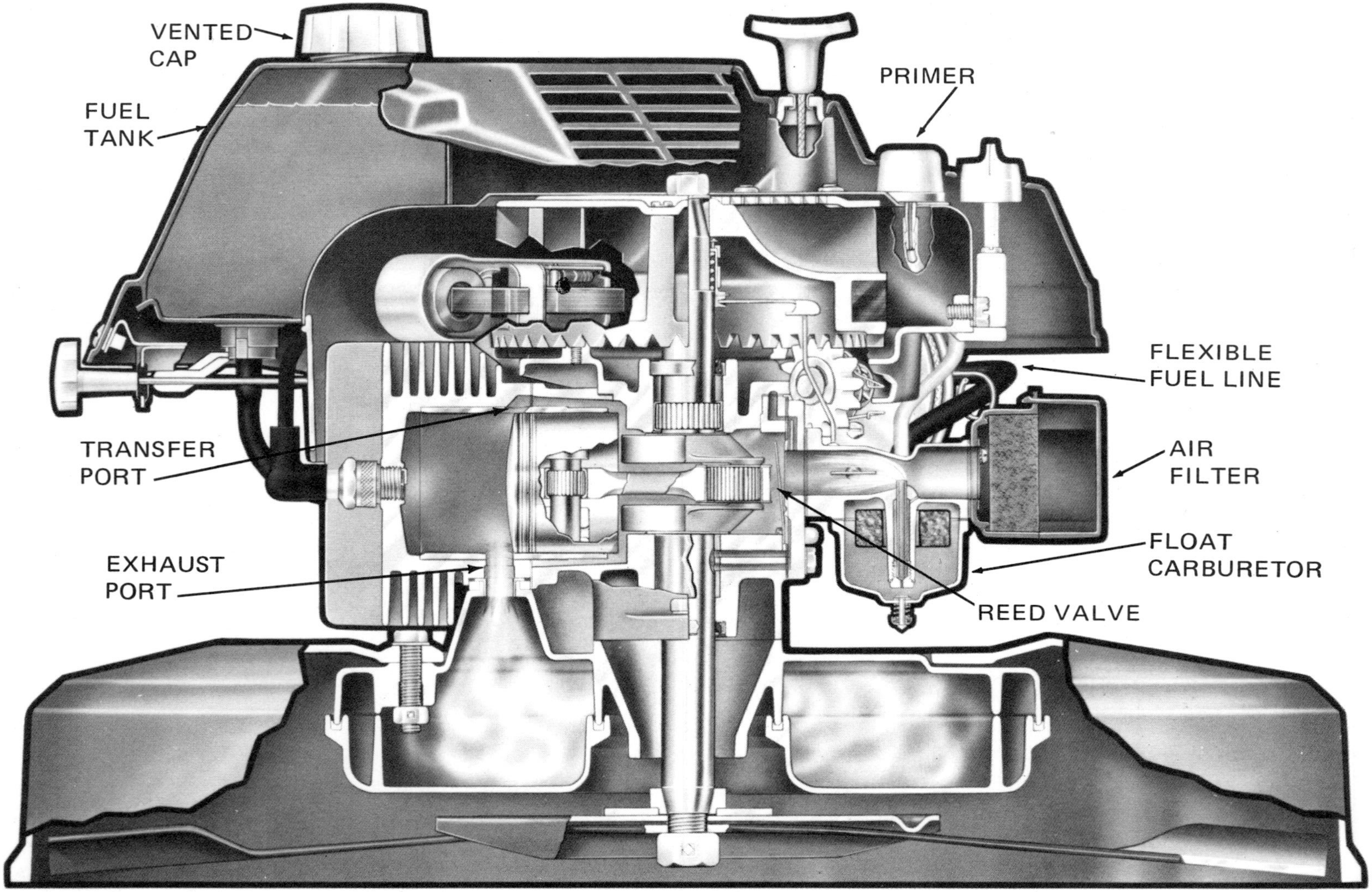

Fig. 3-1. Callouts identify components of a typical fuel system on a two cycle engine used in a power mower application.

PRIMER

FLEXIBLE FUEL LINE

AIR FILTER

FLOAT CARBURETOR

REED VALVE

VENTED CAP

FUEL TANK

TRANSFER PORT

EXHAUST PORT

Chapter 3

FUEL SYSTEMS

Small gas engines can be designed to operate efficiently on gasoline, Liquefied Petroleum Gas (LP-Gas), natural gas, kerosene or diesel fuel. See Fig. 3-1. Gasoline is the most popular of all small engine fuels. *In addition to its power potential, gasoline is readily available, easily transported and tank refilling is simple.* See Fig. 3-2.

Fig. 3-2. Refueling a four cycle engine requires use of an approved fuel can and a fresh fill of regular gasoline.

GASOLINE

For small engines, most manufacturers specify the use of regular grade gasoline with an octane rating around 90. Occasionally, premium fuels are recommended for use in hot climates. This practice may prevent detonation or "dieseling" (after-run). However, a heavier buildup of solid materials in the combustion chamber may be expected because premium fuels contain greater amounts of lead additives.

Gasoline should be clean, free from moisture and reasonably "fresh." After prolonged storage, especially in small quantities, gasoline tends to become "stale." This is caused by a certain amount of oxidation that forms a sticky, gum-like material. The gum may clog small passageways in the carburetor and cause poor engine performance or hard starting.

LP-GAS AND NATURAL GAS

LP-Gas may be propane or butane, or a mixture of both. Properly designed fuel systems will allow the use of LP-Gas with no appreciable loss of horsepower, as compared to a similar engine burning gasoline. LP-Gas burns clean and leaves few combustion chamber deposits. Because of less noxious fumes, these engines often are used in warehouses, factories, etc. LP-Gas also has a high anti-knock rating.

Natural gas generally will cause a horsepower loss of around 20 percent when compared with gasoline. Both LP-Gas and natural gas require a somewhat different fuel system setup than the conventional type used for supplying gasoline. Fig. 3-3 shows the components of one type of LP-Gas system.

COMBUSTION OF LP-GAS

LP-Gas is slower burning than gasoline because it has higher ignition temperatures. For this reason, the timing is often advanced farther on LP-Gas engines.

Due to higher ignition temperatures, greater voltage at the spark plugs may be needed for LP-Gas combustion. "Colder" plugs or smaller spark plug gaps may be used to solve this problem. Check the engine manual for recommendations.

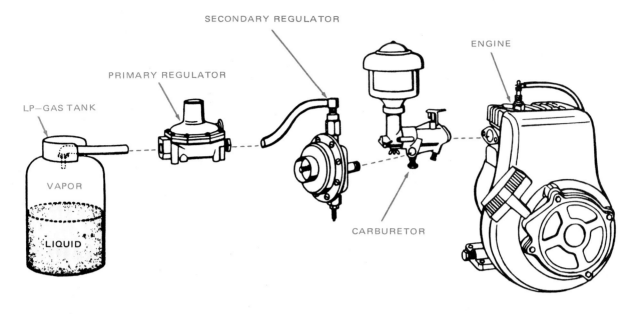

Fig. 3-3. Typical LP-Gas fuel system using vapor withdrawal. (Clinton Engine Corp.)

Less heat is required at the intake manifold to vaporize LP-Gas than gasoline. LP-Gas vaporizes at much lower temperatures than gasoline. It vaporizes at room temperature. This results in less wasted heat and more heat converting into engine power.

ADVANTAGES OF LP-GAS

1. Cheaper, especially when close to source (refinery).
2. Less oil consumption due to engine wear.
3. Reduced maintenance costs - longer engine life between overhauls.
4. Smoother power from slower, more even burning of LP-Gas.
5. Less noxious or poisonous exhaust, such as deadly carbon monoxide gas.

DISADVANTAGES OF LP-GAS

1. Initial equipment costs are high-bulk fuel storage and carburetion equipment are costly.
2. Fewer accessible fuel points (gas stations).
3. Harder to start LP-Gas engines in cold weather − 0 deg. F (−18 C) or below.

KEROSENE AND DIESEL FUELS

Some non-diesel type small gas engines can be converted to operate successfully on kerosene or fuel oil through the installation of a low compression cylinder head and a special carburetor. Engines so equipped are started and operated on gasoline until fully warm, then switched over to the kerosene or fuel oil. These fuel installations generally are confined to heavy-duty engines in industrial service.

The true diesel engine uses diesel fuel injected into the cylinder where it is ignited by the heat of compression. It is not unusual to have compression ratios as high as 20 to 1. Currently, however, diesel application in the small engine field is somewhat limited. It is only practical on applications where continuous use for long periods of time are common.

TWO CYCLE FUEL MIXTURES

Most two cycle engines receive lubrication only from the oil mixed with the gasoline. Because of this, it is important that the correct quantity and proper quality of oil is thoroughly mixed with a specific amount of gasoline. *Always follow the manufacturer's specifications as to type and quantity of oil to use.*

Too little oil could cause the engine to overheat. Overheating, in turn, causes expansion of parts and possible scoring of machined surfaces. Eventually, the pistons may seize (bind, then stick) in the cylinders. Excessive oil, on the other hand, will cause incomplete combustion and rapid buildup of carbon, fouling the spark plugs and adding weight to the pistons.

Fig. 3-4. Portable engine-driven generator with the fuel tank mounted on top side opposite carburetor. (Onan Corp.)

TANKS, LINES AND FITTINGS

Small engine fuel tanks are made of metal or plastic. Some are mounted separately, as in Fig. 3-4. Others are contoured to fit snugly around the engine, Fig. 3-5.

The tank filler cap is vented. If the cap vent becomes clogged, the engine will create enough vacuum in the tank to cause fuel starvation. Most filler caps have baffles and filters. See Fig. 3-6.

The purpose of a fuel tank filler cap with a screw vent is to prevent fuel evaporation when it is closed. This vent should be opened before starting the engine.

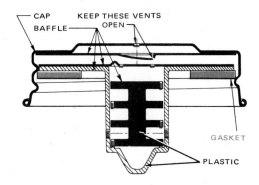

Fig. 3-6. Vented fuel filler caps are baffled to prevent dirt and dust from entering fuel tank. (Clinton Engine Corp.)

A variety of cap styles are shown in Fig. 3-7.

Fuel tanks used in all terrain vehicles (ATVs) and snowmobiles often have the fuel pick-up line inserted from the top of the tank. The pick-up line usually is very flexible and weighted at the bottom, so the line will always be where the fuel is deepest in the tank when the vehicle is at a steep angle.

FUEL FILTERS

Some small engines have a fuel line fitting in the bottom of the tank. A filter screen is placed in the tank fitting or at the end of the pick-up line, Fig. 3-8. A top mounted pick-up line with the filter element at the bottom end is shown in Fig. 3-9.

Other engines have a bottom mounted fuel fitting with a shutoff valve threaded into the tank, Fig. 3-10.

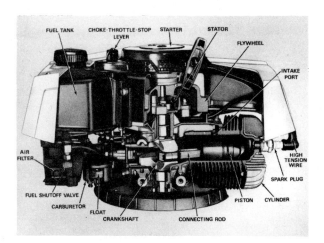

Fig. 3-5. Cutaway view of vertical shaft engine shows a plastic fuel tank contoured to fit snugly around engine. (Jacobsen Mfg. Co.)

Fig. 3-7. Fuel filler caps: A—Three-piece plastic cap showing maze type baffle with fiber gaskets. B—Plastic cap with plastic fabric as a filter with a perforated fiber disc. C—Cap with a threaded screw vent which will seal tank. D—Standard cap with a single vent hole.

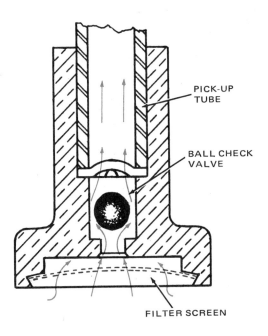

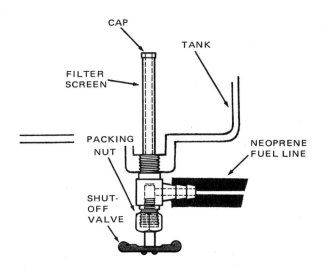

Fig. 3-8. A filter and ball check valve attached to end of a fuel pick-up tube. Ball check valve prevents fuel from draining back into tank while engine is not running. (Clinton Engine Corp.)

Fig. 3-10. Shutoff valve in tank is necessary to stop loss of fuel whenever a part of fuel system is undergoing work. Filter screen is permanent part of valve. Care during installation is necessary. (Lawn-Boy Power Equipment, Gale Products)

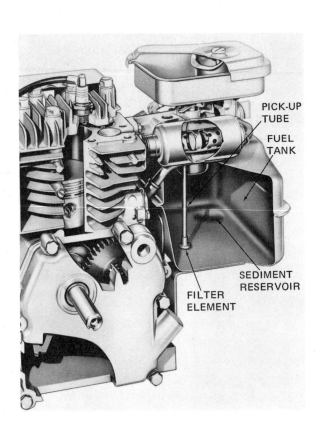

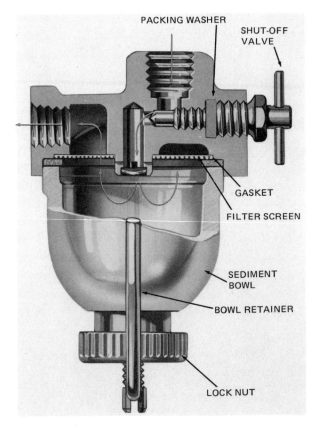

Fig. 3-9. Fuel tank cutaway shows pick-up tube in tank with ball check valve and filter element attached. (Briggs and Stratton Corp.)

Fig. 3-11. Some small engines have remote fuel strainers located somewhere along fuel line. Glass bowl permits visual inspection without dismantling. (Wisconsin Motors Corp.)

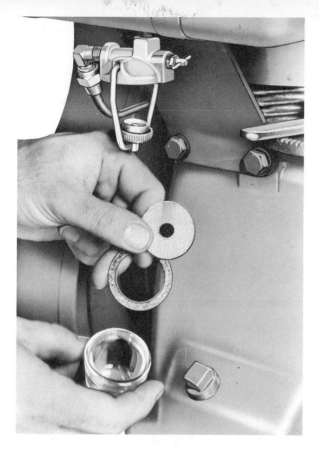

Fig. 3-12. When moisture or dirt is found in sediment bowl of fuel strainer, it can easily and quickly be taken apart for cleaning. Fuel shutoff valve is closed before removing bowl.

They help insure that the engine can always provide quick acceleration and constant full power.

MECHANICAL FUEL PUMPS

The mechanical fuel pump used on small engines is basically the same as the type used on automobile engines. It may or may not have a filter included as part of the pump design. Fig. 3-13 is a cutaway view of a

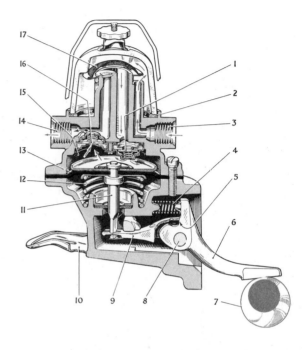

Fig. 3-13. Cutaway of a mechanical fuel pump with a combined fuel strainer on top. Pumps are activated by camshaft (7) as shown. Diaphragm (13) pulsates and forces fuel through check valves (1) and (16).

Older small engines have a more elaborate filter incorporated in a glass sediment bowl, Fig. 3-11. The gasket, screen and bowl can be removed for inspection and cleaning. A similar filter shown in Fig. 3-12 is mounted directly on the fuel tank and has a shut-off valve.

FUEL PUMPS

Fuel pumps are used on engines having the fuel tank mounted in such a way that a gravity fuel supply system will not operate. In these applications, the tank and fuel level is lower than the carburetor, or the fuel level may be above the carburetor at times and below the rest of the time. All terrain vehicles (ATVs) and snowmobiles, for example, may have the fuel tank mounted away from the engine, and the angle of the vehicle is constantly changing.

Fuel pumps provide constant fuel flow under pressure to the carburetor under changing conditions.

combination fuel pump and filter system. Trace the arrows to follow the flow of fuel.

FUEL PUMP OPERATION

The typical mechanical fuel pump shown in Fig. 3-13 operates by means of a diaphragm and atmospheric pressure on the surface of the fuel in the tank. As the engine camshaft revolves, an eccentric (7) actuates the fuel pump rocker arm (6) pivoted at (8) which pulls the pull rod (11) and diaphragm (13) downward against spring pressure (12), creating a depression in the pump chamber (15). Fuel drawn from the tank enters the glass bowl from the pump intake (3). After passing through the filter screen (17) and the inlet valve (1), it enters the pump chamber (15).

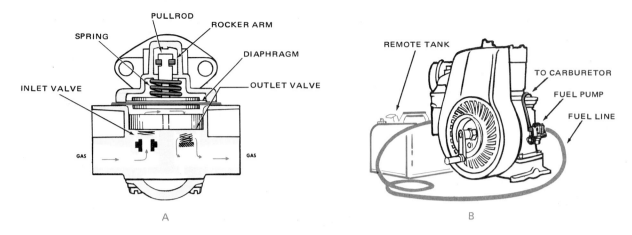

Fig. 3-14. A—Diaphragm fuel pump without a combined filter. B—Pump is cam-operated and can pump fuel from portable fuel tanks if necessary. (Clinton Engine Corp.)

On the return stroke, pressure of the spring (12) pushes the diaphragm (13) upwards, forcing fuel from the chamber (15) through the outlet valve (16) and outlet (14) to the carburetor. When the carburetor bowl is full, the carburetor float will seat the needle valve, preventing any flow from the pump chamber (15). This will hold the diaphragm (13) down against the spring pressure (12), and it will remain in this position until the carburetor requires additional fuel and the needle valve opens. The rocker arm (6) operates the connecting link (9) by making contact at (5). This construction allows idling movement of the rocker arm without moving the fuel pump diaphragm. The spring (4) keeps the rocker arm in constant contact with the eccentric (7) to eliminate noise.

FUEL PUMP HAND PRIMER

The hand primer shown as (10) in Fig. 3-13 is for use when, for some reason, the carburetor float bowl or pump bowl has become empty. By pulling the hand primer upward, the float bowl will fill and insure easy starting without prolonged use of the starter.

Because of the special construction of the pump, it is impossible to overprime the carburetor. After several strokes of the hand primer, its handle will become free acting. This indicates that the float bowl is full.

FUEL PUMP WITHOUT A FILTER SYSTEM

The fuel pump in drawing A in Fig. 3-14 is a mechanical pump without a filter chamber. It is mounted on the crankcase. See B in Fig. 13-14. A

separate filter may be installed somewhere along the fuel line. In other applications, the system may rely solely on the pick-up line filter or a filter in the carburetor or a combination of both. Fig. 3-15 shows a combination fuel pump and filter system. Note shut off needle valve and primer lever.

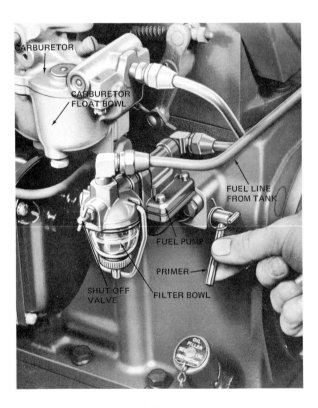

Fig. 3-15. Combination fuel pump and filter mounted in tandem. Fuel pump has a manual primer used to initially get fuel to carburetor. (Wisconsin Motors Corp.)

IMPULSE DIAPHRAGM FUEL PUMPS

One type of diaphragm fuel pump sometimes used on small gas engines is activated by the pulsing vacuum in the intake manifold or crankcase. Four cycle engines use the intake manifold vacuum; two cycle engines use crankcase vacuum.

A typical impulse diaphragm pump is illustrated in Fig. 3-16. When vacuum draws the diaphragm upward

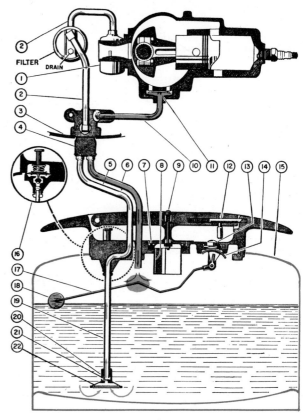

IMPULSE LINE CONNECTOR

SPRING

CHECK VALVE

DIAPHRAGM

FUEL OUTLET

CHECK VALVE

FUEL INLET

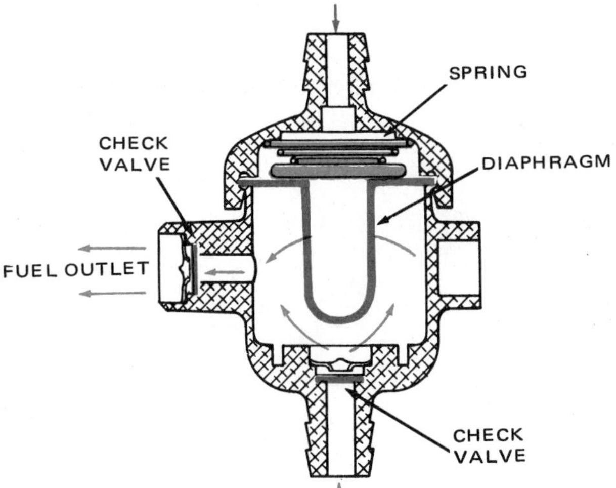

Fig. 3-16. This fuel pump is operated by vacuum pulses transferred from engine crankcase. It can be mounted in any convenient location on engine. (Clinton Engine Corp.)

against spring tension, the inlet check valve opens to allow fuel to flow in. When vacuum is relieved, the spring pushes the diaphragm downward to force fuel through the outlet check valve. This process is repeated as long as the engine is running.

It is common practice today to design the impulse diaphragm fuel pump directly into the carburetor. This design principle will be explained under the section on carburetors.

PRESSURIZED FUEL SYSTEM

Pressurized fuel system, Fig. 3-17 is used when fuel tanks are located a considerable distance below the carburetor. Outboard engines, for example, often operate from portable tanks resting in bottom of boat.

FILTER
DRAIN

1—Carburetor.
2—Fuel line, filter to carburetor.
3—Bottom cowl.
4—Twist connector.
5—Fuel line, tank to twist connector.
6—Air line, tank to twist connector.
7—Seal.
8—Extension tube, filler opening.
9—Pressure relief valve.
10—Air line, crankcase to connector.
11—Pressurized valve, crankcase.
12—Release latch.
13—Magnifying lens.
14—Graduated sector.
15—Remote fuel tank.
16—Priming pump.
17—Float arm.
18—Float.
19—Fuel pick-up tube.
20—Disc filter.
21—Check valve.
22—Strainer.

Fig. 3-17. Pressurized fuel system uses crankcase pressure transferred to fuel tank. Pressure on surface of fuel forces fuel into engine. (Evinrude Motors)

The pressurized fuel system shown in Fig. 3-17 operates as follows:

1. The carburetor (1) is connected to the fuel tank (15) through the fuel filter by fuel lines (2 and 5).
2. Fuel flow from the tank to the carburetor is induced by pressure transmitted from the crankcase to the air space above the fuel line level via the air line (10), running from pressurized valve (11) to the twist connector (4) to the top of the fuel tank (15).
3. For starting, initial flow to the carburetor is induced

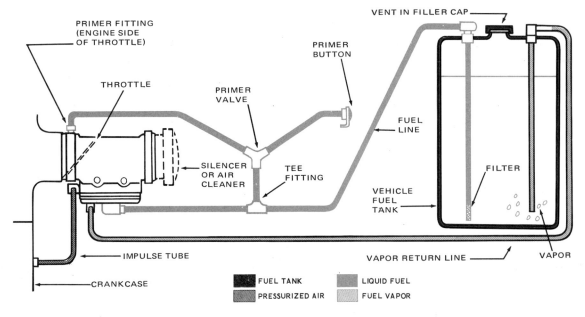

Fig. 3-18. Vapor return fuel system is one of the best methods of preventing a vapor lock. Vapors formed by heat are directed back to fuel tank where they are cooled and condensed to liquid form. (Kohler Co.)

by the hand-operated priming pump (16).

4. A disc filter (20) is incorporated in the bottom of the fuel pick-up tube (19).

5. The fuel level is indicated by a graduated sector (14) actuated by a float (18) attached to arm (17).

6. The pressure relief valve (9) in the center of the carrying handle permits relieving pressure when necessary.

7. The check valve in the twist connector (4) permits disconnecting the air-fuel line without loss of tank pressure.

8. The tank pressure forces fuel up through the pick-up line (10) through the filter to the carburetor (1).

9. The check valve (21) is essential to the operation of the priming pump (16).

VAPOR RETURN FUEL SYSTEMS

If the temperature of the air around or inside a carburetor becomes high enough to vaporize the gasoline, pockets of vapor will stop all flow of fuel. Once this occurs, the engine will remain "vapor locked" and will not run until the temperature drops low enough for the vapor to condense (return to a liquid).

One of the best ways to prevent vapor lock is to use a carburetor with a vapor return line. Then, any vapor that forms is directed back into the fuel tank where the pressure is vented to the atmosphere.

A diagram of a typical vapor return fuel system is shown in Fig. 3-18. The carburetor in this system has a built-in diaphragm fuel pump. The impulse tube operates the pump.

REVIEW QUESTIONS – CHAPTER 3

1. In addition to the power available from gasoline, give three other reasons for its wide acceptance for engine use.

2. Most manufacturers specify regular grade gasoline for small engines. Yes or No?

3. Premium fuels are sometimes recommended for use in hot climates. True or False?

4. A greater build-up of solid materials in the combustion chamber could be expected from using regular grade fuel. Yes or No?

5. Premium fuels contain greater amounts of lead additives. True or False?

6. LP-Gas is either _____ or _____ or a mixture of both.

7. Natural gas used as a small engine fuel is generally accompanied by a horsepower loss of _____ percent.

8. If excessive oil is mixed with the fuel for a two cycle engine:
   a. Overheating may result.
   b. Spark plugs may become overheated.
   c. Incomplete combustion may occur.

d. Seizing will result.

9. Filler caps with screw vents are for the purpose of:
   a. Preventing fuel evaporation when closed.
   b. Preventing fuel starvation when open.
   c. Preventing contamination in the tank.
   d. All of the above are correct.

10. The two types of fuel pumps discussed in this chapter are:
    a. Atmospheric pressure fuel pump.
    b. Impulse diaphragm fuel pump.
    c. Gravity vacuum fuel pump.
    d. Mechanical fuel pump.

11. When a carburetor has been removed and replaced, the engine would be slow starting because of lack of fuel. This problem can be overcome if the engine has a fuel pump with a _____ _____.

12. One satisfactory fuel system which prevents vapor lock is the _____ _____ system.

SUGGESTED ACTIVITIES

1. Collect a variety of tank filler caps. Either cut them in half or disassemble them. Make a display board showing the baffle and filter system.
2. Make a display board of cutaway drawings of fuel tanks with gravity feed fuel lines and top mounted pick-up lines.
3. Obtain and cut away some old fuel pumps so that internal parts can be seen and worked.
4. Cut away parts of an old fuel filter so that the fuel circuit can be traced.

A 7 1/2 hp outboard engine provides a pleasant mixture of fuel "miser" economy, light weight and twin-cylinder "muscle." (Evinrude Motors)

Chapter 4

CARBURETION

A carburetor's primary purpose is to produce a mixture of fuel and air to operate the engine. This function, in itself, is not difficult. It can be done with a simple mixing valve.

The mixing valve, however, is limited in efficiency. It cannot, for example, provide economical fuel consumption and smooth engine operation over a wide range of speeds. Meeting these performance goals requires a much more complex mechanism. This is the main reason why there are so many individual styles and designs of carburetors.

PRINCIPLES OF CARBURETION

Gasoline engines cannot run on "liquid" gasoline. The carburetor must vaporize the fuel and mix it with air in the proper proportion for varying conditions:
1. Cold or hot starting.
2. Idling.
3. Part throttle.
4. Acceleration.
5. High speed operation.

Basically, air enters the top of the carburetor, Fig. 4-1, and is mixed with liquid fuel fed through carburetor passages and sprayed into the airstream. The air-fuel mixture that results is forced into the intake manifold by atmospheric pressure, then burned in the combustion chamber of the engine.

Fig. 4-1 shows how a typical carburetor operates. In this particular small gasoline engine application, the engine is at part throttle operation. Note that the choke valve is open, while the throttle valve is partly closed.

AIR-FUEL MIXTURE

The amount of air needed for combustion is far greater than the amount of fuel required. The usual

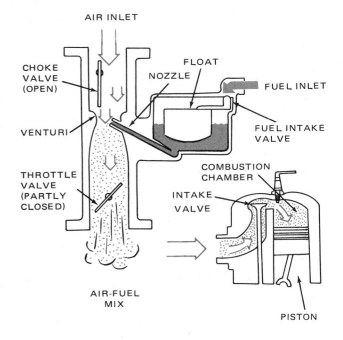

Fig. 4-1. Air entering carburetor mixes with fuel in proper proportion, and mixture flows into combustion chamber. (Deere & Co.)

weight ratio is 15 parts of air to 1 part of fuel, since one pound of air would take up a much greater space than one pound of fuel. Therefore, by volume, one cubic foot of gasoline would have to be mixed with 9000 cubic feet of air to establish a 15 to 1 weight ratio.

Small gasoline engines use varying air-fuel ratios, depending on engine speed and load. The chart in Fig. 4-2 shows how the mixture changes for various operating conditions.

CARBURETOR PRESSURE DIFFERENCES

A carburetor is a device that is operated by pressure differences. When discussing pressure differences, several

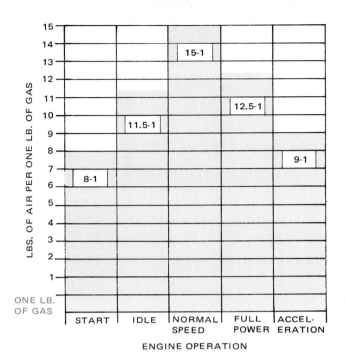

Fig. 4-2. Air-fuel mixture requirements vary, depending upon operating conditions. Chart shows approximate air-fuel ratios for various operating conditions.

MEAN ATMOSPHERIC PRESSURE
at 68 deg. F (20 C)

Altitude (ft.)	Atmospheric Pressure (in. Hg.)
9000	20.92
8000	21.92
7000	22.92
6000	23.92
5000	24.92
4000	25.92
3000	26.92
2000	27.92
1000	28.92
Sea Level	29.92

Fig. 4-3. Weight of air exerted on a given object is determined by height and density of a column of air above object. Air is less dense at higher altitudes. For every 1000 ft. above sea level, mercury column pressure is reduced by 1.0 in.

terms are commonly used. They are VACUUM, ATMOSPHERIC PRESSURE and VENTURI PRINCIPLE.

VACUUM: *An absolute vacuum is any area completely free of air or atmospheric pressure.* This condition is difficult to obtain and is never reached in a small gasoline engine. Therefore, any pressure less than atmospheric pressure generally is referred to as "a vacuum."

ATMOSPHERIC PRESSURE: *The pressure produced by the weight of air molecules above the earth is called atmospheric pressure.* The amount of atmospheric pressure varies with altitude. A person standing on a beach at sea level, for example, would be under a higher vertical column of air than a person standing on a mountain top. Therefore, the total weight of air molecules would be greater at sea level, Fig. 4-3.

Furthermore, any time air molecules are removed from a particular space, a vacuum is created. This space immediately fills with air under atmospheric pressure, if conditions permit.

This effect of atmospheric pressure can be related to small gasoline engines. The downward movement of the piston creates a partial vacuum in the cylinder. As soon as the intake valve opens or the intake port is uncovered, atmospheric pressure forces air through the carburetor and manifold to fill that vacuum.

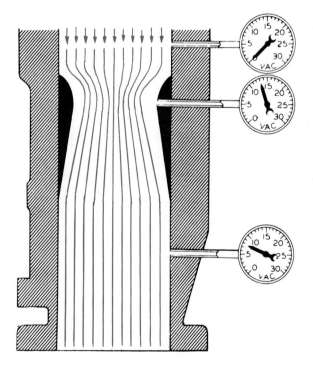

Fig. 4-4. The venturi principle. A restriction in a passage will cause incoming air to increase its velocity, while pressure will be reduced. Reduction in pressure draws fuel into airstream.

VENTURI PRINCIPLE: The carburetor creates a partial vacuum of its own by means of a VENTURI for the purpose of drawing fuel into the airstream. *A venturi is a restriction in a passage which causes air to move faster.* The gage shown at the top in Fig. 4-4 indicates no change in velocity, therefore there is no change in pressure. The area in which the air is moving faster (middle gage) develops a lower pressure.

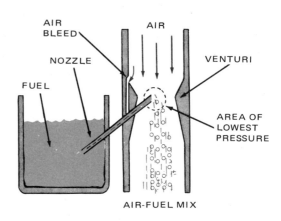

Fig. 4-6. To atomize fuel into finer particles, an air bleed is used. Higher pressure of air horn forces some air to enter at a port midway in nozzle, so fuel is partly atomized before leaving nozzle.

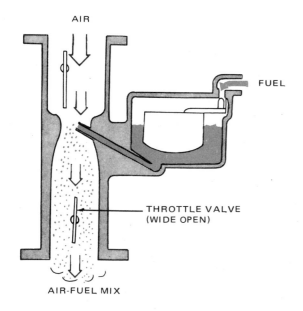

Fig. 4-5. Air flowing through venturi has reduced pressure around nozzle. Fuel is drawn up nozzle by vacuum and mixes in airstream. (Deere & Co.)

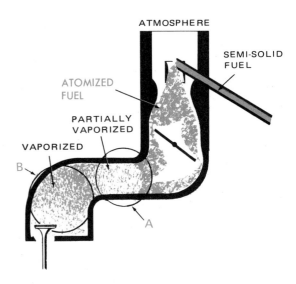

Fig. 4-7. In addition to air bleed and venturi, fuel is further vaporized: A—By vacuum in manifold. B—Engine heat.

Fig. 4-5 shows a simple carburetor with fuel being drawn from the float bowl out the main discharge nozzle. This nozzle is located so that its outer end is in the low pressure area of the venturi section. Fuel coming from the discharge nozzle is still in relatively large liquid droplets that do not burn well. To further atomize the fuel, an air bleed passage is built into the air horn, Fig. 4-6. A small portion of the air rushing through the carburetor is forced through the air bleed passage to the main discharge tube. This air mixes with the stream of fuel, breaking it into small particles before it reaches the venturi. The small particles of fuel are broken into finer particles by the air rushing through the venturi.

When the fuel moves into the intake manifold (which is under partial vacuum), the boiling point of the gasoline is lowered. This causes many of the atomized particles to "boil" or "flash" into a vapor. See

A in Fig. 4-7. As the partially vaporized fuel moves through the manifold, it is warmed by the heat of the manifold walls. This causes further vaporization, shown at B in Fig. 4-7. When the mixture enters the combustion chamber, both the swirling motion and the sudden increase in temperature due to the compression stroke completes the vaporization of the fuel.

TYPES OF CARBURETORS

The three basic types of carburetors are named according to the direction in which the air flows from their outlets to the engine manifold:
1. "Natural draft" or "side draft," Fig. 4-8.

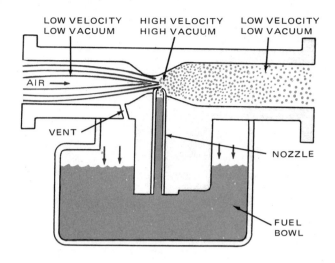

Fig. 4-8. Natural draft carburetor (cross-draft) has horizontal air flow through it.
(Deere & Co.)

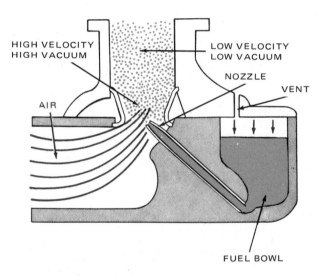

Fig. 4-9. Air flowing through updraft carburetor moves vertically upward into venturi. Passages must be comparatively smaller to increase air velocity so it will carry fuel upward.

2. "Updraft," Fig. 4-9.
3. "Downdraft," Fig. 4-10.

The natural draft carburetor is used where there is little space on top of the engine. The air flows horizontally into the manifold.

Updraft carburetors are placed low on the engine and use a gravity feed fuel supply. However, the air-fuel mixture must be forced upward into the engine. The air velocity must be high, so smaller passages must be used in the carburetor and manifold.

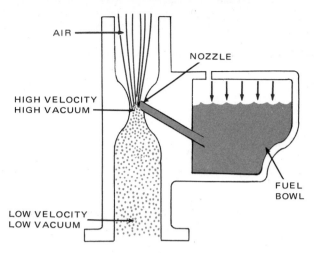

Fig. 4-10. Downdraft carburetor has downward flow of air through venturi. Since it can operate with lower velocities, it has larger passages.

Downdraft carburetors operate with lower air velocities and larger passages. This is because gravity assists the air-fuel mixture flow to the cylinder. The downdraft carburetor can provide larger volumes of fuel when needed for high speed and high power output.

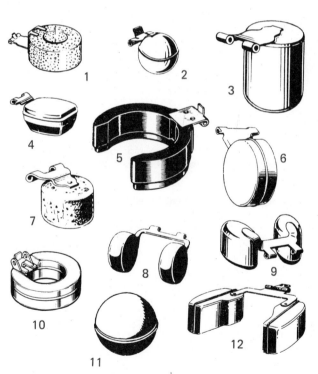

Fig. 4-11. Various float designs. 1—Doughnut-shaped cork. 2—Ball-shaped metal. 3—Cylindrical metal. 4—Rectangular metal. 5—Horseshoe-shaped plastic. 6—Cylindrical metal. 7—Round cork. 8 and 9—Twin type metal. 10—Doughnut-shaped metal. 11—Ball-shaped metal. 12—Twin type plastic.

FLOAT TYPE FUEL SUPPLY

The carburetor float is a small sealed vessel made of brass or plastic. Currently, some floats are made of solid flotation materials that eliminate the possibility of leakage. See Fig. 4-11.

The purpose of the carburetor float is to maintain a constant level of fuel in the float bowl. The float rises and falls with the fuel level. As fuel is used from the float bowl, the float lowers and unseats a needle valve, which lets fuel enter the bowl. This, in turn, raises the float, seating the needle and shutting off fuel supply to the bowl.

The closed position of the needle valve is illustrated in Fig. 4-12. The needle valve illustrated in Fig. 4-13

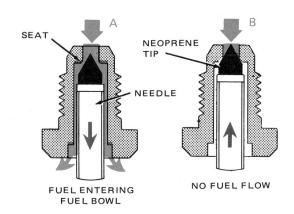

Fig. 4-13. Needle in float bowl opens and closes fuel passage into chamber. Needle is operated by hinged arm of float. (Evinrude Motors)

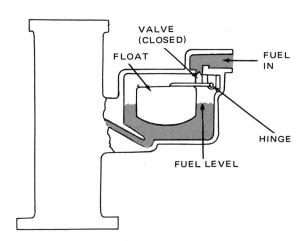

Fig. 4-12. Float in float bowl maintains a constant fuel level: When fuel level rises, float closes needle valve, stopping incoming fuel. When fuel level lowers, float unseats needle and lets more fuel in. (Deere & Co.)

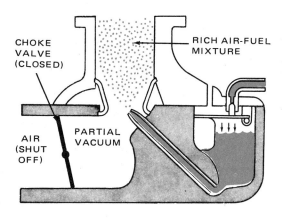

Fig. 4-14. Choke valve is closed and vacuum is high in carburetor. Fuel mixture entering intake manifold is extremely rich.

shows valve action in greater detail. The neoprene needle point is soft and seats well in the valve. Also, it is not as likely to wear out as readily as the harder brass needle points.

FLOAT BOWL VENTILATION

Most carburetors are sealed and balanced to maintain equal air pressure. The air pressure above the fuel in the bowl and the air pressure entering the carburetor are equalized by a vent in the float bowl. Refer to Fig. 4-9. This vent assures a continuous free flow of fuel.

CHOKE SYSTEM

The carburetor choke is a round disc mounted on a shaft located at the intake end of the carburetor, Fig. 4-14. When closed, the choke provides a rich air-fuel mixture needed during starting of a cold engine. It allows less air to enter the carburetor. The manifold vacuum draws harder on the fuel nozzle. Therefore, more fuel and less air enters the combustion chamber.

THROTTLE SYSTEM

The throttle valve also is a round disc mounted on a shaft like the choke. This valve is located beyond the main fuel nozzle, Fig. 4-15.

The main purpose of the throttle valve is to regulate the amount of air-fuel mixture entering the cylinders. It

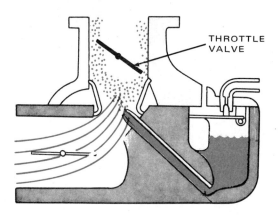

Fig. 4-15. Throttle valve is located beyond main fuel nozzle. Throttle regulates amount of air-fuel mixture entering engine.

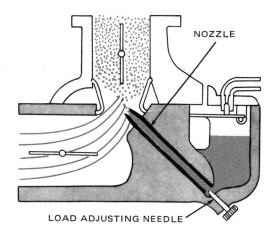

Fig. 4-16. A load adjusting needle, located as shown, regulates amount of fuel entering main nozzle.

also permits the operator to vary engine speed to suit conditions or to maintain a uniform speed when the load varies.

On many engines, linkage connects the throttle valve to a governor. The governor, in turn, is connected to a speed control lever. When the speed control lever is set for a given speed, the governor will maintain that speed until the engine reaches its limit of power.

When the load on the engine increases, the governor automatically opens the throttle valve. This permits more air-fuel mixture to enter the engine, providing increased power to maintain a uniform speed. When the load decreases, the governor closes the throttle to reduce engine power. More details on governors is presented later in this chapter.

LOAD ADJUSTMENT

The amount of fuel entering the main discharge nozzle is sometimes regulated by a load adjusting needle, Fig. 4-16. Many carburetors have a fixed jet or orifice which is preset to allow proper fuel for maximum power and economy. These carburetors are nonadjustable types.

ACCELERATION SYSTEM

When the throttle valve is opened quickly for acceleration, a large amount of air is allowed to enter. Unless some method is used to provide additional fuel to maintain a satisfactory air-fuel ratio, the engine will slow down and possibly stop. On larger engines and multi-cylinder engines, a mechanical plunger type pump is connected to the throttle linkage. When the throttle

valve is opened on acceleration, the pump automatically depresses and forces fuel into the carburetor.

ACCELERATION WELL

An acceleration well is a reservoir of fuel. During idling (when load nozzle is inactive), fuel rises inside the nozzle. The fuel flows through holes in the side of the nozzle and into the accelerating well, Fig. 4-17.

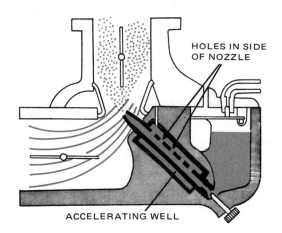

Fig. 4-17. Acceleration well stores fuel for use during rapid acceleration. When fuel has been used from acceleration well, nozzle holes act as air bleeds. (Deere & Co.)

When the throttle valve is opened quickly, the stored fuel rushes through the holes in the nozzle without being metered by the adjusting needle. This fuel combines with the fuel in the nozzle, and the double charge enters the airstream. This provides a much richer air-fuel mixture when there is a sudden need for more power. As the fuel supply decreases in the accelerating well and the holes are uncovered, they become air bleeds for the main nozzle.

ECONOMIZER CIRCUIT

The economizer circuit is designed to retard fuel flow to the engine at part throttle. During part throttle, the full capacity of the main nozzle is not required. The basic process is the same. To reduce capacity, some carburetors of all three types have economizer circuits.

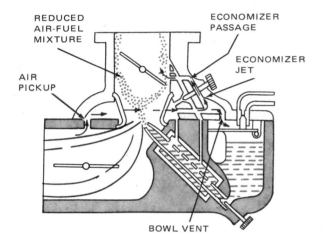

Fig. 4-18. Economizer system creates a reduced pressure in float bowl during part throttle operation, which retards amount of fuel discharged from main nozzle.

The basic "economizing" process is the same for all carburetors. Fig. 4-18 shows an updraft carburetor with the bowl vent passage extended to a point near the throttle valve. When the throttle valve is partially open, the economizer passage is on the engine side of the plate. This permits the engine to draw air through the passage, reducing air pressure in the bowl and cutting down on fuel flow from the nozzle.

IDLING CIRCUIT

During idling operation, the throttle valve is closed. In this condition, the idling system of any type of carburetor supplies just enough air-fuel mixture to keep the engine running. However, actual idling system operation varies in updraft, downdraft and natural draft carburetors.

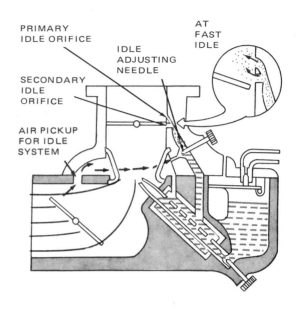

Fig. 4-19. During idling, some incoming air is directed through a passage around venturi. This air mixes with fuel and is drawn out primary and secondary idle orifices. Throttle valve is closed for idle, slightly open for fast idle.

The UPDRAFT CARBURETOR in Fig. 4-19 is in the idling mode of operation. The choke is partially closed, directing air flow through the pickup. Since the throttle valve is closed, the air moves through a passage outside of the venturi to the idle orifice. At this point, the idle adjusting needle regulates the amount of air mixing with the fuel in the idle orifice. Less air provides a richer mixture, more air produces a leaner mixture.

At slow idle, the throttle valve is closed. Only the primary orifice is exposed to allow fuel to enter into the manifold. At fast idle, the throttle valve opens slightly to expose both primary and secondary orifices. Remember, the speed and power of the engine is directly related to the amount of air-fuel mixture allowed to enter the cylinder. Take note that, at idling speed, the main discharge nozzle is inoperative due to lack of air flow through the venturi.

The DOWNDRAFT CARBURETOR in Fig. 4-20 is in the idling mode. The air bleed is located above the venturi and serves both the idling ports and main discharge nozzle. NOTE: Main discharge nozzle is not shown in Fig. 4-20 for purpose of clarity. It would be located as shown in Fig. 4-10. The idle adjustment screw in this carburetor, regulates flow of air-fuel mixture.

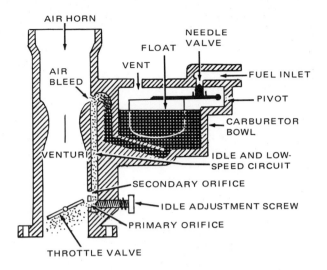

Fig. 4-20. In downdraft carburetor, incoming air enters in air bleed above venturi and travels with fuel to idle orifice. Carburetor is in idling state, since throttle valve has uncovered primary orifice only.   (Deere & Co.)

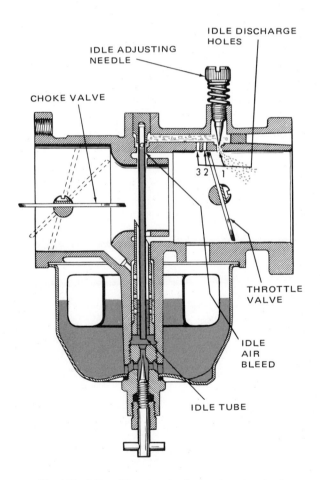

Fig. 4-21. Idling. Throttle valve is closed, and engine is operating from primary idle orifice.
(Zenith Div., Bendix Corp.)

The NATURAL DRAFT CARBURETOR in Fig. 4-21 is in the idling mode. The throttle valve is closed, and the engine is running from the primary idle discharge hole. The choke valve is wide open. The engine is idling.

PART THROTTLE, FULL THROTTLE SEQUENCE

Beyond idling speed, the carburetor has other circuits for part throttle and full throttle operation. In Fig. 4-22, the throttle valve in this natural draft carburetor is partly open. Primary and secondary discharge holes are open, allowing more air-fuel mixture to enter. The engine is running at part throttle.

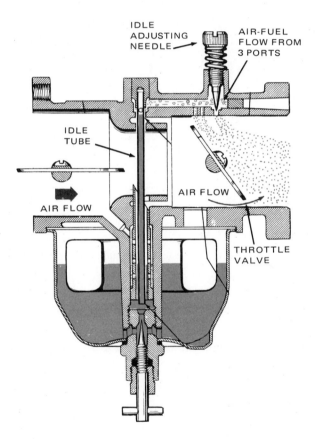

Fig. 4-22. Part throttle. Engine is running from primary and secondary orifices.

Refer to Fig. 4-23 for full throttle mode of operation. The throttle is wide open, and the maximum amount of air is flowing through the venturi. The main discharge nozzle is operating because of high vacuum in the nozzle area. The maximum air-fuel mixture is

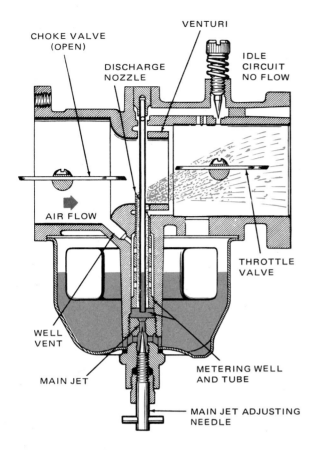

Fig. 4-23. Full throttle. Idle orifices have stopped feeding fuel due to reduced vacuum in that part of carburetor. A full flow of fuel is being drawn from main nozzle.

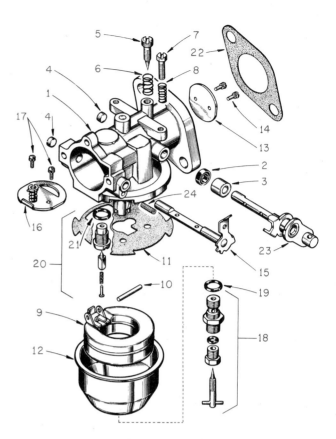

Fig. 4-24. An exploded view of natural draft carburetor: 1—Throttle body. 2—Seal. 3—Retainer. 4—Cup rings. 5—Idle adjustment needle. 6—Spring. 7—Throttle stop screw. 8—Spring. 9—Float and hinge assembly. 10—Float pin. 11—Gasket. 12—Fuel bowl. 13—Throttle valve. 14—Screw. 15—Lever and shaft assemble for choke. 16—Choke valve. 17—Screw. 18—Main jet and adjustment assembly. 19—Washer. 20—Fuel valve and seat assembly. 21—Gasket. 22—Flange gasket. 23—Throttle shaft and lever assembly. 24—Spring.
(Zenith Div., Bendix Corp.)

entering the cylinders, and the engine is developing full speed and power.

Fig. 4-24 is an exploded view of the natural draft carburetor shown in Figs. 4-21 through 4-23.

THE PRIMER

Some carburetors are equipped with primers. The primer is a hand-operated plunger which, when depressed, forces additional fuel through the main nozzle prior to starting a cold engine.

In operation, primer pumps air pressure into float bowl, forcing fuel up the nozzle. A primer mounted on a float type carburetor is shown in Fig. 4-25.

DIAPHRAGM TYPE CARBURETORS

The diaphragm carburetor does not have a float system. Instead, the difference between atmospheric pressure and the vacuum created in the engine pulsates

Fig. 4-25. A primer plunger mounted on the carburetor.
(Deere & Co.)

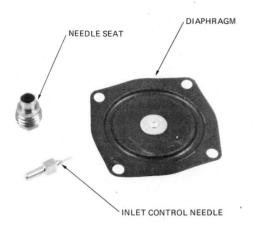

Fig. 4-26. A diaphragm, control needle and needle seat of type used in diaphragm carburetor system. (Deere & Co.)

a flexible diaphragm. A diaphragm, control needle, and needle seat are shown in Fig. 4-26. The diaphragm draws fuel into a chamber of the carburetor from which it is readily drawn into the venturi.

The carburetor shown at A in Fig. 4-27 is a diaphragm type natural draft carburetor. Views B and C illustrate the operating system of the carburetor.

In diagram B in Fig. 4-27, vacuum created in the manifold draws fuel from the upper chamber through the check valve into the venturi. Then, reduced pressure in the upper chamber allows atmospheric pressure to lift the diaphragm, compressing the inlet tension spring. Finally, movement of the diaphragm opens the fuel valve, permitting fuel to flow into the upper chamber. Remember, this action takes place on the intake stroke of the piston.

In diagram C in Fig. 4-27, manifold pressure increases to equal atmospheric pressure when the piston rises on the compression stroke. Since there is no difference in pressure between the upper chamber and the lower chamber, the inlet tension spring closes the fuel valve and returns the diaphragm to a neutral position. The check valve closes immediately when the pressure is equalized. Therefore, the retracting diaphragm draws fuel into the upper chamber before the fuel valve closes completely.

The pulsation of the diaphragm takes place on every intake and compression stroke, regardless of the number of engine cylinders. On four cycle engines, fuel is drawn into the cylinder on the downstroke of the piston. On two cycle engines, the fuel is drawn into the crankcase during the upstroke of the piston.

In some applications, the diaphragm spring is adjustable (adjustment screw not shown) for the purpose of balancing the force of the inlet tension spring.

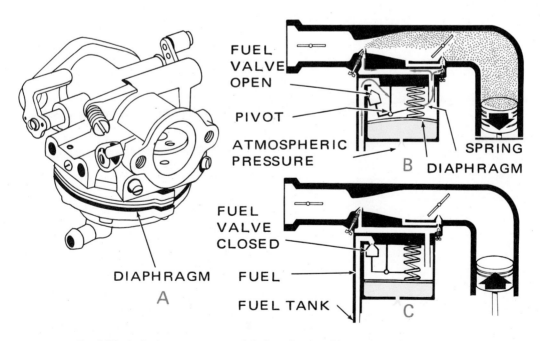

Fig. 4-27. A diaphragm type natural draft carburetor. Diaphragm is lifed by manifold vacuum while fuel is being drawn from jets. When vacuum is reduced, diaphragm returns to normal, drawing new fuel into upper fuel chamber.

STARTING (CHOKE) OPERATION

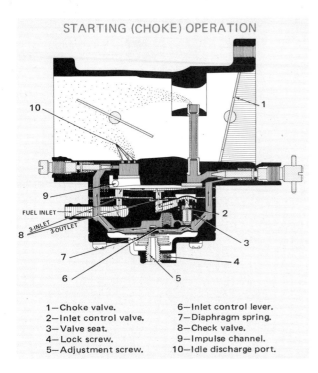

1—Choke valve.
2—Inlet control valve.
3—Valve seat.
4—Lock screw.
5—Adjustment screw.
6—Inlet control lever.
7—Diaphragm spring.
8—Check valve.
9—Impulse channel.
10—Idle discharge port.

Fig. 4-28. With choke plate closed, a very strong vacuum is formed in air horn. A large quantity of gasoline is "sucked" out of idle jets and main nozzle. A rich mixture results which can support cold engine operation.

IDLING OPERATION

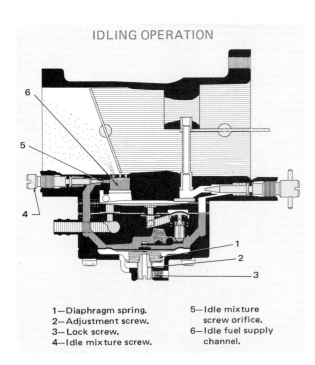

1—Diaphragm spring.
2—Adjustment screw.
3—Lock screw.
4—Idle mixture screw.
5—Idle mixture screw orifice.
6—Idle fuel supply channel.

Fig. 4-29. During idling or very slow speed operation, only the primary orifice (passage) is feeding fuel to engine. Air velocity is not high enough to draw fuel out of the high speed circuit. Remember, this circuit controls fuel mixture when an engine is idling.

INTERMEDIATE OPERATION

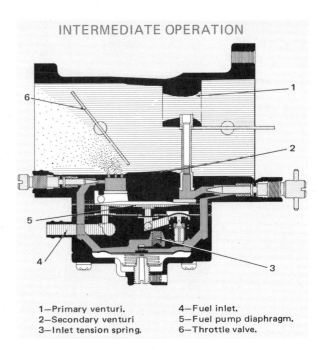

1—Primary venturi.
2—Secondary venturi
3—Inlet tension spring.
4—Fuel inlet.
5—Fuel pump diaphragm.
6—Throttle valve.

Fig. 4-30. Fuel feeding from idling discharge ports provides intermediate speed operation. (Rupp Industries, Inc.)

DIAPHRAGM CARBURETOR OPERATION

A study of the various circuits of a typical diaphragm type carburetor will help clarify operating principles. The carburetor shown in Fig. 4-28 is in starting condition with the choke valve closed. Follow the arrows that indicate direction of fuel flow.

Fuel is drawn from the idle discharge ports and main nozzle because manifold vacuum is high. The carburetor diaphragm is drawn upward during the intake stroke of the engine piston, unseating the fuel inlet needle to allow fuel to flow. Note in Fig. 4-30 that this natural draft carburetor also incorporates a fuel pump diaphragm in its body.

Fig. 4-29 illustrates idling operation with the choke valve open and the throttle valve closed. Vacuum is in effect on the engine side of the throttle valve. Since only one idle discharge port is exposed, a small quantity of fuel is being used, and the engine runs slowly.

Fig. 4-30 shows the throttle partially open for intermediate speed. Air flow through the primary venturi is still not great enough to draw fuel up the main nozzle. Three idle discharge ports are feeding fuel for medium speed. These extra idle discharge ports are termed off-idle ports. They must supply more fuel than the single idle port, yet not as much as the main discharge port. The intermediate circuit must provide fuel for transition from idle to high speed operation.

Fig. 4-31 illustrates high speed operation with maximum air and fuel flowing through the carburetor. All idle ports and the main nozzle are feeding fuel. NOTE: Choke and throttle valves are open.

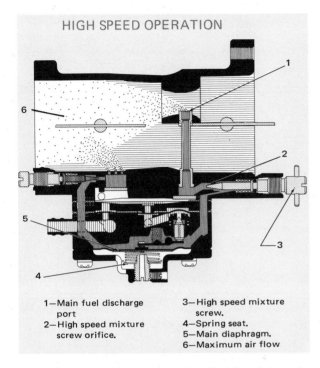

HIGH SPEED OPERATION

1—Main fuel discharge port
2—High speed mixture screw orifice.
3—High speed mixture screw.
4—Spring seat.
5—Main diaphragm.
6—Maximum air flow

Fig. 4-31. During high speed operation, fuel flow from main nozzle and idle jets combines with maximum airflow.

MANUAL THROTTLE CONTROLS

A basic manual throttle control consists of either mechanical linkage or a flexible cable. One end of the control is attached to the throttle shaft lever. The other end is connected to a lever, slide or dial that is operated manually to open or close the throttle valve.

The manual throttle can be used as the sole control for positioning the throttle valve. Typical applications of this type are chain saws, motorcycles, snowmobiles and outboard engines. In other installations, the manual control is used in conjunction with a governor. This setup permits governed speed to be changed when desired.

Fig. 4-32 pictures a manual remote control using a flexible cable to transmit motion from the speed

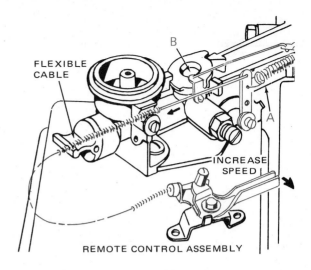

Fig. 4-32. This manual throttle control uses a flexible cable to transmit motion from hand lever to governor spring lever. (Briggs and Stratton Corp.)

control lever to the throttle valve. In this case: A—Motion changes tension on the governor spring. B—Tension on governor spring changes the throttle valve position.

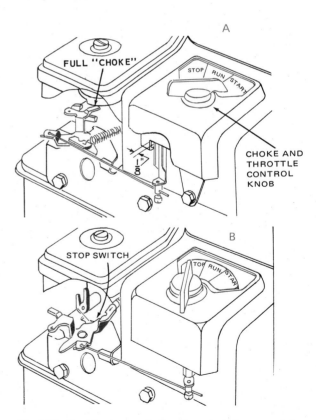

Fig. 4-33. Combined manual throttle and choke control A—Control knob turned to START activates choke. B—Knob turned to STOP closes stop switch.

Fig. 4-33 shows a throttle control that varies governor spring tension, positions the throttle valve and actuates the choke for starting. In diagram A in Fig. 4-33, the control knob is turned to START and rotates the choke valve shaft to the choked position. When the engine is started, the control knob is turned to RUN.

To stop the engine, the knob is turned to STOP, as shown in B in Fig. 4-33. The stop switch grounds the ignition system, cutting off the flow of electricity to the engine. NOTE: This switch is illustrated in detail in the chapter on ignition systems.

GOVERNOR THROTTLE CONTROLS

In many small gasoline engine applications (lawn mowers, generators and garden tractors), the load on

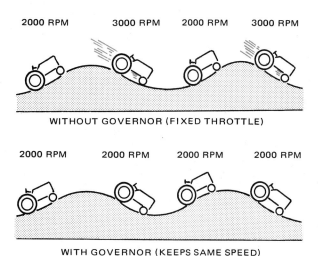

Fig. 4-34. Notice what the governor does for the engine. (Deere & Co.)

the engine can change instantly. The change in load would require constant throttle changes on the part of the operator. Instead, governors are used to provide a smooth, constant speed, regardless of engine loading.

WHAT AN ENGINE GOVERNOR DOES

Governors can be designed to serve three basic functions:
1. Maintain a speed selected by operator which is within range of governor.
2. Prevent overspeeding that may cause engine damage.
3. Limit both high and low speeds.

In Fig. 4-34, observe how tractor speed varies without a governor but stays constant with a governor.

Small engine governors generally are used to maintain a fixed speed not readily adjustable by the operator, or to maintain a speed selected by means of a throttle control lever. In either case, the governor protects against overspeeding. If the load is removed, the governor immediately closes the throttle. If the engine load is increased, the throttle will be opened to prevent engine speed from being reduced.

For example, a lawnmower normally has a governor. When mowing through a large clump of grass, engine load increases suddenly. This tends to reduce engine speed. The governor reacts by opening the carburetor throttle valve. Engine power output increases to maintain cutting blade speed. When the mower is pushed over a sidewalk (no grass or engine load), engine speed tends to go up. The governor reacts by closing the carburetor throttle valve. This limits maximum cutting blade speed. As a result, mower engine and cutting blade speeds stay relatively constant.

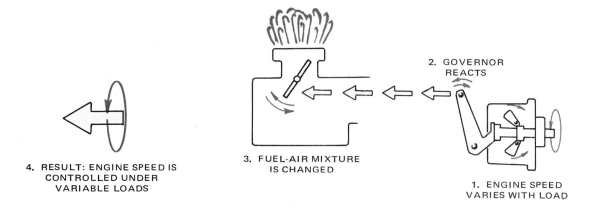

Fig. 4-35. Centrifugal governor controls engine speed by varying fuel mixture. (Deere & Co.)

TYPES OF GOVERNORS

There are several types of engine governors: AIR VANE (also called PNEUMATIC), CENTRIFUGAL, Fig. 4-35, (also called MECHANICAL) and VACUUM. Most modern governors are air vane types which adjust fuel intake according to engine demands, Fig. 4-36. However, centrifugal governors are also fairly common. Vacuum governors are usually found on farm and industrial engines. Basically, all accomplish the same purpose—to protect the engine from overspeeding and to maintain a constant speed, independent of load. However, different speed sensing devices are used.

AIR VANE GOVERNOR

The air vane governor is operated by the stream of air created by the flywheel cooling fins. The force developed by the airstream is in direct proportion to the speed of the engine.

A lightweight, thin strip of metal called an air vane is placed in the direct path of the airstream. It is pivoted on a pin or shaft set near one end. The vane is connected, with linkage, to the throttle shaft lever. When the engine is running, the airstream pivots the vane and attempts to close the throttle valve, Fig. 4-36.

The governor spring is attached to the throttle lever or to linkage from the vane. This spring is designed to pull the throttle valve to wide open position.

Note in Fig. 4-36 how the airflow pivots the vane, causing it to exert rotary pressure on the upper throttle shaft lever while the governor spring tries to pull the throttle valve open. The ratio of pressure developed by the vane, as opposed to the tension of the governor spring, determines throttle valve position. When the engine is stopped, the airstream ceases and the throttle valve is pulled to wide open position by the governor spring.

The governor spring bracket can be moved to vary the amount of tension exerted by the spring. This will alter vane spring pressure balance and establish a new throttle setting, Fig. 4-36.

Governor spring tension is carefully calibrated by the manufacturer. If the spring is stretched or altered in any way, it should be replaced with a new spring designed for that make and model engine. If the

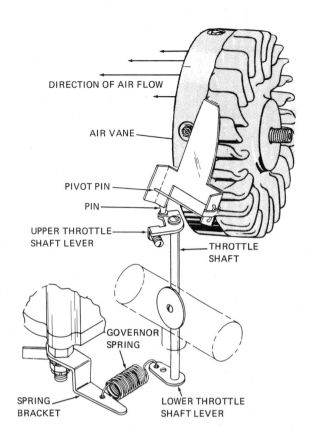

Fig. 4-36. Schematic illustrates operation of an air vane governor. Vane tries to close throttle valve, while governor spring tries to open it. Balance between these two forces determines throttle position. (Tecumseh Products Co.)

linkage is bent, worn, or damaged, it, too, should be straightened, repaired, or replaced. Look at Fig. 4-37.

If it is necessary to replace any governor components on an air vane governor, the top no-load rpm should be checked with an accurate tachometer. The top speed must NOT exceed the amount recommended for the implement being driven.

In the case of lawnmowers, blade tip speed should not exceed 19,000 feet per minute in a no-load condition. If necessary, change the governor spring or adjust the top speed limit device so the engine stops accelerating at the recommended rpm, based on blade length. See Fig. 4-38.

Since blade tip speed is a function of blade length and engine rpm, longer blades require lower engine speeds. It is suggested that top governed engine speed be adjusted at least 200 rpm lower than the speeds shown in Fig. 4-38 to account for tachometer inaccuracy.

Fig. 4-37. Parts of a typical governor system for actual small engine.
Note location of idle speed screw and needle valve screw.

FIXED SPEED

If the engine is designed to run at only one specific rpm, the tension of the governor spring is carefully adjusted until the speed is correct. Then it is left at this

BLADE LENGTH IN INCHES (MILLIMETRES)	MAXIMUM ROTATIONAL RPM
18 (460)	4032
19 (485)	3820
20 (510)	3629
21 (535)	3456
22 (560)	3299
23 (585)	3155
24 (610)	3024
25 (635)	2903
26 (660)	2791

Fig. 4-38. Chart lists various lengths of lawnmower blades and maximum rotational speeds which produce blade tip speeds of 19,000 feet per minute. It is recommended that top speeds be set 200 rpm less than shown.

setting, Fig. 4-39. Fixed speed engines of this type have a limited range of governor spring adjustment. When the engine is started, the force of the airstream on the vane closes the throttle until that force is equal to spring tension.

VARIABLE SPEED

It is often desirable to have an engine operate at many different speeds that may be quickly and easily set by the operator. In this case, a variable speed air vane governor is used, Fig. 4-40. Engine rpm is changed merely by pivoting the governor spring bracket. Remember, the throttle control adjusts governor spring tension. The spring is not connected directly to the throttle lever.

CENTRIFUGAL OR MECHANICAL GOVERNOR

Like the air vane governor, the centrifugal (mechanical) governor also controls engine speed. The

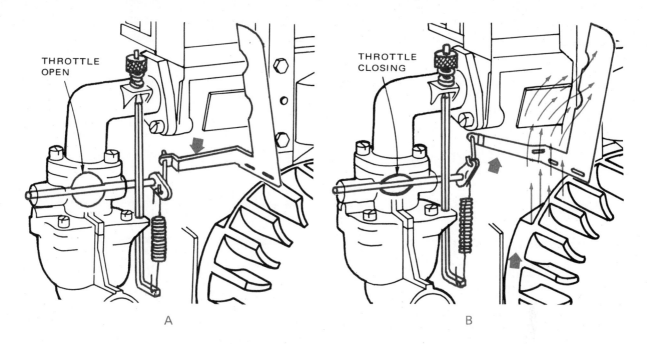

Fig. 4-39. Air vane governor. A—Engine stopped. Spring holds throttle open. B—Engine running. Air pressure pivots vane of fixed speed governor and shuts throttle valve until spring pressure and vane pressure are balanced. Knurled nut alters spring tension and adjusts speed. (Briggs and Stratton Corp.)

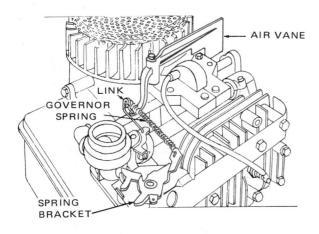

Fig. 4-40. Variable speed air vane governor. To change rpm, operator alters governor spring tension by rotating spring bracket.

centrifugal governor, however, utilizes pivoted fly-weights that are attached to a revolving shaft or gear driven by the engine. With this setup, governor rpm is always directly proportional to engine rpm.

Diagrams A and B in Fig. 4-41 show how centrifugal governor action operates. When the engine is stopped, the heavy ends of the flyweights are held close

to the shaft by the governor spring. The throttle valve is held fully open as illustrated at A in Fig. 4-41.

When the engine is started, the governor is rotated. As its speed increases, centrifugal force goes up and causes the flyweights to pivot outward. This forces the spool upward, raising the governor lever until spring tension equals the centrifugal force on the weight. This action partially closes the throttle valve shown in view B in Fig. 4-41.

If the engine is subjected to a sudden load that reduces rpm, the reduction in speed lessens centrifugal force on the flyweights. The weights move inward, lowering the spool and governor lever. This series of actions opens the throttle valve and lost rpm is regained.

CHANGING GOVERNOR SPEED SETTING

The centrifugal governor speed setting can be changed by turning a knurled adjusting nut on the end of the tension rod, Fig. 4-42. This system is used when the engine is expected to run at a constant speed setting for long periods of time.

A remote hand control cable that alters spring tension also can be used to set governor speed, Fig. 4-43.

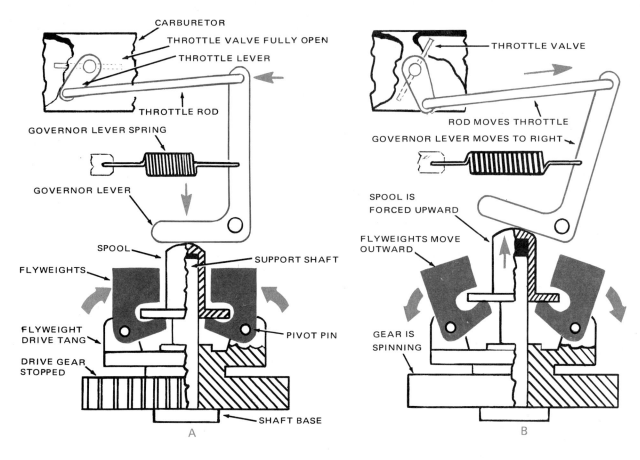

Fig. 4-41. A centrifugal type governor. Centrifugal force causes flyweights to pivot outward, raising spool. Spool rotates governor lever which closes throttle valve. Balance between centrifugal force and governor spring tension determines throttle valve setting.

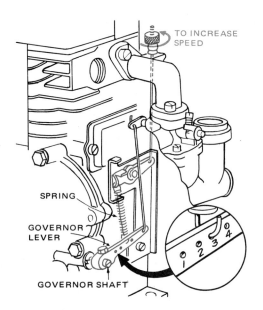

Fig. 4-42. Single speed setting, using knurled nut to provide a limited governor speed range. (Briggs and Stratton Corp.)

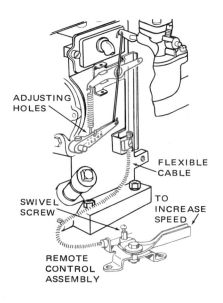

Fig. 4-43. Operator-controlled governor speed setting device allows quick, wide, governed speed changes. (Briggs and Stratton Corp.)

Movement of the control handle increases or decreases spring tension, which speeds up or slows down the engine. The operator can quickly select any speed within the range of the governor.

Another type of governor speed adjusting arrangement is shown in Fig. 4-44. Movement of the governor adjusting lever changes spring tension and engine rpm.

HUNTING OF CENTRIFUGAL GOVERNORS

Frequently, when an engine is first started or is working under load, its speed becomes erratic or oscillates. The engine speeds up rapidly; the governor responds and engine speed drops quickly. The governor stops functioning and engine speed again increases. The governor responds and this action is repeated over and over. This condition is known as HUNTING.

Hunting is usually a result of improper carburetor adjustment. Leaning or richening the fuel mixture can often correct the problem. Also, the governor may cause hunting if it is too stiff or binds at some point. It must work freely.

VACUUM GOVERNORS

Farm and industrial engines are often equipped with a vacuum governor for regulating maximum engine speed. The vacuum governor, Fig. 4-45, is located between the carburetor and the intake manifold. It senses changes in intake manifold pressure (vacuum). There is no other mechanical connection between governor and other parts of the engine.

As engine speed and suction (vacuum) increase, the

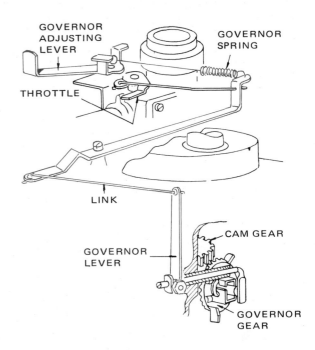

Fig. 4-44. Method of setting governed speed by using an adjusting lever to control governor spring tension.

governor unit closes the throttle butterfly valve. This causes a decrease in fuel flow and engine speed.

When engine speed and vacuum increase, the spring opens the throttle valve. Fuel flow and engine speed increases. An adjustment of spring tension is used to set the desired speed range.

CARBURETOR ADJUSTMENTS

1. Turn idle speed screw in until it touches stop. Then turn it several (about three) more times, Fig. 4-37.

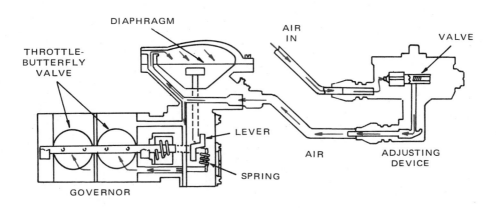

Fig. 4-45. A vacuum governor must maintain a preset maximum engine speed, independent of engine load.

2. Turn carburetor needle valve screw all the way in. Seat it lightly. Back the needle valve screw out 1-1/2 turns. Close choke.

3. Start engine. If engine will not start, back out needle valve screw by 1/2 turn increments until engine does start.

4. When engine starts, open choke and let engine warm up. Adjust idle speed screw so that engine is running slowly, but is not lugging.

5. Slowly close needle valve until engine begins to slow. Open needle valve until engine resumes speed.

6. Readjust idle speed screw to between 800 and 1000 rpm.

7. Open throttle and check maximum rpm with a reliable tachometer.

8. Adjust the maximum speed in accordance with the type of governor on the engine. You may need to refer to the owner's manual or service manual for the proper method.

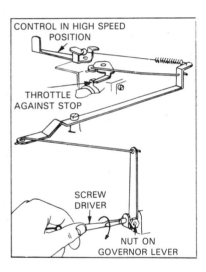

Fig. 4-46. A vertical shaft engine governor. Throttle in high speed position for making governor lever adjustments.

GOVERNOR ADJUSTMENTS

Fig. 4-36 shows a governor which requires moving the spring bracket to increase or decrease governor spring tension. Loosen the bracket nut to move the bracket; then retighten nut.

Fig. 4-42 shows a knurled thumbscrew used to adjust spring tension, thus changing rpm. The governor spring should be in hole number three for standard operation. In general, the closer the spring is to the pivot end of lever,the smaller the difference between load and no-load engine speed.

If the spring is brought too close to the pivot point, the engine will begin to "hunt" (engine speed increase and decrease). The further from the pivot end, the less tendency to "hunt" but there will be a greater speed drop under load. If the governed speed is lowered, the spring can usually be moved closer to the pivot.

If the governor shaft has been removed or loosened, the proper adjustment is:
1. Loosen screw holding governor lever.
2. Place throttle in high speed position and hold.
3. With screwdriver, turn governor shaft COUNTER-CLOCKWISE as far as it will go.
4. Tighten screw holding governor lever to governor shaft to specifications (typically 35-45 inch pounds or 4.0-5.0 N·m).
5. Before starting engine, move linkage manually to check for binding.

Fig. 4-43 utilizes a remote control lever assembly to adjust spring tension in place of knurled thumb screw. Governor shaft adjustment same as B.

Fig. 4-44 is a governor used on a vertical shaft engine. It has a manually operated governor adjusting lever. If the governor lever has been removed or loosened, the proper adjustment is:
1. Loosen screw holding governor lever to governor shaft, Fig. 4-46.
2. Place throttle in high speed position and hold.
3. With a screwdriver, turn governor shaft CLOCK-WISE as far as it will go.
4. Tighten screw holding governor lever to governor

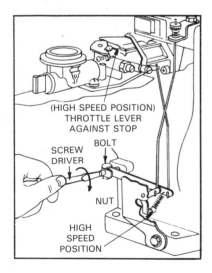

Fig. 4-47. A horizontal shaft engine governor.

shaft to specifications (about 35-45 inch pounds or 4.0-5.0 N·m).

5. Check linkage for freedom of movement.

Fig. 4-47 is a governor used on a horizontal shaft engine. Adjustment of the governor lever is the same as Fig. 4-46.

Some governors require bending the spring attach arm, as shown in Fig. 4-48.

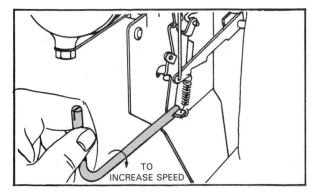

Fig. 4-48. Bending governor spring attach arm to increase spring tension and engine speed.

GOVERNOR MAINTENANCE

Two major requirements for a trouble-free governor are a vibration free drive mechanism and cleanliness of all governor parts.

Governors are very sensitive devices requiring a steady drive without vibration. Many governors are driven through a spring or rubber coupling to dampen vibrations and roughness. This coupling should always be inspected periodically or when problems in governor operation are encountered.

Cleanliness of governor parts and linkage reduces friction, the possibility of hunting as well as the work load on the governor. Keep everything clean.

GOVERNOR FEATURES

The operating principles of the governor mechanism are quite simple and reliable. Governors provide accuracy and efficiency of operation combined with convenience and comfort for the operator. Some of the operating features of governing engine speed and power output are as follows:

STABILITY: Ability to maintain a desired engine speed without fluctuating. Instability results in HUNTING or oscillating due to over-correction. Excessive stability results in a DEAD-BEAT governor, one that does not correct sufficiently for load changes.

SENSITIVITY: The percent of speed change required to produce a corrective movement of the fuel control mechanism. High governor sensitivity would help keep the engine operating at a constant speed.

AIR CLEANERS, AIR FILTERS

An engine breathes a tremendous quantity of air during its normal service life. If the incoming air is not thoroughly cleaned by passing it through a filtering device, dirt and grit entering the cylinder would cause rapid wear and scoring of machined parts throughout the engine. Engine life, under severe dust conditions, actually could be reduced to minutes.

Three types of air cleaners widely used in small gasoline engines are the OIL BATH, OIL WETTED and DRY TYPES.

OIL BATH AIR CLEANER

Oil bath air cleaners usually consist of two housings,

Fig. 4-49. Oil bath air cleaner. Oil sump traps heavier particles of dirt as air flow changes direction. Much of remaining contaminants are trapped as air passes through filter element. (Kohler Co.)

Fig. 4-49. The upper housing has a built-in filtering element. The lower housing is constructed with an oil reservoir at the bottom.

Oil bath air cleaners, Fig. 4-50, draw air down between the inner and outer housings. The air is moving at high speed and must suddenly reverse its direction when it strikes the bottom, which is covered with a pool of engine oil. When the air reverses direction, airborne dust and grit (heavier than air) is thrown into the oil, where it sinks to the bottom.

A small amount of oil is drawn into the filter element by the airstream. Dirt small enough to make the high speed change of direction is trapped by the oil-

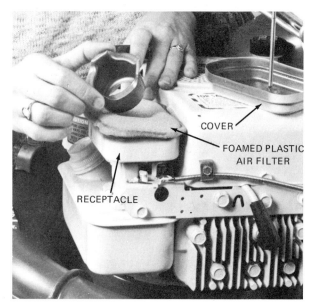

Fig. 4-51. An oil wetted air cleaner. Polyurethane foam is dampened with oil and contained in a vented case attached to carburetor. Filter can be cleaned and re-oiled.

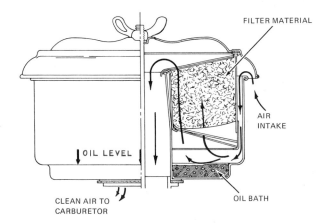

Fig. 4-50. Oil bath air cleaner cut away. Note oil level. If it is too low, poor filtering results. Too high, engine draws excess oil through carburetor, causing carbon deposits in engine. (Wisconsin Motors Corp.)

moistened filter. When the engine is stopped, the oil in the filter element drains back into the oil sump, carrying much of the dirt with it.

The construction of a typical oil bath air cleaner is shown in Fig. 4-50. Note how incoming air changes direction above the oil sump, then passes up through the filter pack and down into the carburetor. The filter material must be rinsed and dried, and the oil in the sump must be changed periodically.

OIL WETTED AIR CLEANER

The oil wetted air cleaner utilizes a filtering element (crushed aluminum, polyurethane foam, etc.) dampened with engine oil. Fig. 4-51 shows a polyurethane foam oil wetted element, reservoir and cover. In operation, the air is drawn directly through the oil wetted ele-

ment where the damp material effectively filters out contaminants. This type element can be re-used by rinsing in cleaning solvent, drying and then re-oiling.

DRY TYPE AIR CLEANER

Dry type air cleaners pass the airstream through treated paper, felt, fiber or flocked screen. Some filter elements (flocked screen) can be cleaned, but most are designed to be thrown away when they become dirty.

A typical treated paper air cleaner element is shown in Fig. 4-52. You can clean this filter by tapping it on a flat surface to dislodge light accumulations of dirt. However, when it will not tap clean, a treated paper filter must be replaced with a new element designed for the given engine application.

REVIEW QUESTIONS—CHAPTER 4

1. Give five different engine running conditions that must be met by the carburetor.
2. Normal air-fuel mixture by weight is:
   a. 12 to 1.
   b. 13 to 1.
   c. 14 to 1.
   d. 15 to 1.
3. If the barometric pressure on a standard day at 1500 ft. MSL was 29.95, then at 3500 ft. MSL the

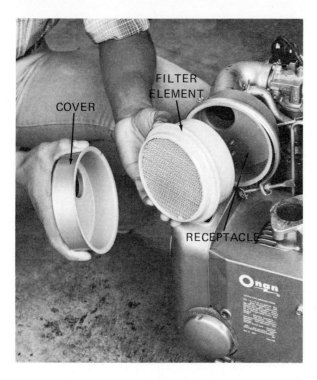

Fig. 4-52. Dry type filter element can be partially cleaned by tapping gently. New element should be installed as needed.

barometric pressure would be: (Refer to Fig. 4-3.)
a. 30.95 in. Hg.
b. 31.95 in. Hg.
c. 29.95 in. Hg.
d. 27.95 in. Hg.

4. In a venturi:
a. Air pressure is greatest where velocity is greatest.
b. Air pressure is least where velocity is greatest.
c. Velocity is least where air pressure is least.
d. The volume of air entering is slightly greater than the air leaving due to the restriction to flow.

5. Name the three basic types of carburetors. (Consider direction of air flow.)

6. Which type of carburetor would normally require a smaller air passage than the other two types?

7. Needle valve points in the carburetor float chamber are usually made from one of two materials. The two materials are _____ and _____.

8. The choke valve in the carburetor is always located:
a. Nearest to the intake end of the carburetor.
b. Nearest the manifold end of the carburetor.
c. In the center of the carburetor.
d. Above the float chamber level.

9. The richest air-fuel mixture takes place during:
a. Full throttle.
b. Half throttle.
c. Idle.
d. Starting.

10. On some carburetors, the amount of fuel entering the main discharge nozzle is regulated by:
a. The float level.
b. A load adjusting needle.
c. The idle adjustment needle.
d. A spray bar needle valve.

11. The acceleration well fills when the engine:
a. Is running at steady high speed.
b. Is running at half throttle.
c. Is under heavy load.
d. Is idling.

12. The primary purpose of air bleeds is to:
a. Increase the air-fuel ratio.
b. Improve atomization of the fuel.
c. Remove air bubbles that may be mixed with the fuel.
d. Prevent vapor lock.

13. During idle and fast idle conditions, the main discharge nozzle is:
a. Discharging a small amount of fuel.
b. Inoperative.
c. Acting as an air bleed.
d. Providing most of the fuel.

14. The carburetor economizer:
a. Reduces float bowl pressure.
b. Reduces the amount of fuel discharged into the venturi.
c. Operates only after the engine reaches part throttle.
d. All of the above are true.

15. One main advantage of a diaphragm carburetor as compared to a float type carburetor is its ability to _____.

16. The diaphragm in the carburetor:
a. Forces fuel through the main discharge nozzle.
b. Operates only at high speed.
c. Draws fuel under spring pressure.
d. Draws fuel during a vacuum pulse from the manifold or crankcase.

17. Governors serve three basic functions. What are they?

18. Name two basic types of small engine governors.

19. On a governor installation, the governor spring is attached to the throttle lever. The governor spring is intended to:
a. Have no effect on the throttle valve.
b. Close the throttle valve.

c. Open the throttle valve.

d. Return the throttle lever to the off position.

20. Give three types of air cleaners.

SUGGESTED ACTIVITIES

1. Make a venturi tube. Provide a connection so that air can be forced through the venturi. Install one pressure gage before the restriction and one gage in the restriction. Demonstrate what happens when the air is applied to the venturi.

2. Working with the same venturi used in activity number one, remove the pressure gage in the restriction. Place a pickup tube in the restriction and draw water out of a beaker. Demonstrate the atomization of the water particles.

3. Make a cutaway of a float type carburetor so that the float, needle, throttle valve, choke and internal passages can be seen.

4. Make a working mock-up model of a variable speed centrifugal governor system that demonstrates governor principles.

5. Make a large cross section of a float type carburetor mounted on a board. Make the choke, throttle valve, float and needle movable from the back. Paint the various parts, passages and ports with bright colors. Give a demonstration to the class of choked, idle, part throttle and full throttle carburetor functions.

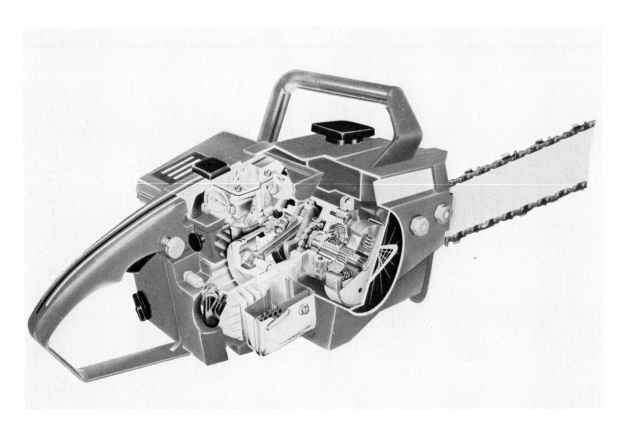

This is a phantom view of a modern two-stroke cycle chain saw engine. Study part construction and locations.

Cutaway of a four-stroke cycle, horizontal shaft engine. Can you find the valves, carburetor, air cleaner, and ignition coil? (Kohler)

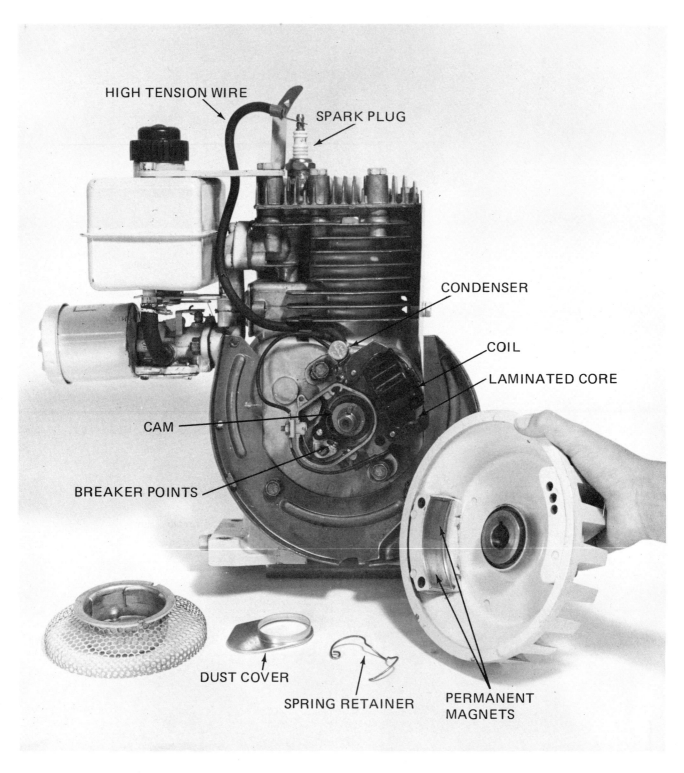

HIGH TENSION WIRE

SPARK PLUG

CONDENSER

COIL

LAMINATED CORE

CAM

BREAKER POINTS

DUST COVER

SPRING RETAINER

PERMANENT MAGNETS

Fig. 5-1. Major parts of small engine magneto are breaker points, condenser, coil, flywheel magnets and spark plug.

Chapter 5

IGNITION SYSTEMS

Fig. 5-2. Ignition system of small engine works hard to produce enough voltage to force electrons to jump spark plug gap.

The primary purpose of the ignition system, Fig. 5-1, of a small gasoline engine is to provide sufficient electrical voltage to discharge a spark between the electrodes of the spark plug, Fig. 5-2. Secondly, the spark must occur at exactly the right time to ignite the highly compressed air-fuel mixture in the combustion chamber of the engine.

The ignition system must be capable of producing as much as 20,000 volts to force the electrical current (electrons) across the spark plug gap. Then the intense heat created by the electrons jumping the gap ignites the air-fuel mixture surrounding the electrodes.

The rate, or times per minute, at which the spark must be delivered is very high. For example, a single cylinder, four cycle engine operating at 3600 rpm re-

quires 1800 ignition sparks per minute. A two cycle engine running at the same speed requires 3600 sparks per minute. In multi-cylinder engines, the number of sparks per minute is multiplied by the number of cylinders.

Every spark must take place when the piston is at exactly the right place in the cylinder and during the correct stroke of the power cycle (see Chapter 2). Considering the high voltage required, the precise degree of timing and the high rate of discharges, the ignition system has a remarkable job to do.

Most small gasoline engines use magnetos, Fig. 5-1, to supply ignition spark. *Magnetos are self-continued units that produce electrical current for ignition without any outside primary source of electricity.* It serves as a simple and reliable ignition system. Basic parts of the magneto are:
1. Permanent magnets.
2. High tension coil with laminated iron core and primary and secondary windings.
3. Breaker points and breaker cam.
4. Condenser.
5. High tension spark plug wire.
6. Spark plug, Figs. 5-1 and 5-2.

Today there are several basic kinds of magnetos in use on small engines. The one described in the preceding paragraph is referred to as a "mechanical breaker ignition" (MBI) system. This type has been used exclusively until the recent development of the "transistor controlled ignition" (TCI) system, and the "capacitor discharge ignition" (CDI) system. These are sometimes referred to as "solid state" ignition systems. They will be described in detail later in this chapter.

To make it easier to understand how the various magneto parts function, a review of some basis electrical principles follows.

THE ELECTRON THEORY

All matter is composed of atoms. An atom is extremely small, so small that it cannot be seen with the most powerful microscope. It is the smallest particle of an element that can exist, alone or in combination, yet it consists of electrons, protons and neutrons.

Atoms can be broken down into types, determined by the number and arrangement of the electrons, protons and neutrons. A few types of atoms are hydrogen, oxygen, carbon, iron, copper and lead. There are many others, about 100 in all.

The structure of the atom determines the weight, color, density and other properties of an element. The electrons, though varying in number, are identical in all of the elements. An electron from silver would be the same as an electron from copper, tin, or any other substance.

Electrons travel in orbits around the center of the atom. They are very light and their number per atom varies from one element to another. Electrons have negative ($-$) electrical charges, as shown in Fig. 5-3.

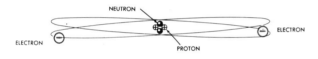

Fig. 5-3. All atoms consist of electrons, neutrons and protons. Neutrons and protons form the nucleus. Neutrons have no electrical charge, but each proton carries a positive ($+$) charge. Electrons orbit the nucleus and carry negative ($-$) charges.

Protons are large, heavy particles when compared with the electrons. One or more protons help form the nucleus (center) of the atom and are positively ($+$) charged.

A neutron is made up of an electron and proton bound tightly together. Neutrons are electrically neutral and are also located in the nucleus of the atom, Fig. 5-3. The number of electrons is equal to the number of protons in any atom. Normally, then, atoms are electrically neutral because the negative electrons cancel the positive force of the protons.

Actually, an atom is held together because unlike electrical charges attract each other. The positive

charged protons hold the negative charged electrons in their orbits. Since like electrical charges repel each other, the negative electrons will not collide as they spin.

The ease with which an electron from one atom can move to another atom determines whether a material is an electrical conductor or nonconductor. In order to have electric current, electrons must move from atom to atom. Materials allowing electrons to move in this way are called conductors. Examples are copper, aluminum and silver.

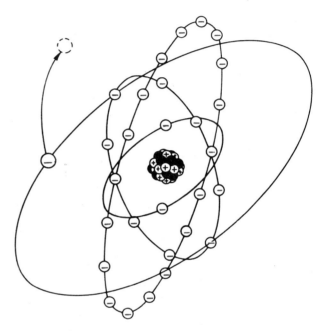

Fig. 5-4. Conductors are materials having electrons that can easily leave orbit or one atom and move to orbit of another atom. When many electrons do this, electricity is produced.

In nonconductors, it is difficult, if not impossible, for electrons to leave their orbits. Nonconductors are called insulators. Some examples are glass, mica, rubber, plastic and paper.

Electron flow in a conductor follows a path similar to that shown in Fig. 5-4. The flow of electrons will take place only when there is a complete circuit and a difference in electrical potential. A difference in potential exists when the source of electricity lacks electrons, or is positively ($+$) charged. Since electrons are negatively ($-$) charged, and unlike charges attract, the electrons move toward the positive source, Fig. 5-5.

An electrical potential is produced in three ways:
1. Mechanically.
2. Chemically.
3. Statically.

The electrical generator is a mechanical producer of electricity and can be run by water power, steam turbines or internal combustion engines. A magneto is a type of generator. Mechanical energy from the crankshaft is used to rotate the permanent magnet.

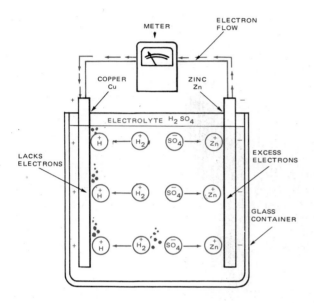

Fig. 5-5. A difference in potential exists if source of electricity lacks electrons and, therefore, is positively (+) charged. Electrons, being negatively (−) charged, are attracted to positive source.

Electricity used in homes and factories is produced mechanically. Batteries are chemical producers of electricity. Lightning is a result of static electricity.

ELECTRICAL UNITS OF MEASUREMENT

The three basic units of electrical measurement are:
1. Amperes — Rate of electron flow.
2. Volts — Force that causes electrons to flow.
3. Ohms — Resistance to electron flow.

An ampere is a measurement of the number of electrons flowing past any given point in a specific length of time. One ampere of current is equal to 6,240,000,000,000,000,000 (6.24×10^{18}) electrons per second. Since electricity generally is transmitted through wires, the greater the number of electrons flowing, the larger the wire size must be.

The difference in electrical potential between two points in a circuit is measured in volts. Voltage is the force, or potential, that causes the electrons to flow.

Resistance to electron flow is measured in ohms. Some materials produce a strong resistance to electron flow, others produce little resistance. If a wire is too small for the amount of current produced at the source, the wire will create excessive resistance and it will then get hot.

The air gap between spark plug electrodes is highly resistant to electron flow, creating the need for high voltage to cause the electrons to jump the gap. The high resistance also creates heat which, in this case, ignites the fuel in the cylinder.

OHM'S LAW

Every electrical circuit operates with an exact relationship of volts, amps and ohms. It is possible to work out their mathematical relationship through application of Ohm's Law.

Briefly, the formula is $I = \dfrac{E}{R}$ where:

I = amperes
E = volts
R = ohms

If circuit voltage is 12 and resistance is 8 ohms, the current would be:

$$I = \frac{E}{R} \qquad I = \frac{12}{8} \qquad I = 1.5 \text{ amperes}$$

If amperage is 15 and voltage is 6, resistance would be:

$$R = \frac{E}{I} \qquad R = \frac{6}{15} \qquad R = .4 \text{ ohms}$$

If amperage is 3, resistance 10 ohms, the voltage would be:

$$E = I \times R \qquad E = 3 \times 10 \qquad E = 30 \text{ volts}$$

MAGNETISM

The molecular theory of magnetism is the one most widely accepted by scientists. Molecules are the smallest particles of matter which are recognizable as being that matter. For example, a molecule of aluminum oxide will contain atoms of aluminum and atoms of oxygen.

In most materials, the magnetic poles of adjoining molecules are arranged in a random pattern, so there is no magnetic force, Fig. 5-6. Iron, nickel and cobalt molecules, however, are able to align themselves so that all their north poles point in one direction and south poles in the opposite direction as in Fig. 5-7. The individual magnetic forces of each molecule combine to produce one strong magnetic force. In magnets, opposite poles attract each other and like poles repel each other, much in the same way that like and unlike electrical charges react.

Certain materials have good magnetic retention. That is, they retain their molecular alignment. These materials are suitable as permanent magnets. Some other materials maintain their molecular alignment only when they are located within a magnetic field. When the field is removed, the molecules disarrange themselves into random patterns and the magnetism is lost. When permanent magnets are cut into pieces, each piece takes on the polarity of the parent magnet, as illustrated in Fig. 5-8.

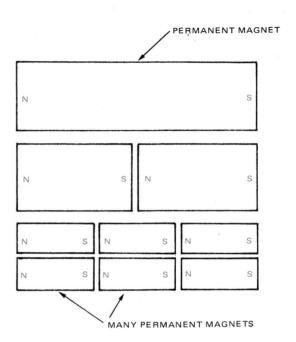

Fig. 5-8. If a permanent bar magnet is broken into sub parts, each sub part has a north and south pole like parent magnet. If parts could be further broken into individual molecules, each molecule would be an individual magnet.

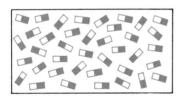

UNMAGNETIZED IRON

Fig. 5-6. An unmagnetized substance is made up of molecules whose poles are not aligned. Molecules have north and south poles, like bar magnets.

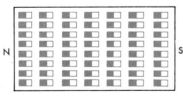

MAGNETIZED IRON

Fig. 5-7. A magnetized substance has all molecules in alignment, north to south. Individual molecules combine magnetic forces to produce a strong overall magnetic force.

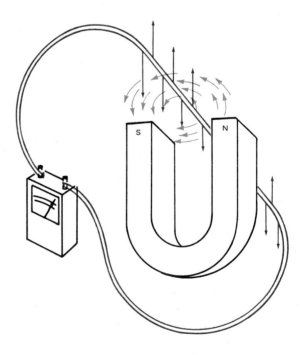

Fig. 5-9. If a conductor, such as copper wire, is moved so that it cuts magnetic lines of force, an electron flow is induced in conductor. Flow of electrons (electricity) can be measured with a sensitive meter.

MAGNETS AND ELECTRICITY

The fact that there is a close relationship between electricity and magnetism serves as the basis for making a workable magneto. Over 100 years ago, Michael Faraday discovered that electricity could be produced from magnetism. One of his experiments showed that if a wire is moved past a magnet, the magnetic field is cut by the wire and current will flow. See Fig. 5-9. When movement of the wire is stopped, the current also stops, Fig. 5-10. Therefore, electricity will flow when the magnetic lines of force are being cut by the wire.

An important principle used in magneto construction is that a magnetic field is developed when electrons flow through a coil of wire. Fig. 5-11 shows a simple coil of wire with current flowing through it. In this illustration, the magnetic field is indicated by lines around the coil.

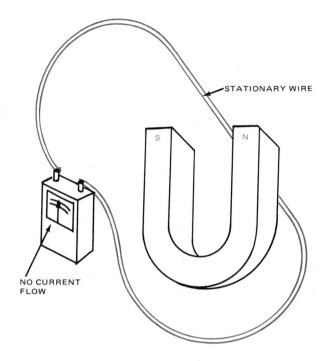

Fig. 5-10. A conductor that is not moving and not cutting magnetic lines of force will not induce electrical current.

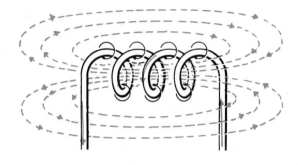

Fig. 5-11. A coil of wire with current flowing through it will produce a magnetic field around itself and around each turn of wire in coil.

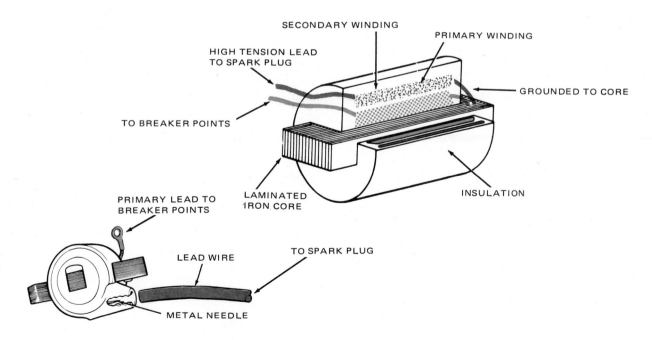

Fig. 5-12. Ignition coil consists of two windings, one inside the other. Coil functions as a step-up transformer to produce high voltage and low amperage from low voltage and high amperage.

The lines of force in Fig. 5-11 come from the north pole and return to the south pole. If the direction of current is changed, the polarity also changes. In magneto operation, an electric current is passed through a coil which develops a magnetic field.

IGNITION COIL

The ignition coil used in a magneto operates like a transformer. The coil contains two separate windings of wire insulated from each other and wound around a common laminated iron core. See Fig. 5-12. The primary winding is heavy gage wire with fewer turns than the secondary winding, which has many turns of light gage wire.

When electrical current is passed through the primary winding, a magnetic field is created around the iron core. When the current is stopped, the magnetic field collapses rapidly, cutting through the secondary windings. This rapid cutting of the field by the wire in the coil induces high voltage in the secondary circuit. The high secondary voltage, in turn causes a spark to jump the spark plug gap and ignite the air-fuel mixture.

SPARK PLUGS

At first glance, an assortment of spark plugs may look very much alike. Actually, there are many variations. Using the correct spark plug for a given engine application can greatly increase the efficiency, economy and service life of the engine.

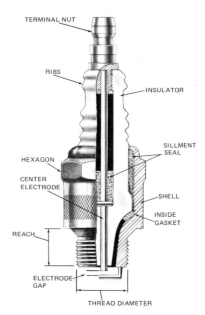

Fig. 5-13. Spark plug carries high voltage current produced by ignition sytem. It also must withstand the high temperatures and shock of combustion, insulate center electrode against current loss, and seal against compression leakage.
(Deere & Co.)

Fig. 5-13 shows major parts of a typical spark plug. The terminal nut is the external contact with the high tension coil. Some terminal nuts are removable, others are not.

Two common methods of high tension lead connections are shown in Fig. 5-14. Application A uses the exposed clip, which is satisfactory in uses where

Fig. 5-14. Two common high tension lead connectors. A—Exposed clip type. B—Neoprene boot type. Exposed clip connector is used in conjunction with a metal strip stop switch.

moisture, oil or dirt will not get on the plug or can easily be wiped off. The boot type shown at B provides better plug protection.

The spark plug insulator usually is an aluminum oxide ceramic material, which has excellent insulating properties. The insulator must have high mechanical strength, good heat conducting quality and resistance to heat shock. Generally, ribs on the insulator extend from the terminal nut to the shell of the plug to prevent "flashover." Flashover is the tendency for current to travel down the outside of the spark plug instead of through the center electrode. See Fig. 5-15.

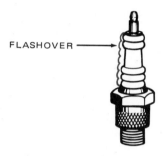

Fig. 5-15. Flashover is caused by moisture or dirt, or by a worn out terminal boot, which allows voltage to short across outside of ceramic insulator.

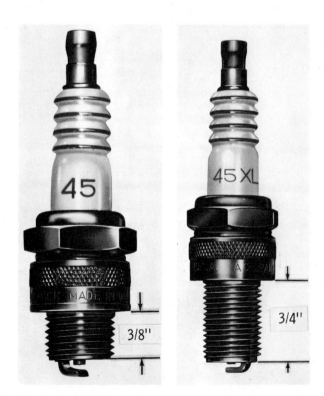

Fig. 5-16. Spark plug reach, (length of thread) can vary considerably from one plug to another. Too long a reach can damage a piston. Too short a reach provides poor combustion. (AC Spark Plug Div., GMC)

The center electrode carries the high voltage current to the spark gap. If the electrical potential is great enough to cause the current to jump the plug gap, the grounded electrode will complete the circuit to ground. NOTE: Always refer to manufacturer's specifications for correct electrode gap, Fig. 5-16A.

Fig. 5-16A. Mechanic is setting correct electrode gap from manufacturer's specifications. (Champion Spark Plugs)

The sillment seal is a compacted powder that helps insure permanent assembly and eliminates compression leakage under all operating conditions. The inside gasket also acts as a seal between the insulator and the steel shell.

Spark plug "reach" varies with type of spark plug. Some are long, others quite short, Fig. 5-16. Never use a spark plug that has a longer reach than specified, Fig. 5-17. Serious engine damage can result if the piston hits the plug.

Threads on the spark plugs are metric sizes, usually 14 mm. Several standard sizes are in common use.

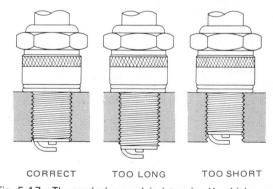

CORRECT   TOO LONG   TOO SHORT

Fig. 5-17. The spark plug reach is determined by thickness of cylinder head.

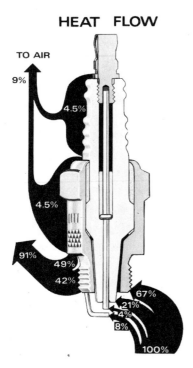

HEAT FLOW

TO AIR

9%

4.5%

4.5%

91%

49%

42%

67%

21%

4%

8%

100%

Fig. 5-18. Heat of ignition and combustion must be conducted away from critical parts of spark plug to prevent preignition and burning of electrodes.

Fig. 5-20. Operating temperature of spark plug can be studied with a special spark plug having a thermocouple (heat sensor) installed in it. (Champion Spark Plug Co.)

SPARK PLUG HEAT TRANSFER

Heat transfer in spark plugs is an important consideration. The heat of combustion is conducted through the plug as shown in Fig. 5-18. Spark plugs are manufactured in various heat ranges from "HOT" to "COLD," Fig. 5-19. Cold running spark plugs are those which transfer heat readily from the firing end. They are used to avoid overheating in engines having high combustion temperatures.

In figuring spark plug heat range, the length of the insulator nose determines how well and how far the heat travels. The spark plug at A in Fig. 5-19, for example, is a hot plug because the heat must travel a greater distance to the cylinder head. Spark plug at D in Fig. 5-19 is comparatively colder than A. A cold plug

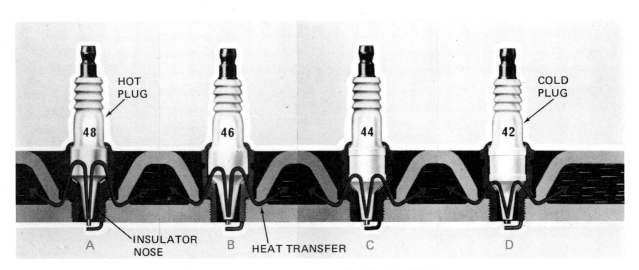

Fig. 5-19. Spark plug heat transfer determines whether plug is "hot" or "cold." Heat is controlled by insulator nose. (AC Spark Plug Div., GMC)

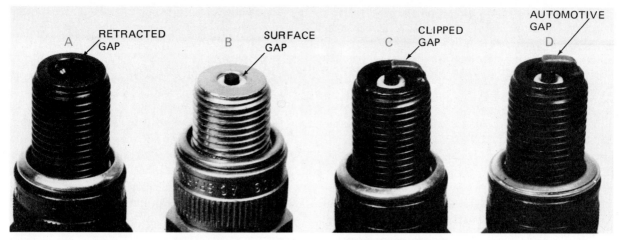

Fig. 5-21. Many electrode designs are available for engine use. Surface gap spark plug is extremely ''cold'' and is finding application with capacitor discharge ignition systems. (AC Spark Plug Div., GMC)

installed in a cool running engine would tend to foul. Cool running usually occurs at low power levels, continuous idling or in start/stop operation.

The tip of the insulator is the hottest part of the spark plug and its temperature can be related to preignition (firing of fuel charge prior to normal ignition) or plug fouling. Experiments show that if combustion chamber temperature exceeds 1750 deg. F (954 C) in a four cycle engine, preignition is likely to occur. If insulator tip temperature drops below 700 deg. F (371 C), fouling or shorting of the plug due to carbon is likely to occur.

MEASURING SPARK PLUG TEMPERATURE

Specially assembled thermocouple spark plugs, Fig. 5-20, are used to accurately determine spark plug temperatures during actual engine operation. These plugs have a small temperature sensing element imbedded in the insulator tip and serve as a valuable aid to engineers in gathering spark plug and combustion data.

TYPES OF ELECTRODES

Electrode configurations vary considerably. In Fig. 5-21, plug A is a retracted gap type used in some engines where clearance is a problem or protection of the firing tip is desirable. Plug B is a surface gap type, which is extremely cold. Surface gap spark plugs are sometimes used in engines with capacitive discharge ignition systems. Plug C is a clipped gap type, in which the side electrode extends only part way across the center electrode. Plug D is a standard gap automotive spark plug.

BREAKER POINTS AND CONDENSER

The breaker points generally consist of two tungsten contacts. One contact point is stationary, the other is movable. Each contact is fastened to a bracket.

Tungsten is a hard metal with a high melting temperature. These characteristics are needed to withstand the continual opening and closing that takes place and the eroding effect of the arc that occurs when the points "break" (start to open). Fig. 5-22 shows the breaker point assembly with the dust cover removed.

Fig. 5-22. Magneto breaker point assembly. A—Dust cover. B—Stationary point. C—Movable point. D—Pivot pin. E—Gap adjustment screw. F—Wear block. G—Breaker point spring. H—Felt lubricator. K—Cam block. L—Point housing. M—Advance adjustment screw.

The breaker point assembly is an electrical switch. When the points are closed, the magnetic field created by the flywheel magnets is being cut by the primary coil winding. This induces current in the primary circuit, and the primary winding builds its own magnetic field around the coil.

When the breaker opens, current flow in the primary circuit stops and the magnetic field collapses. Immediately upon collapse, a surge of high voltage induced in the secondary winding of the coil forces the electrons to jump the spark plug gap. This process is repeated every time the breaker points close, then open, and the spark plug sparks only at the instant the breaker points open.

CONDENSER CONSTRUCTION AND OPERATION

The condenser plays an important part in ignition system operation. Its primary purpose is to prevent current from arcing across the breaker point gap as the points open. If arcing would occur, it would burn the points and absorb most of the magnetic energy stored in the ignition coil. Not enough energy would be left in the coil to produce the necessary high voltage surge in the secondary.

The condenser absorbs some current during the first instant breaker points begin to separate. Since the condenser absorbs some current, little current is left to form an arc between the points. A condenser must be selected that has just the right capacitance to absorb the amount of energy required to produce an arc.

The construction of the condenser is quite simple. Fig. 5-23 is a condenser partially opened up to show laminations of aluminum foil and an insulating strip. Two strips of aluminum foil of specific length are wound together with the insulator strip between them. One foil strip is grounded, while the other is connected across the breaker points.

THE MAGNETO

Magnetos supply the ignition spark on most small engines. A magneto will produce current for ignition without any outside primary source of electricity.

Major components of a typical flywheel magneto ignition system are illustrated in Fig. 5-24. Except for the

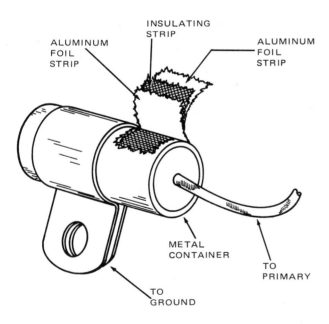

Fig. 5-23. Condenser consists of two strips of aluminum foil separated by an insulating material (dielectric). One foil strip is connected to metal container, other is attached to primary lead. (Kohler Co.)

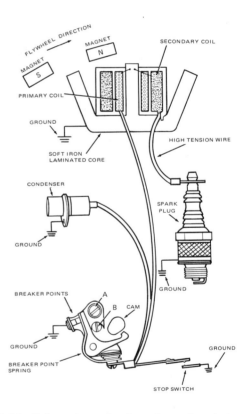

Fig. 5-24. Coil, condenser, breaker points and spark plug make up primary and secondary circuits of magneto. Magnets create current flow in primary winding of coil, which induces current in secondary winding.

spark plug, the coil, condenser and breaker points may be found inside or outside of the flywheel. This varies with type of engine, but the principles of operation remain basically the same.

Magnets usually are cast into the flywheel and cannot be removed. They are strong permanent magnets made of "Alnico" (aluminum, nickel, cobalt alloy) or of a newer ceramic magnetic material.

The coil in Fig. 5-24 is cut away to show primary and secondary windings. The primary winding usually has about 150 turns of relatively heavy copper wire. The secondary winding has approximatley 20,000 turns of very fine copper wire. One end of primary and one end of secondary is grounded to the soft iron laminated core which, in turn, is grounded to engine.

The odd shape of the core of the coil is designed to effiently direct the magnetic lines of force. The spacing (air gap) between the magnets and the core ends is critical and can greatly affect the whole system. This gap can be checked with special gages or standard feeler gages. The high tension wire is heavily insulated because it carries high voltage. If the insulation deteriorates, much of the voltage can be lost by arcing to nearby metallic parts of the engine.

The breaker points are mechanically actuated, opened by the cam and closed by the breaker point spring. Referring to Fig. 5-24, the breaker point gap is adjusted by loosening clamp screw A and turning eccentric screw B to move the stationary point.

THE MAGNETO CYCLE

As the flywheel turns, the magnets pass the legs of the laminated core of the coil. When the north pole of the magnet is over the center leg of the lamination, the magnetic lines of force move down the center leg through the coil, across the bottom of the lamination and up the side leg to the south pole. See Fig. 5-25.

As the flywheel continues to turn, Fig. 5-26, the north pole of the magnet comes over the side leg and the south pole is over the center leg of the lamination. Now the lines of force move from the north pole down through the side leg and up through the center leg and

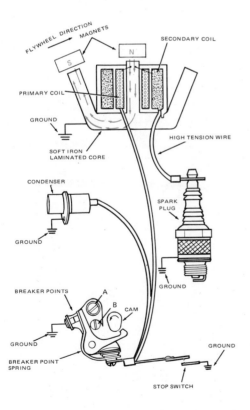

Fig. 5-25. As flywheel turns and magnets align with legs of laminated core of coil, a magnetic field is conducted through primary winding.

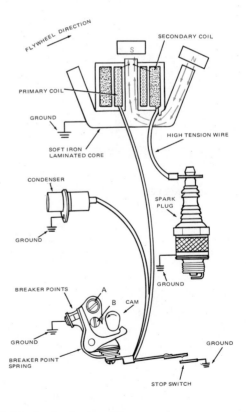

Fig. 5-26. As flywheel continues to turn, magnets realign with center and outside leg of core, causing magnetic field to reverse and induce low voltage in primary coil.

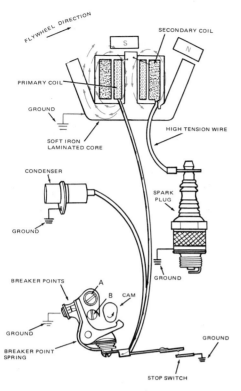

Fig. 5-27. Magnetic field reversal takes place as magnets pass from left to right. Breaker points are closed.

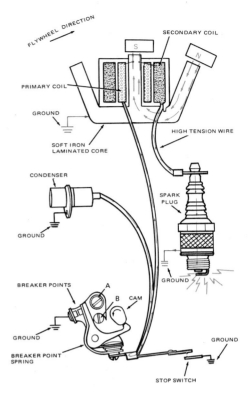

Fig. 5-29. Spark plug fires and condenser discharges voltage back into primary circuit.

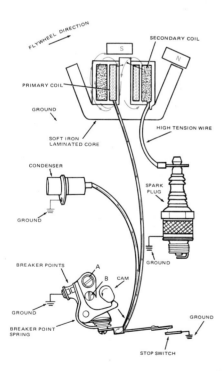

Fig. 5-28. Breaker points open, causing primary magnetic field to collapse at an extremely high rate through secondary winding. This induces high voltage required to fire spark plug. Field collapse also cuts through windings of primary coil, inducing moderate voltage that is absorbed by condenser.

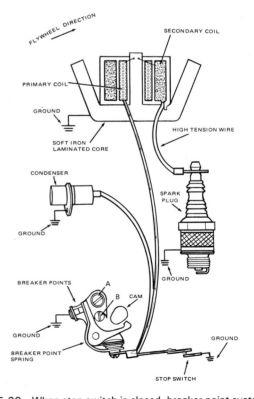

Fig. 5-30. When stop switch is closed, breaker point system is grounded, stopping the engine.

the coil to the south pole. At this point, the lines of force have reversed direction.

Fig. 5-27 shows the field reversal taking place in the center leg of the lamination and coil. The reversal induces low voltage current in the primary circuit through the breaker points. Current flowing in the primary winding of the coil creates a primary magnetic field of its own, which reinforces and helps maintain the direction of the lines of force in the center leg of the lamination. It does this until the magnets' pole move into a position where they can force the existing lines of force to change direction in the center leg of the lamination. Just before this happens the breaker points are opened by the cam.

Opening of the points breaks the primary circuit and the primary magnetic field collapses through the turns of the secondary winding. See Fig. 5-28. The condenser makes the breaking of the primary current as instantaneous as possible by absorbing the surge of primary current to prevent arcing between the breaker points.

As the magnetic field collapses through the secondary winding of the coil, high voltage current is induced in the secondary winding. At exactly the same time, the charge stored in the condenser surges back into the primary winding, Fig. 5-29, and reverses the direction of current in the primary windings. This change in direction sets up a reversal in direction of the magnetic field cutting through the secondary and helps increase the voltage in the secondary circuit. The potential of the high voltage causes secondary current to arc across the spark plug gap.

THE STOP SWITCH

The spark plug can only fire when the ignition points open the primary circuit. Using this as a basis for a stop switch, the switch is designed to ground the movable breaker point so that, in effect, the points never open. See Fig. 5-30. Therefore, the engine stops running.

Another common method of stopping single cylinder engines is by means of a strip of metal fastened to one of the cylinder head bolts. When the engine is running, the strip is suspended about 1/2 in. from the spark plug wire terminal. By depressing the strip against the plug wire, the current flows down the strip to the cylinder head to prevent a spark at the plug. There is no danger of shock to the operator. CAUTION: Do not touch the spark plug directly.

IGNITION ADVANCE SYSTEMS

Some small engines have mechanical systems that retard occurrence of spark for starting. For intermediate and high speed operation, the advance mechanism causes spark to occur earlier in the cycle.

One type of ignition advance system is illustrated in Fig. 5-31. Two different spark timings are provided, one for starting and one for running. For starting, the spark-advance flyweight holds the cam in a position so that the ignition spark occurs at 6 deg. of crankshaft rotation before the piston reaches top dead center (TDC). See A in Fig. 5-31.

When the engine reaches a speed of nearly 1000 rpm, centrifugal force moves the flyweight out, forcing the cam to rotate. This position of the cam causes the points to open and a spark to occur at 26 deg. before top dead center, as shown at B in Fig. 5-31.

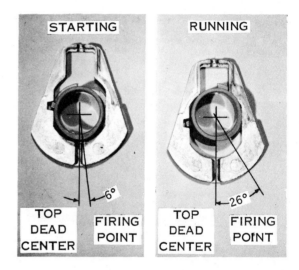

Fig. 5-31. For starting engine, flyweight is held in retarded position by a spring. When engine reaches 1000 rpm, flyweight overcomes spring and rotates cam to advanced position. (Lawn-Boy Power Equipment, Gale Products)

Fig. 5-32 shows the flyweight and cam assembled on the crankshaft in starting (solid lines) and running (dashed lines) positions.

DWELL AND CAM ANGLE

Dwell (cam angle) is the time the breaker points stay closed during one revolution of the cam. Dwell is the

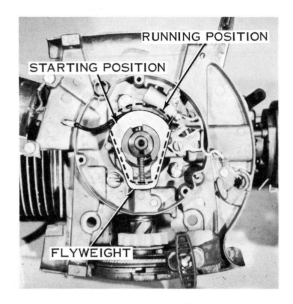

Fig. 5-32. Flyweight, spring and cam rotate together as a single assembly. Outer position of flyweight is for advanced operation.

number of degrees measured around the cam from the point of closing to the point of opening. Fig. 5-33 shows the direction relationship between the breaker point gap setting and dwell time.

Illustration A in Fig. 5-33 indicates what might be considered normal dwell time. Illustration B shows a large point gap with a correspondingly short dwell. Note that the wear block must travel quite a distance to a lower position on the lobe before the points make contact. Then, they open with just the slightest rise in the lobe.

The narrow point gap illustrated at C in Fig. 5-33 causes the breaker points to close after only a slight travel down the lobe, but more distance is required up

the lobe to open the points. Remember that the cam is driven directly from the crankshaft. When the breaker points open, the spark plug fires.

Obviously, then, changing the point setting can also change spark timing. The engine manufacturer specifies which gap setting is best (usually between .020 to .030 in.) and the number of degrees before top dead center (BTDC) that the spark should occur.

SOLID STATE IGNITION

Solid state is a broad term applied to any ignition system which uses electronic semi-conductors (diodes, transistors, silicon controlled rectifiers, etc.) in place of one or more standard ignition components.

The capacitive discharge ignition (CDI) system is a solid state ignition system. It is one of the newest for use in an internal combustion engine. It is standard equipment in many applications and has improved the reliability of modern small gasoline engines.

Electronic components in solid state systems are extremely small. Since there are no moving parts, mechanical adjustments are not required. Solid state ignition systems provide many advantages:
1. Elimination of ignition system maintenance.
2. No breaker points to burn, pit or replace.
3. Increased spark plug life.
4. Easy starting, even with fouled plugs.
5. A flooded engine will start easily.
6. Higher spark output and faster voltage rise.
7. Spark advance is electronic and automatic. It never needs adjusting.
8. Electronic unit is hermetically sealed and unaffected by dust, dirt, oil, or moisture.

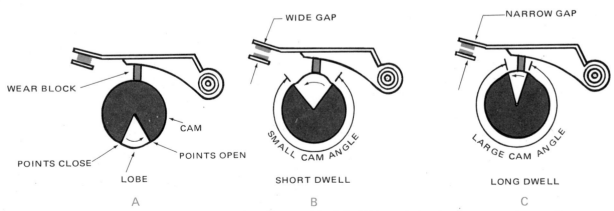

Fig. 5-33. Relationship of point gap and dwell. A—Normal gap and dwell. B—With a wider breaker point gap setting, dwell decreases. C—A narrower gap increases dwell.

9. System delivers uniform performance throughout component life and under adverse operating conditions.
10. Improves idling and provides smoother power under load.

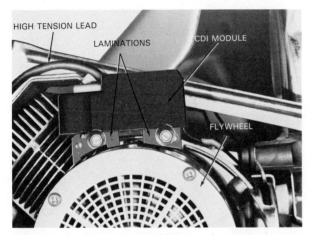

Fig. 5-34. Solid state, CDI ignition module is compact and maintenance free. Only moving parts are flywheel magnets.

The CDI ignition system is breakerless. The mechanical points and accessories are replaced with electronic components. The only moving parts are the permanent magnets in the flywheel. Fig. 5-34 shows the CDI module installed on a small gasoline engine.

The CDI module can be tested to see if it is producing a spark by grounding a known good spark plug to the engine. Turn the flywheel and watch the plug electrode gap for a spark. If there is no spark, the ignition switch or switch lead may be faulty; or the air gap may be incorrect (spacing between flywheel and CDI module). No other troubleshooting is necessary.

The ignition switch is the most vulnerable part of the solid state ignition system. The switch must be kept dry and clean.

OPERATION OF CDI SYSTEM

Refer to the following illustrations to progressively trace current flow through the various electronic components of a typical CDI system.

In Fig. 5-35, flywheel magnets rotate across the CDI module laminations, inducing a low voltage alternating current (ac) in charge coil. The ac passes through a rectifier and changes to direct current (dc), which travels to the capacitor (condenser) where it is stored.

In Fig. 5-36: The flywheel magnets rotate approximately 351 deg. before passing the CDI module laminations and induce a small electrical charge in trigger coil. At starting speeds, this charge is just great enough to turn on the silicon controlled rectifier (SCR), a solid state switch in a retarded firing position (9 deg. BTDC) for easy starting.

Fig. 5-35. Flywheel operation. 1A—Magnets induce low voltage alternating current into charge coil at 2. 3—Rectifier changes alternating to direct current. 4—Direct current from rectifier is stored in capacitor (condenser). See Fig. 5-38.

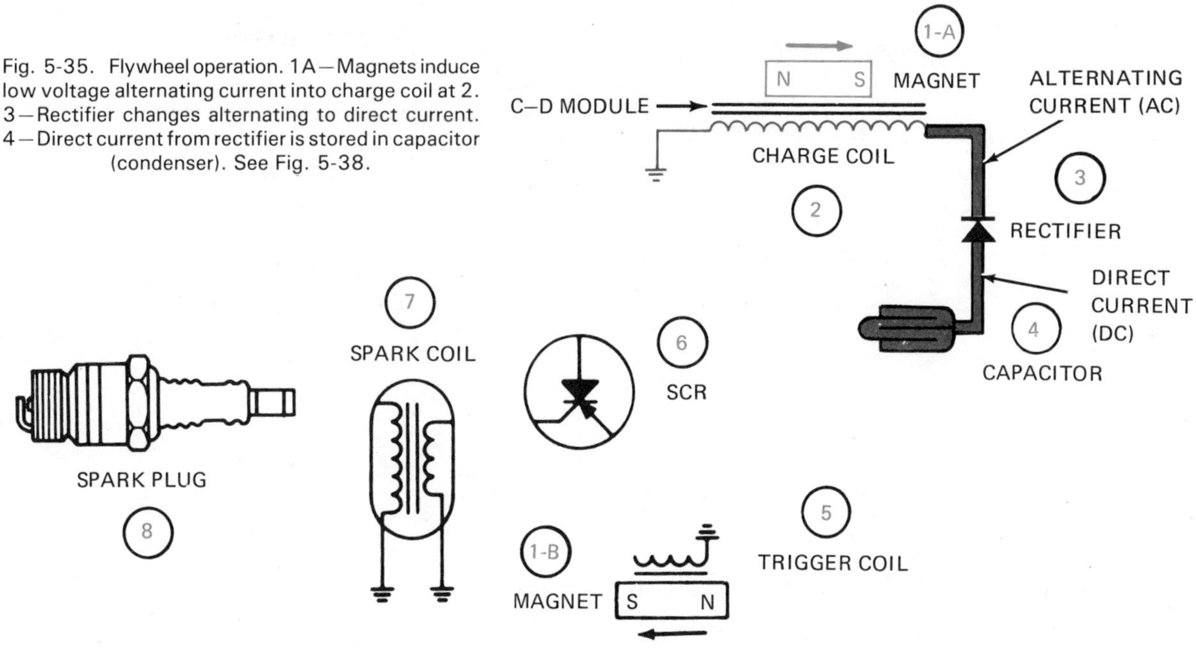

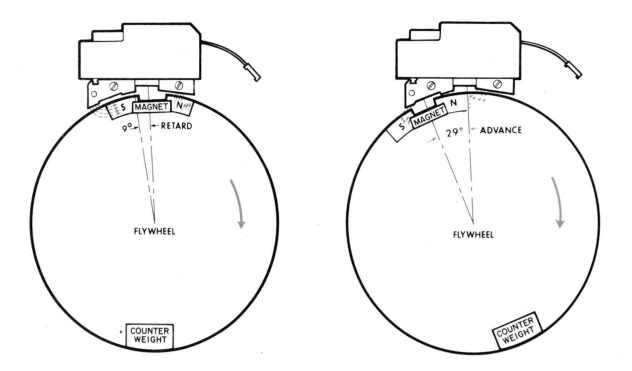

Fig. 5-36. At low speed, flywheel magnets induce a small current in trigger coil, which turns on silicon rectifier at 9 deg. BTDC for easy starting.

Fig. 5-37. At 800 rpm, stronger trigger coil current turns on silicon rectifier at 29 deg. BTDC for satisfactory ignition during normal engine operation.

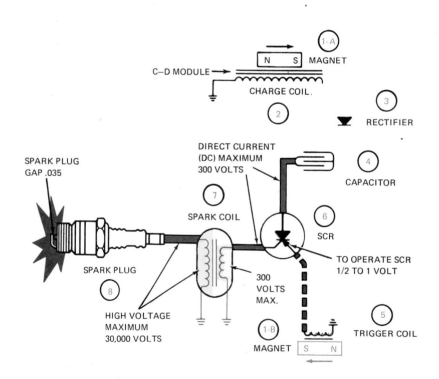

Fig. 5-38. 1B—Magnet induces a small current in trigger coil. 5—Trigger coil switches on silicon controlled rectifier. 6—Rectifier permits capacitor to discharge 300V into primary winding of spark coil. 7—Spark coil steps up voltage in secondary winding. 8—Spark plug fires.

In Fig. 5-37: When the engine reaches approximately 800 rpm, advanced firing begins. The flywheel magnets travel approximately 331 deg., at which time enough voltage is induced in the trigger coil to energize the silicon controlled rectifier in the advanced firing position (29 deg. BTDC).

In Fig. 5-38: When the silicon controlled rectifier is triggered, the 300V dc stored in the capacitor travels to the spark coil, where the voltage is stepped up instantly to a maximum of 30,000V. This high voltage current is discharged across the spark plug gap.

OPERATION OF TRANSISTOR CONTROLLED IGNITION (TCI) SYSTEM

The individual components that make up the transistor controlled ignition system are given in a chart in Fig. 5-39. Study the function of each part carefully.

There are a variety of transistor controlled circuits. Each has its own unique characteristics and modifications. Fig. 5-40 illustrates a typical circuit for a tran-

sistor controlled ignition. Refer to this circuit as its principles are described.

As the engine flywheel rotates, the magnets on the flywheel pass by the ignition coil. The magnetic field around the magnets induces current in the primary windings of the ignition coil.

The base circuit of the ignition system has current flow from the coil primary windings, common grounds, resistor (R1), base of the transistor (T1), the emitter of the transistor (T1), and back to the primary windings of the ignition coil.

Current flow for the collector circuit in Fig. 5-40 is from the primary windings of the coil, common grounds, collector of transistor (T1), emitter of transistor (T1), and back to the primary windings.

When the flywheel rotates further, the induced current in the coil primary increases. When the current is high enough, the control circuit turns on and begins to conduct current. This causes transistor (T2) to turn on

Diode (D1, D2)	 A ──▶── K	Allows one way current only from Anode "A" to Cathode "K" as rectifier.
Flywheel		Provides magnetic flux to primary windings of ignition coil.
High-tension lead		Conducts high voltage current in secondary windings to spark plug.
Ignition coil		Generates primary current, and transforms primary low voltage to secondary high voltage.
Ignition switch		No spark across gap of spark plug when switch is at "STOP" position.
Resistor (R1, R2)	 ──⋀⋀⋀──	Resists current flow.
Spark plug		Ignites fuel-air mixture in cylinder.
Thyristor (S)	 A ──▶◀── K / G	Switches from blocking state to conducting state when trigger current/voltage is on gate "G."
Transistor (T, T1, T2)	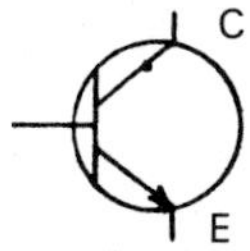 C / B / E	Very small current in the base circuit (B to E) controls and amplifies very large current in the collector circuit (C to E). When the base current is cut, the collector current is also cut completely.

Fig. 5-39. Study components of transistor controlled ignition system.

and conduct. A strong magnetic field forms around the primary winding of the ignition coil, Fig. 5-40.

The trigger circuit for this ignition system consists of the primary windings, common grounds, control circuit, base of transistor (T2), emitter of transistor (T2).

When transistor (T2) begins to conduct current, the base current flow is cut. This causes the collector circuit to shut off and transistor (T1) stops conducting current.

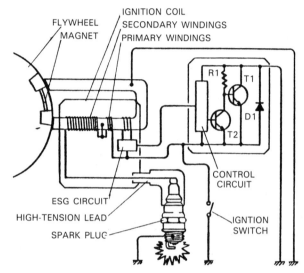

Fig. 5-40. Study how the transistor circuit is used to operate the ignition coil.

When transistor (T1) stops conducting, current stops flowing through the primary of the ignition coil. This causes the primary magnetic field to collapse across the secondary windings of the ignition coil. High voltage is then induced into the secondary to "fire" the spark plug.

The secondary circuit includes the coil secondary windings, high tension lead, spark plug, and common grounds returning to the coil secondary.

With the ignition switch at stop, Fig. 5-40, the primary circuit is grounded to prevent the plug from firing.

Diode (D1) is installed in the circuit to protect the TCI module from damage.

The ESG circuit in Fig. 5-40 is used to retard the ignition timing. At high engine rpm, the ESG circuit conducts. This bypasses the trigger circuit and delays when current reaches the base of transistor (T2).

MAGNETO IGNITION SYSTEMS COMPARED

The following are definitions for the three general classifications of magneto ignition systems.
1. A mechanical breaker ignition (MBI) system is a flywheel magneto inductive system commonly used for internal combustion engines. It employs mechanical breaker contacts to time or trigger the system.

COMPARISONS	MECHANICAL BREAKER IGNITION SYSTEM	TRANSISTOR CONTROLLED IGNITION SYSTEM	CAPACITOR DISCHARGE IGNITION SYSTEM
Abbreviation	MBI	TCI	CDI
Circuit type	Conventional	Solid state	Solid state
Energy source	Primary current of ignition coil	Primary current of ignition coil	Stored in capacitor
Trigger switch	Breaker contacts	Power transistor	Thyristor
Secondary voltage	Standard	Standard	Higher
Spark duration	Standard	Standard	Shorter
Rise time*	Standard	Standard	Shorter
Maximum operating speed	Standard	Higher	Higher
Maintenance	Regap and retime	None	None

*RISE TIME—time required for maximum voltage to occur.

Fig. 5-41. Chart compares breaker point, transistor controlled, and capacitor discharge systems. Not differences and similarities.

2. A transistor controlled ignition (TCI) system is an inductive system that does NOT use mechanical breaker contacts. It utilizes semiconductors (transistors, diodes, etc.) for switching purposes.
3. A capacitor discharge ignition (CDI) system is a solid state (no moving parts) system which stores its primary energy in a capacitor and uses semiconductors for timing or triggering the system.

Look at Fig. 5-41. It compares these three types of magneto ignition systems. Study them!

BATTERY IGNITION SYSTEMS

The battery ignition system has a low voltage primary circuit and a high voltage secondary circuit.

Like the magneto, it consists of a coil, condenser, breaker points and spark plug. The basic difference is that the source of current for the primary circuit is supplied by a lead-acid battery. See Fig. 5-42.

When the ignition switch is turned on, current flows from the positive post of the battery to the ignition coil. Current traveling through the primary windings of the coil builds up a magnetic field, Fig. 5-43. During this time, the breaker points are closed. Ignition at the plug is not required, so the current returns to the battery through the common ground.

Then, at the exact time when ignition at the plug is required, the breaker points are opened by the cam. Current flow stops abruptly, causing the magnetic field

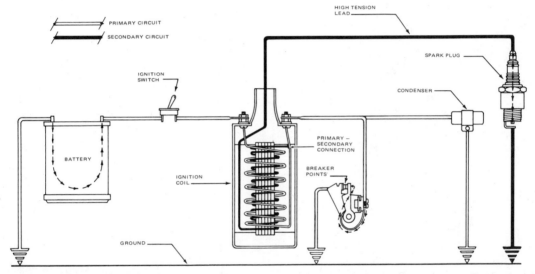

Fig. 5-42. Above. Battery ignition system is similar to a magneto system, except that battery replaces flywheel magnets.
Fig. 5-43. Below. When points close in battery ignition system, primary current builds a magnetic field around coil. (Kohler Co.)

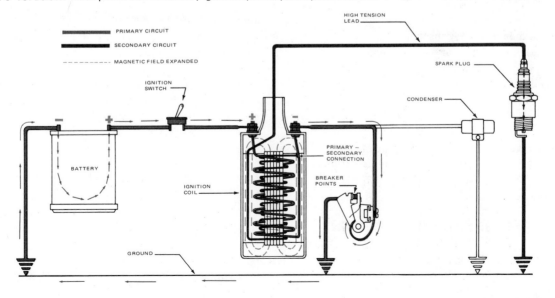

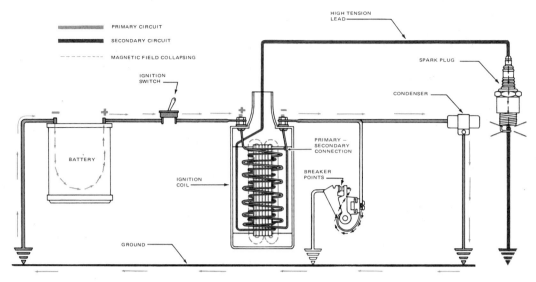

Fig. 5-44. Above. Breaker points open and field collapses inducing high voltage in secondary winding of coil. Condenser absorbs voltage surge in primary winding. Fig. 5-45. Below. Condenser discharges back into primary circuit.

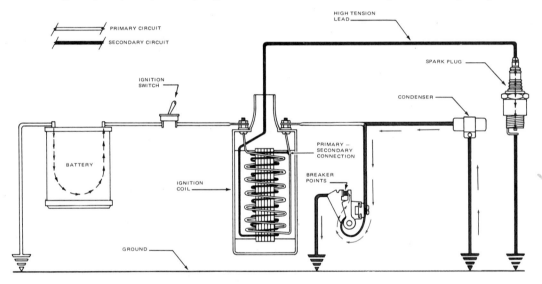

surrounding the coil to collapse, Fig. 5-44. This rapid change of magnetic flux causes voltage to be induced in every turn of the primary and secondary windings. Voltage in the primary winding (about 250 volts) is quickly absorbed by the condenser.

Without the condenser, the current would arc at the breaker point gap, burning the points. The condenser acts as a reservoir for the sudden surge of power in the primary windings of the coil. The condenser holds the current for an instant, then releases it to the primary circuit, as shown in Fig. 5-45.

HIGH VOLTAGE SECONDARY CURRENT

The voltage built up in the secondary winding of the coil could become as high as 25,000 volts. It has approximately 100 times as many turns of wire as the primary, Fig. 5-46. Normally, the voltage does not reach this value. Once it becomes great enough to jump the spark plug gap, the voltage drops. Usually, the amount required to jump the gap is between 6000 and 20,000 volts. The actual amount of voltage required depends upon variables such as compression, engine speed, shape and condition of electrodes, spark plug gap, etc.

AUTO-TRANSFORMER TYPE
IGNITION COIL

The auto-transformer type ignition coil used on some small gasoline engines serves as a step-up transformer. It increases low voltage primary current to the high voltage required to bridge the spark plug gap. The

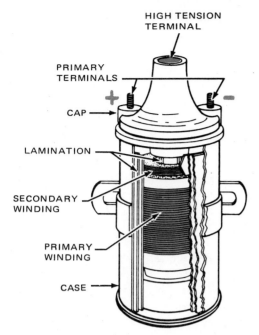

HIGH TENSION
TERMINAL

PRIMARY
TERMINALS

CAP

LAMINATION

SECONDARY
WINDING

PRIMARY
WINDING

CASE

Fig. 5-46. Auto-type ignition coil has about 100 times as many turns in secondary winding than in primary winding. Coil can produce 25,000V, if needed.

primary and secondary windings are connected, and the common ground of the battery and primary circuit is used to complete the secondary circuit.

With this type of coil, very little primary current can flow into the secondary circuit because, normally the secondary circuit is open at the spark plug gap. Primary current is just not great enough to jump that gap. Therefore, the two circuits function separately.

The primary winding of the coil consists of about 200 turns of heavy copper wire. The secondary winding has approximately 20,000 turns of very fine copper wire. Because the magnetic field collapses through such a great number of conductors in the secondary, a very high voltage, or electrical potential is developed. The amperage (rate of electron flow), however, is proportionately low. This is a typical characteristic of any transformer.

Laminated iron is used as the center core of the coil, Fig. 5-46. It also forms the outer shell of the inner assembly, providing maximum concentration of the magnetic field. The inner assembly is sealed in a coil case, and the remaining space inside is filled with a special oil to minimize the effects of heat, moisture and vibration.

The top of the coil is provided with two primary terminals. They are marked positive (+) and negative

(−). The positive terminal must be connected to the positive side of the battery. The negative terminal connects to the breaker points, Fig. 5-45. The center tower of the coil contains the high tension terminal.

THE LEAD-ACID BATTERY

The battery is the sole source of energy for the battery ignition system of a small gasoline engine. A generator is used to replenish energy in the battery. However, the generator does not supply energy directly to the ignition system.

Lead-acid type batteries are used. The cell plates are made of lead, and a sulfuric acid and water solution serves as the electrolyte. "Wet" or "dry charged" types are available. Wet batteries are supplied with the electrolyte in them, ready for use if the charge has been kept up. Dry charged batteries must have electrolyte installed after purchase. Both types of batteries function in the same way.

BATTERY CONSTRUCTION

The typical 12V battery is constructed with a hard rubber case and six separate compartments called cells, Fig. 5-47. There is a specific number of negative and positive plates in each cell. The greater the number of plates per cell, the higher the ampere-hour rating (capacity to provide current for a specific length of time) of the battery. The positive plates have a lead oxide covering. The negative plates have a porous or spongy surface.

BATTERY VOLTAGE

A chemical reaction causes each negative plate to lose electrons and each positive plate to gain electrons when surrounded by electrolyte. The plates, therefore, develop an electrical potential between them. All plates of a like charge are electrically connected, causing accumulative charges to be present at the positive and negative bettery terminals.

Each cell of a battery in good condition contributes approximately 1.95 to 2.08V. Six fully charged cells will produce at least 12V. If not, the battery must be recharged or a replacement battery installed.

A DISCHARGING BATTERY

When a battery discharges without replacement of energy, the sulfuric acid is chemically withdrawn from the electrolyte, specific gravity goes down and lead

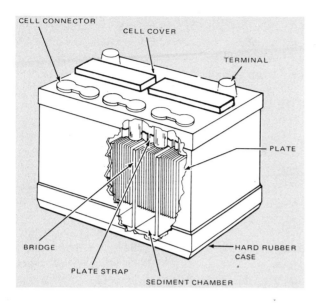

CELL CONNECTOR
CELL COVER
TERMINAL
PLATE
BRIDGE
HARD RUBBER CASE
PLATE STRAP
SEDIMENT CHAMBER

Fig. 5-47. Battery provides all current for battery ignition system. A 12V battery has six cells, producing 2V each.

sulfate deposits accumulte on the plates. If sulfate deposits become too great, or the level of the electrolyte falls lower than the top of the plates, permanent damage may be done to the battery.

When a battery is recharged, a controlled direct current is passed through the battery in reverse direction from normal operation. This causes a reversal in chemical action and restores the plates and electrolyte to active condition.

REVIEW QUESTIONS-CHAPTER 5

1. The two major tasks performed by an ignition system are to _____ and _____.
2. If a four cycle engine runs at 3600 rpm, the number of sparks per minute required at the spark plug would be _____ per minute.
3. Name the six main electrical components that make up the magneto ignition system.
4. Regarding atoms:
   a. Electrons and protons are tightly bonded together.
   b. Electrons are positively charged particles.
   c. Neutrons are negatively charged, protons are positively charged.
   d. Electrons do not collide because they are negatively charged.
5. The electron theory states that:
   a. Like charges attract each other.
   b. Like charges repel each other.
   c. Only negative charges repel each other.

d. Electrons orbit the nucleus because the protons repel the electrons.
6. Substances that have electrons which can move freely from atom to atom are said to be good:
   a. Conductors.
   b. Dielectrics.
   c. Insulators.
   d. Nonconductors.
7. List three good conductors of electrical current.
8. List three good insulating materials.
9. A source of electricity:
   a. Has an excess of electrons and is positively charged.
   b. Has an excess of electrons and is negatively charged.
   c. Lacks electrons and is negatively charged.
   d. Lacks electrons and is positively charged.
10. Electrical current in a battery flows from:
    a. Negative to positive.
    b. Positive to negative.
11. Electrical current in the primary circuit of a battery ignition system flows from:
    a. Negative to positive.
    b. Positive to negative.
12. What are the three ways in which an electrical potential can be produced?_____ _____ and _____.
13. Match the following units of electrical measurement:
    a. _____ Rate of electron flow.   Ohms.
    b. _____ Resistance to electron flow. Volts.
    c. _____ Electrical potential.   Amperes.
14. In a circuit of 5 amperes and 2 ohms, the voltage would be _____.
15. In a circuit with 12 volts and 5 amperes, the resistance would be _____ ohms.
16. Soft iron used in the laminations of a coil or transformer has:
    a. Good magnetic retention.
    b. Poor magnetic retention.
    c. Non-magnetic properties.
17. What happens when a coil of wire is passed through a magnetic field?
18. What happens when electric current is passed through a coil of wire?
19. In the ignition coil, the primary winding has:
    a. Many turns of fine wire.
    b. Few turns of fine wire.
    c. Few turns of heavy wire.
    d. Many turns of heavy wire.
20. The coil acts as a transformer which:
    a. Steps down the voltage and increases the output amperage.

b. Steps up the voltage and amperage.

c. Steps down the voltage and amperage.

d. Steps up the voltage and decreases the output amperage.

21. The greatest resistance to the highest voltage in the ignition circuit occurs at:

a. Spark plug.

b. Breaker points.

c. Condenser.

d. Secondary of the coil.

22. Would a "cool" spark plug have a short or long insulator nose? _____.

23. Spark plugs of the "cool" type should always be matched with cool running engines. True or False?

24. The device used by engineers and technicians to measure spark plug temperatures under running conditions is a _____ _____.

25. Breaker point contacts are made of a very hard material called _____.

26. When the breaker points in the magneto are closed:

a. Current is induced in the primary circuit by the flywheel magnets and coil.

b. The spark plug fires.

c. A high voltage is induced in the secondary circuit.

d. The condenser absorbs current from the primary circuit for later use in the secondary.

e. All of the above are correct.

27. The spark plug fires when:

a. Magnetic field collapses through the secondary windings of the coil.

b. Piston is nearing top of compression stroke.

c. Breaker points open.

d. Voltage potential is high enough to bridge the electrodes.

e. All of the above are correct.

28. The condenser acts as a voltage reservoir or shock absorber for the primary circuit. Its primary purpose is to prevent _____.

29. When the breaker points open, voltage in the primary circuit may surge to as much as _____.

30. The STOP SWITCH grounds the _____.

31. When connecting the auto-type ignition coil in the circuit, the positive terminal of the battery must be connected to:

a. The positive terminal of the coil.

b. The negative terminal of the coil.

c. Either terminal of the coil.

32. When breaker points are set with a wider gap, the dwell:

a. Becomes greater.

b. Becomes less.

c. Does not change.

33. Name five advantages of a solid state ignition system.

SUGGESTED ACTIVITIES

1. Experiment with a coil of wire and a magnet to induce a current. Use a galvanometer to show the current flow.

2. With a dry cell, iron filings, copper wire and a piece of paper, demonstrate the magnetic flux produced when the current flows through the wire.

3. Using copper wire, form it into a loose coil. Use a dry cell to pass a current through the wire and determine the polarity with a permanent magnet. Reverse the polarity.

4. Using the coil of wire from No. 3, wrap insulating tape around an iron bar and place it in the coil. Demonstrate how this can improve the magnetic strength of the coil when current is flowing.

5. Make a visible magneto mounted on a display board or built into a clear acrylic box so that it can be manually turned with a crank. Old, but usable, engine parts can be used.

6. A workable battery ignition system can be built and mounted as a display board. Demonstrate the operation and principles involved in this system.

7. Make a large Ohm's law triangle and display it until it has been learned. Make up some problems that require use of the formulas.

8. Make a collection of various kinds of spark plugs.

9. Section a spark plug, so that the inner construction can be studied.

10. Cutaway a "dry" type battery that has not had electrolyte added to it and discuss how it works.

11. Section an old ignition coil to show the primary and secondary windings around the core.

12. Carefully open a condenser to display the lamination of aluminum foil and insulation.

13. Collect several engines with different types of ignition advance mechanisms. Demonstrate them to the class.

14. Disassemble a magneto and demonstrate how it works.

Small four cycle engines powering mobile equipment generally call for oil level checks every 5 hr. of operation, oil changes every 25 hr. Stationary engines (shown) should be level-checked daily, with oil changes every 50 hr.

Chapter 6

ENGINE LUBRICATION

Lubrication is the process of reducing friction between sliding surfaces, Fig. 6-1, by introducing a slippery or smooth substance between them. These lubricants come in dry (powdered), semi-dry (grease) and liquid (oil) forms. Oil is the most important lubricant for small engine use, simply because it is often the only lubrication the engine needs, Fig. 6-1A.

FRICTION

Friction is the resistance to motion created when one dry surface rubs against another. Even highly polished metal surfaces have irregularities (when studied under a microscope) that would create much friction if rubbed together. The microscopic roughness would resist movement and create heat.

As the relatively rough projections on the contact surfaces rub across each other, they eventually would break off and become loose particles. The particles, in turn, would work between the surfaces and gouge grooves in the metal. Then, as the friction and heat increases, the metal parts would expand, causing greater pressure between the surfaces and creating even greater friction. This condition of wear exists until the parts either weld themselves together or seize (expand so much that mating parts cannot move).

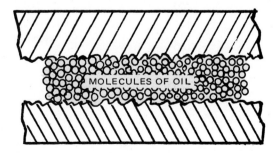

Fig. 6-1. A film of engine oil serves to separate and lubricate machined surfaces. When parts are in motion, oil molecules roll over one another like microscopic ball bearings.

Fig. 6-1A. Oils, grease, and graphite are commonly used in engine lubrication.

In some cases, the excessively worn parts lose so much material from their contact surfaces, they become too loose to function properly. When this happens, the scored part should be replaced with a new one.

PREVENTING WEAR DUE TO FRICTION

In designing a small gasoline engine, the manufacturer selects suitable materials for parts that will be in moving contact with one another. For example, precision insert bearing shells, Fig. 6-2, are used in connecting rods and caps, and in main bearing saddles and caps. The steel backing has a cast babbitt surface. Babbitt is an alloy of tin, copper and antimony, which has good anti-friction qualities.

Rod and main bearing inserts must withstand reciprocating and rotational forces while fitting closely to the crank journal or throw. Notice the oil hole in insert in Fig. 6-2 permits oil to enter oil groove. With crankshaft in motion, the oil groove distributes oil all the way around the insert. Then, oil travels outward to bearing edges and back into crankcase.

Some bearing inserts have laminations of other metals to back up the babbitt. Refer to Fig. 6-3. This

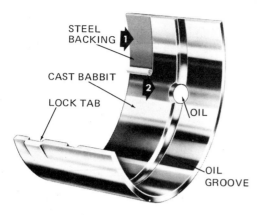

Fig. 6-2. Precision insert type of engine bearing has a steel shell with a cast babbitt inner surface. Oil hole and groove carries pressurized oil to bearing surfaces.

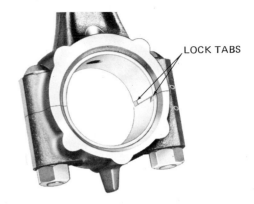

Fig. 6-4. Lock tabs prevent bearing insert from rotating in connecting rod. Tabs must be seated in recesses provided in rod. (Wisconsin Motors Corp.)

construction enhances the properties of the insert, making it more heat resistant and durable.

The lock tab illustrated in Fig. 6-3 is used to prevent the bearing insert from rotating in the connecting rod or main bearing cap. An insert with the tab in place in a rod assembly is shown in Fig. 6-4.

Regardless of the quality of the material used, all bearing surfaces in small gasoline engines must have oil separating moving parts that are in close contact. See Fig. 6-5. A thin film of engine oil must coat the area between: piston/piston rings and cylinder wall, piston pin and piston or connecting rod, valve stem and guide, valve tappet and guide, main and rod bearing inserts and crankshaft journals, etc.

The thin film of oil between the close fitting parts may be only a few molecules in thickness, yet this is

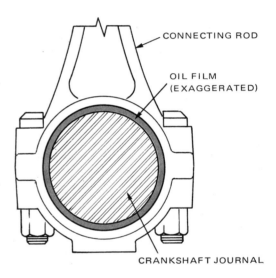

Fig. 6-5. Oil film between close fitting metallic parts prevents actual contact. Oil provides a relatively frictionless movement of parts.

enough to prevent the two metal surfaces from actually touching. The molecules of oil then roll over one another acting like microscopic ball bearings between the surfaces. See Fig. 6-1.

LUBRICATING OIL

Oil performs a number of important jobs:
1. Oil lubricates. It reduces wear of moving parts by preventing metal-to-metal contact. A film of oil .0005 to .002 in. thick separates moving parts and acts like minute liquid ball bearings to reduce friction. The lubricated parts turn or slide freely, Fig. 6-1.
2. Oil helps cool the engine. It absorbs heat while cir-

Fig. 6-3. Some precision inserts have several layers of material between steel shell and babbitt. (Clevite Corp.)

culating through and around the various parts. The absorbed heat is transferred to the air or to the reservoir of oil in the crankcase, which is cooler than the oil returning from the engine parts.

3. Oil cleans the engine. As it circulates, oil washes the internal parts and picks up impurities resulting from combustion. The impurities are removed from the oil as it circulates through a filter, Fig. 6-6, or removed with the oil when it is drained from the engine.

4. Oil provides a seal between the piston rings and cylinder wall.

OIL SELECTION

The cost of oil for a small gasoline engine is relatively low. However, the particular oil selected for use in a small engine is extremely important to the life of the engine, Fig. 6-6A.

The oil recommended for use in a given engine may be shown on the engine nameplate or on a special label attached to the engine. The operator's manual also carries the manufacturer's recommendation, as does the lubrication guide provided by major oil companies.

Specifications for engine oils are given in two ratings:
1. API (Americal Petroleum Institute) Service classifi-

Fig. 6-6A. Use a sealing cap to keep engine oil clean during storage when a full can is not used.

cation, often referred to as "type" of oil.
2. SAE (Society of Automotive Engineers) Viscosity, referred to as "viscosity grade."

SAE VISCOSITY GRADE

SAE viscosity is an important consideration when selecting engine oil. Viscosity, refers to resistance of a fluid to flow, and it is rated in SAE numbers such as "SAE 10." Because oils become thicker at low temperatures and thinner at high temperatures, various viscosities are recommended corresponding to surrounding air temperatures.

Fig. 6-7 shows a comparison of viscosity recommendations by five manufacturers for their four cycle engines at various operating temperatures. Note that the higher viscosity oils (more resistant to flow) are recommended for higher temperatures.

On the other hand, a thick oil used in low temperature operation makes a cold engine very difficult to start. Upon starting, certain critical parts may not receive adequate lubrication for some time while the oil is gaining heat from combustion. "Cold running" can result in scored cylinder walls and engine bearings.

MULTI-VISCOSITY OILS

Some of the oils listed in Fig. 6-7 are single viscosity grade oils, such as "SAE 20." Others are multi-viscosity grade oils, such as "SAE 5 W-20." The "W" represents

Fig. 6-6. An oil filter traps contaminants picked up by engine oil. Most disposable filters are located on side of engine. (Wisconsin Motors Corp.)

FOUR-CYCLE CRANKCASE LUBRICATION (VISCOSITY-GRADE)
RECOMMENDATIONS OF MANUFACTURERS

MANUFACTURER	Above 40°F.	Above 32°F.	Below 5°F.	Below 0°F.	Below -10°F.
Briggs and Stratton	SAE 30 or 10W-30	5W-20 or 10W			
Clinton	SAE 30		10W		5W
Kohler	SAE 30		SAE 10W	5W or 5W 20	
Wisconsin	SAE 30	SAE 20 or SAE 20W	10W		
Tecumseh	SAE 30		10W-30		

Fig. 6-7. A comparison of viscosity-grade recommendations by five engine manufacturers. Recommendations are for specific models only, not for full line coverage.

a winter type oil. The multi-viscosity grade oils can be substituted for single viscosity grades in four cycle engines, but not in two cycle engines.

ENGINE LUBRICATION

The way in which moving parts are lubricated differs in two cycle and four cycle small gasoline engines. In preparing a two cycle engine for use, a specified amount of two cycle engine oil is mixed with each gallon of gasoline to provide fuel for the engine. This fuel is thoroughly mixed in a separate container, Fig. 6-8, then it is poured into the fuel tank. In operation, an oil mist is created that lubricates the cylinder wall and all internal engine parts.

In readying a four cycle engine for use, the fuel tank is filled with fresh gasoline. In addition, a specified amount of four cycle engine oil of the type and viscosity class recommended by the manufacturer is poured into the crankcase of the engine. In operation, the gasoline-air mixture is "fired" in the combustion chamber. At the same time, the oil sump in the crankcase supplies lubrication for the cylinder wall and all internal engine parts.

TWO CYCLE ENGINE LUBRICATION

Air-cooled engine operation covers a wider range of varying speeds with much higher combustion chamber temperatures than water-cooled engines. Therefore, automotive engine oils are NOT suitable for two cycle air-cooled engines. Likewise, multi-viscosity and/or detergent type oils may contain additives not intended

for two cycle engine operation and should not be used. *Always use the type and viscosity grade of oil recommended by the engine manufacturer.*

Certain manufacturers recommend the use of a specific SAE viscosity, diluted, two cycle engine oil. The diluent is added to make the oil pour more freely, particularly at low temperatures. As engine oil gets colder, one of its ingredients becomes an interlocking crystalline structure which binds together until the oil finally solidifies. The diluent blocks the formation of these crystals and the oil stays fluid.

Special additives are often recommended for two cy-

Fig. 6-8. Fuel for small gasoline engines should be stored in a clean, properly marked can with a vent cap and pouring spout.

cle engine use. Since two cycle engines are lubricated by mixing oil with the fuel, the oil eventually enters the combustion chamber and is burned. Some regular engine oils have additives that do not burn completely and leave a residue that fouls spark plugs and clogs exhaust ports.

The spark plugs shown in Fig. 6-9 were used in four test engines. Each engine ran for the same length of time, and identical preventive maintenance and adjustments were performed. The only exception in the test procedures was in the brand of engine oil used. Spark plug A in Fig. 6-9 was taken from an engine that used the oil recommended by the manufacturer. The other three spark plugs were removed from test engines operating with other brands of engine oil.

Not all engine oils will produce deposits of the type shown in Fig. 6-9, but the importance of using the recommended oil should be quite clear from this comparison. Special additives for two cycle oils must be selected to avoid or prevent unburned deposits. Oils containing these additives generally are sold under the brand name of the engine manufacturer.

FOUR CYCLE ENGINE LUBRICATION

Four cycle engines must be operated with the proper "type" of oil. For this reason, engine oil service classifications have been established by the Society of Automotive Engineers and the American Petroleum Institute. In Fig. 6-10, the "S" stands for service, while A, B, C, D and E designate the type of service for which the oil is best suited.

Most four cycle engine manufacturers recommend oils supplemented by additives. These chemicals are added to improve the quality of the oil. They may prevent corrosion, provide a better cushioning effect between moving parts and/or help prevent scuffing and reduce wear.

ENGINE OIL SERVICE CLASSIFICATIONS

OLD	NEW	SERVICE CONDITIONS
ML (Motor Light)	SA	Service typical of gasoline and other spark-ignition engines used under light and favorable operating conditions, the engines having no special lubrication requirements.
MM (Motor Moderate)	SB	Service typical of gasoline and other spark-ignition engines used under moderate to severe operating conditions, but presenting problems of deposit or bearing-corrosion control when crankcase oil temperatures are high.
MS (Motor Severe)	SC, SD SE, SF	Service typical of gasoline and other spark-ignition engines used under unfavorable, or severe, types of operating conditions, and where there are special lubrication requirements for deposits, wear, or bearing corrosion control.

Fig. 6-10. American Petroleum Institute (API) has published oil service classifications in which oils are recommended for specific service conditions.

Detergent-dispersants, often called "detergents," are added to some oils. Detergents suspend the dirt and sludge in the oil, where the contaminants can be trapped by a filter or readily drained before fresh oil is put in. Basically, if the recommended oil is used in an engine, special oil treatments should not be required.

SPLASH LUBRICATION SYSTEM

Small gasoline engines generally use some type of "splash system" to lubricate internal machined surfaces. The splash lubrication system shown in Fig. 6-11 features an oil dipper arm on the connecting rod cap. The dipper is designed to pick up oil from the crankcase on every revolution of the crankshaft, splashing oil on the various moving parts as it is carried around by the crank throw.

With the splash system, the cylinder wall receives a generous amount of oil. To avoid oil-burning problems, the oil control ring on the piston removes excess oil, returning it to the crankcase as illustrated in Fig. 6-12. The connecting rod bearings and piston pin receive lubrication through oil passage holes, Fig. 6-13.

CONSTANT LEVEL SPLASH SYSTEM

The constant level splash system provides three major improvements over the simple splash system:

A   B   C   D

Fig. 6-9. Spark plugs A, B, C, D were used in identical test engines using different oils. Plug A was taken from engine using oil recommended by engine manufacturer.

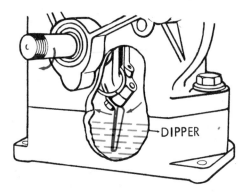

Fig. 6-11. With splash system, some oil dippers are cast onto connecting rod, others are bolted on. Oil level must be high enough for dipping action. (Briggs and Stratton Corp.)

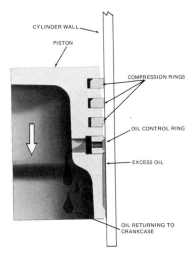

Fig. 6-12. Oil splashed on cylinder wall lubricates piston and piston rings. Excess oil is scraped from wall by oil control ring.

1. A lubrication oil pump.
2. A splash trough.
3. A strainer.

The cam-operated pump supplies oil to the trough where the oil dipper dips it up and distributes it to the cylinder wall and moving parts. The term "constant level" is used because the pump can supply more oil than the dipper can remove. Therefore, the trough is always full. Oil returning to the crankcase must pass through the filter before it can be pumped back to the trough. This keeps large contaminants in the crankcase.

The constant level splash system will provide adequate lubrication as long as there is enough oil to supply the pump. However, if the oil level is low, the cooling effect of the oil is reduced.

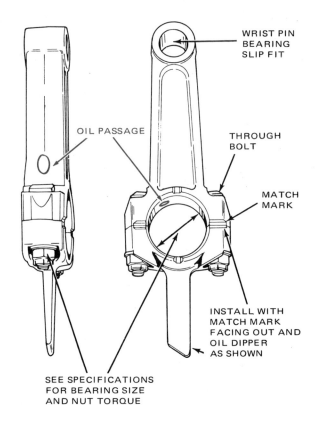

Fig. 6-13. Holes drilled in connecting rod, bearing insert and piston pin boss provide passageways for lubricating oil.

EJECTION AND BARREL PUMPS

The ejection pump forces oil under pressure against the rotating connecting rod. Some oil enters the connecting rod bearings, while the remaining oil is deflected to other parts in the crankcase. The ejection pump system is similar to the splash system, but it provides a more forceful spray of oil.

The barrel pump, Fig. 6-14, is a cylinder and plunger type of lubrication pump. By design, an eccentric on the camshaft moves the plunger in and out of the pump cylinder. The camshaft is hollow and has holes from the center of the shaft to the eccentric.

In operation, the plunger is drawn out until a hole in the eccentric aligns with a hole in the plunger. This allows the cylinder to fill with oil. When the plunger is forced in, a different hole in the eccentric aligns with the plunger, and oil is forced through passages to the main bearings and crankshaft connecting rod journal.

POSITIVE DISPLACEMENT PUMPS

Several types of positive displacement oil pumps are used in pressurized lubrication systems. See Fig. 6-15.

BARREL
PUMP

ECCENTRIC
SHAFT

Fig. 6-14. Plunger of barrel pump is actuated by an eccentric shaft driven by valve camshaft. Ball-shaped plunger end is held in a socket so that it can pivot as shaft turns. (Deere & Co.)

One common type is the gear pump shown in Fig. 6-16 with the end cover removed to expose two meshed gears. One gear is shaft-driven from the engine. It drives the second gear.

Note in Fig. 6-16 that the driving gear is keyed to the driving shaft. As the gears turn, oil fills the spaces between the teeth and is carried around to the oil outlet. No oil passes between the gears where the teeth are meshed, because of the tight fit.

If, for some reason, oil flow is restricted somewhere in the engine, the increase in pressure would raise the ball against the spring in the pressure relief valve. When this happens, oil will pass through the valve and recirculate through the pump. Recirculation of the engine oil continues until the restriction to flow ceases and pressure declines, allowing the ball to seat and relief valve to close. Without a pressure relief valve in the

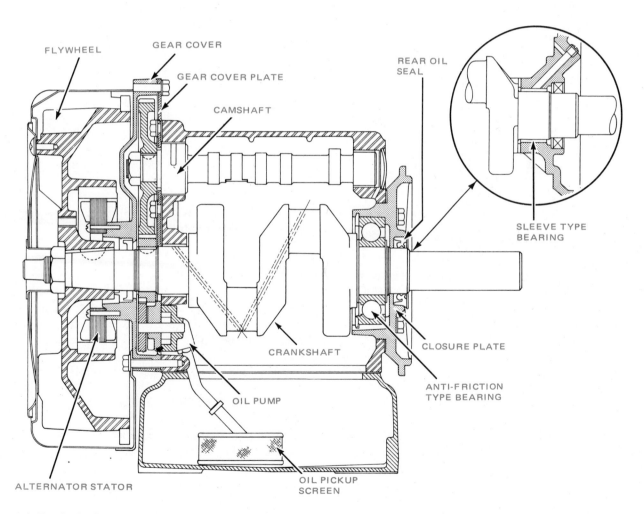

FLYWHEEL

GEAR COVER

GEAR COVER PLATE

CAMSHAFT

REAR OIL
SEAL

SLEEVE TYPE
BEARING

CLOSURE PLATE

ANTI-FRICTION
TYPE BEARING

CRANKSHAFT

OIL PUMP

ALTERNATOR STATOR

OIL PICKUP
SCREEN

Fig. 6-15. Oil pump used in two cylinder engine supplies oil to moving internal engine parts. Note drilled crankshaft.

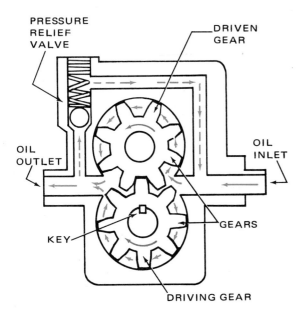

Fig. 6-16. In gear type pump operation, oil is carried between teeth of matching gears. If oil pressure is too high, relief valve recirculates oil through pump.

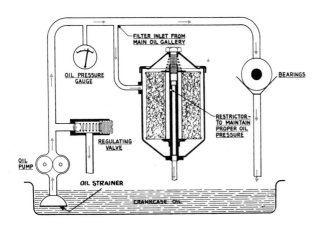

Fig. 6-17. Bypass filter system pumps some engine oil through filter. Remaining oil goes to engine bearings. (Wix Filters)

system, pressures would become excessively high during high engine speeds.

FULL PRESSURE LUBRICATION SYSTEM

A full pressure lubrication system is the type used in automobile engines. On some of the larger small engines, an almost completely pressurized system is used, including a positive displacement gear or rotor pump. Passages for oil flow are drilled to all critical points, such as camshaft bearings, main bearings, connecting rod bearings and piston pins. A splash system

is used in conjunction with the pressure system, particularly for lubricating cylinder walls.

OIL FILTER SYSTEMS

Oil filters similar to the one shown in Fig. 6-6 are used on some small engines. *Filters trap dirt, carbon and other harmful materials and prevent them from circulating through the engine.* The oil filter prevents very fine particles from circulating. The oil strainer, Fig. 6-17 (usually attached to intake side of oil pump), prevents large particles from entering the filter.

Three basic types of oil filter systems are in common use: bypass, shunt and full-flow. Generally, the filter element is replaceable and can be discarded when dirty.

BYPASS SYSTEM

The bypass filter system, Fig. 6-17, pumps part of the oil through the filter, while the remaining oil is pumped to the engine bearings. Oil pumped through the filter is returned directly to the crankcase. The primary purpose of the filter in the bypass system is to keep a clean supply of oil in the crankcase.

The pressure relief valve in the bypass system controls the maximum allowable pressure in the system. If there is a restriction to oil flow, pressure buildup will overcome spring tension and open the valve. When this occurs, oil pressure will be relieved and oil will flow through the valve, back to the crankcase.

Pressure relief valves are installed in all pressurized lubrication systems. Often they are an integral part of the oil pump.

SHUNT FILTER SYSTEM

In the shunt-type filter system, part of the oil delivered by the pump is filtered and directed to the engine bearings. Some of the oil is shunted past the filter. The remaining oil is circulated through the pressure relief valve, back to the crankcase.

FULL-FLOW FILTER SYSTEM

The full-flow system, Fig. 6-18, directs the entire volume of pumped oil through the filter to the bearings. If the filter element becomes clogged with dirt, oil pressure will increase. The added pressure will open the relief valve permitting the oil to flow. If the filter did not have a relief valve and the filter became clogged, serious damage to the engine would result.

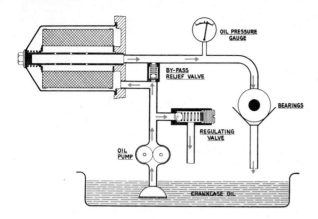

Fig. 6-18. Full-flow filter system directs all engine oil through filter. Relief valve opens if filter becomes clogged.

The filter cartridge must correspond to the filter system on the engine. For example, a full-flow cartridge used in a partial-flow system will give longer service life, but initial efficiency will be poor due to its high flow rate. One the other hand, a partial-flow cartridge used in a full-flow system would drastically reduce oil pressure. Proper oil filtration in modern gasoline engines cannot be overemphasized.

REVIEW QUESTIONS—CHAPTER 6

1. Give three general types of lubricants.
2. Name four important jobs performed by a good engine lubricant.
3. Which engine lubrication system utilizes a trough in the crankcase?
   a. Barrel type lubrication system.
   b. Ejection pump system.
   c. Constant level system.
   d. Splash system.
4. Which engine lubrication system used on some small engines is most similar to the system used on automobiles?
   a. Ejection pump system.
   b. Constant level splash system.
   c. Barrel type pump system.
   d. Full pressure system.
5. Babbitt metal used as a bearing material is an alloy of three metals. Name them.
6. The type of bearing shell shown in Fig. 6-2 is referred to as a _____ _____ bearing.
7. The lock tabs on the bearing in Fig. 6-3 are for the purpose of _____.
8. The lubricating oil recommendation for a particular make and model engine would be found on a special label on the engine. True or False?
9. API stands for _____.

10. SAE stands for _____.
11. "Type" of oil refers to its:
    a. Service classification.
    b. Viscosity.
    c. Grade.
    d. Maximum operating temperature.
12. Most four cycle engine manufacturers recommend oils supplemented by additives. True or False?
13. What is the purpose of detergent-dispersants in oils?
14. Why are some oils with additives not suitable for two cycle engine use?
15. A high viscosity oil would:
    a. Be thin.
    b. Be suitable in relatively high temperatures.
16. Several types of positive displacement oil pumps are used in pressurized lubrication systems. Yes or No?
17. Name three types of oil filter systems in use.
18. Generally, the oil filter element is replaceable. True or False?

SUGGESTED ACTIVITIES

1. Demonstrate friction to the class by first rubbing two sheets of wasted paper together, then rub two sheets of coarse abrasive together. Discuss how a lubricant could reduce friction in each case.
2. Make a viscosity measurement of an engine oil selected by your instructor, using a viscosimeter.
3. Compare a multi-viscosity oil with single viscosity oils, using a viscosimeter. Test the oils at various temperatures and make a chart of your results.
4. Compare a detergent oil with a non-detergent oil by placing equal amounts of carbon black in each. Shake them up, then let them set for several days. Explain any observed differences in the oils.
5. Produce condensation in a glass beaker partly filled with gasoline. Place the beaker loosely covered in a warm place or in direct sunlight. When the air in the beaker is warm, pack ice around the outside. Repeat the process until a quantity of water can be seen in the bottom of the beaker. A hot humid day will produce best results.
6. Create heat with friction. Place a dull, unwanted drill in a drill press. Try to drill a bar of metal. Show what happens to drill as a result of friction.
7. Demonstrate to the class how oil cleans. With engine grease and grime on your hands, wipe them clean in a container of clean engine oil. Compare the cleaning power with soap or a detergent.
8. Remove the end cover from a gear type oil pump and demonstrate how it works. Reinstall the cover, place the pump in oil and operate it by hand.

Air-cooled engines rely on flywheel fan for air circulation and on cooling fins of cylinder and cylinder head for heat dissipation.

Chapter 7
ENGINE COOLING

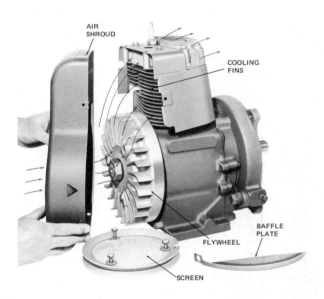

Fig. 7-1. Air cooling system of small gasoline engine consists of shroud, screen, flywheel, baffles and cooling fins. Air flow follows path shown by arrows. (Wisconsin Motors Corp.)

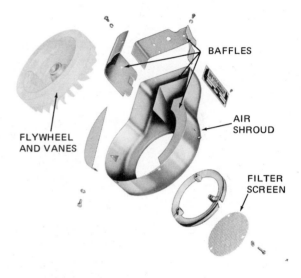

Fig. 7-2. A number of sheet metal parts enclose flywheel and cylinder. Baffles direct flow of air to critical areas.

The efficiency and life of an engine depends upon how well it is cooled. The average temperature of burned gases in the combustion chamber of an air-cooled engine is about 3600 deg. F (1982 C).

About a third of the heat is carried away by the cooling system, Fig. 7-1. Various sheet metal parts surrounding the air-cooled engine direct the flow of cooling air as shown in Fig. 7-2. The exhaust system carries away another third of the heat. That which remains is used to produce engine power.

Loss of heat through the cylinder walls to the cooling system reduces the temperature from 1200 deg. F (649 C) to approximately 350 deg. F (177 C). This drops to 100 deg. F (38 C) by the time it reaches the outer edges of the cooling fins. Fig. 7-3 illustrates cylinder wall temperatures at various locations.

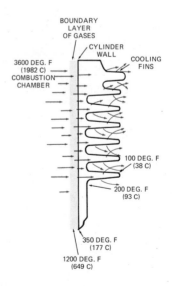

Fig. 7-3. High combustion chamber temperature is reduced by exhaust gases and cooling system. Large area of cooling fins controls heat dissipation from cylinder.

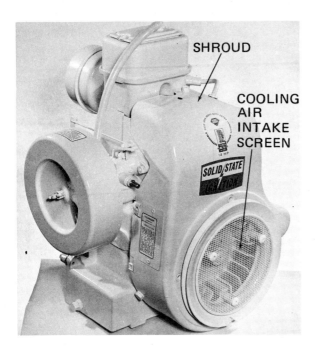

SHROUD

COOLING AIR INTAKE SCREEN

Fig. 7-4. Cooling air intake screen must be kept clean for unrestricted air flow. Screen must be kept in place to prevent clogging of cooling fins. (Tecumseh Products Co.)

There is a reason for the comparatively low temperature along the inner cylinder wall. A "boundary" layer of stagnant gas lies next to it and acts as an insulator. Therefore, less heat travels through the cylinder wall than you might expect from the 3600 deg. F (1982 C) produced by the burning fuel.

HOW AIR COOLING WORKS

Operating temperature is lowered to about 200 deg. F (93 C) as heat passes through cylinder wall to outer surface. Air forced over the cooling fins and directed by the sheet metal baffles rapidly dissipates it. The cooling system, then, carries heat away from the engine.

The heat of combustion (rapid burning and expansion of gas) travels from the cylinder through the cylinder walls by "conduction." Conduction is heat transfer through a solid material.

When the heat reaches the outer surfaces of the cylinder, air forced over the surface carries it away by "convection." *Convection occurs when heat transfers through movement of gas — in this case, air.*

The flywheel has fins which blow air around the cylinder wall and cooling fins. The flow of air is controlled and directed by a sheet metal shroud and baffles surrounding the flywheel and cylinder. An engine should never be run without the shroud in place or it will quickly overheat. The flywheel is covered for safety, Fig. 7-4, and a screen or perforated plate allows air to be drawn in. *The screen should be kept clean to permit unrestricted air flow.*

Thin cooling fins increase the surface area around the outside of the cylinder, Fig. 7-5. The greater the surface area in contact with cool air, the more rapidly the heat can be carried away. Cooling fins are necessary on air-cooled engines but not on water-cooled engines. Water is four times more effective than air for engine cooling.

Fig. 7-5. Cooling fins are designed and placed to provide adequate cooling for each part of cylinder. Thickness, surface area and spacing are important considerations

HOW EXHAUST COOLING WORKS

As noted earlier, the exhaust system carries away approximately a third of the engine heat. If the exhaust system is restricted in any way, part of the 1200 deg. F (649 C) temperature will remain in the metals of the engine. This heat buildup will increase friction. In turn maintenance costs increase on moving parts and engine life is shortened. Dirt, grass clippings, leaves, straw or other materials lodged between the cooling fins will tend to insulate the cylinder and cause "hot spots" and engine overheating. *Keeping an engine clean, cool and properly lubricated will significantly extend its life.*

HOW WATER COOLING WORKS

Water is an excellent medium for cooling engines. It is inexpensive, readily available and absorbs heat well.

Fig. 7-6. Outboard engines are frequently water cooled. They are identified by absence of cooling fins. (Evinrude Motors)

Some small engines are water-cooled because they are used in or around a water source. The outboard engine in Fig. 7-6 is typical of a relatively small water-cooled engine. Notice the absence of cooling fins found on air-cooled engines.

Water-cooled engines, in general, are made with a double wall surrounding the cylinder, Fig. 7-7. This is called a water jacket. A small pump keeps the water in circulation. The water circulating through the jacket absorbs the heat of combustion and carries it away from the engine. In cold weather antifreeze solutions, added to the water in the proper amounts, prevent freezing.

WATER PUMPS

Water pumps of many different designs are used to move the water through the cooling system. A sliding vane type pump is illustrated in Fig. 7-8.

In outboard engines the pump is generally located in the lower unit near or below the water line. The main pump member is driven by the vertical drive shaft or horizontal propeller shaft. The pump must have an opening for water to come in and one for it to go out. These are called the inlet and outlet.

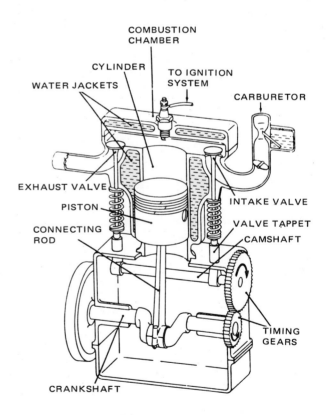

Fig. 7-7. Water is circulated around cylinders through water jacket, where it absorbs some combustion heat.

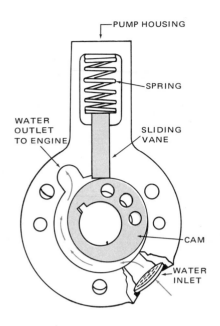

Fig. 7-8. Revolving cam in sliding vane water pump fills large volume of space, and water is pushed through outlet to engine. Sliding vane prevents water from revolving with cam.

As the eccentric cam (an off-center enlargement on a shaft) is rotated in the sliding vane pump, the volume of space between the cam and pump housing is constantly changing. Water enters the inlet and fills the space. The cam rotates and closes the inlet, pushing the water ahead of it toward the pump outlet.

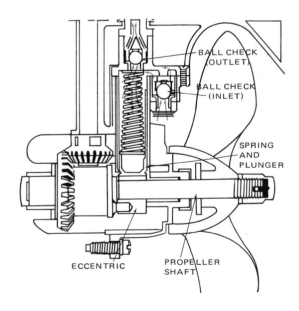

Fig. 7-9. Plunger pump draws water in through one ball check valve, then forces it out through a second valve. Plunger is operated by an eccentric.

The outlet is connected to the water jacket by suitable tubing. A second tube carries the heated water out of the engine. The sliding vane is kept in close contact with the cam by a spring. The vane provides a seal so that water cannot continue past the outlet. However, during high-speed operation, the water pressure may become high enough to lift the vane against the spring, allowing some water to recirculate through the pump. This action prevents too much pressure from building up in the cooling system.

The rotor type pump operates much like the sliding vane pump. The vane and rotor are one piece and the eccentric gyrates the rotor (rotates with a bobbing motion), causing a pumping action.

The plunger pump, Fig. 7-9, has a cylinder and plunger. The plunger is raised and lowered in the cylinder by an eccentric on the propeller shaft. The spring keeps the plunger close to the eccentric.

When the spring forces the plunger down in the cylinder, water is drawn in through the inlet ball check valve while the outlet check valve is closed. As the eccentric lifts the plunger, the inlet check valve is closed by the increasing pressure in the cylinder and the outlet check valve is opened. Water in the cylinder is forced through tubing to the engine.

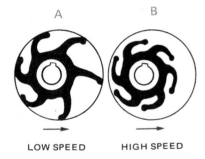

Fig. 7-10. Vari-volume pump has a flexible synthetic rubber impeller offset from center of impeller housing. A—Space between impeller arms varies as they rotate. B—At high speed, increased water pressure forces impeller arms inward to prevent overcooling of engine.

Vari-volume pumps use a synthetic rubber impeller. See A in Fig. 7-10. Since the impeller housing is off-center with the drive shaft, the impeller arms must flex as they revolve. The volume between the arms increases and decreases with rotation.

The inlet port or opening is located where the volume is increasing, and the water is drawn into the housing. The outlet is on the side where volume is decreasing, and the water is forced out into the water jacket.

When a vari-volume pump is driven at high speed, the back pressure of the water in the system becomes great enough to force the impeller arms inward. See B in Fig. 7-10. The pump loses some of its effectiveness and moves only enough water through the engine to maintain proper operating temperature in the cylinders. Overcooling of the cylinders of an outboard engine can happen at high engine speeds unless some sort of flow control, such as this, is designed into the cooling system.

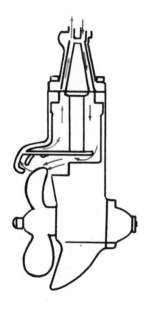

Fig. 7-11. Pressure-vacuum cooling system utilizes thrust of propeller tips and venturi (narrowed section of a passage) vacuum principle created by forward motion of engine to circulate water.

OUTBOARD WATER CIRCULATION SYSTEMS

Some small outboard engines use a simple cooling device called a pressure-vacuum water flow system, Fig. 7-11. The water flow from the propeller tips creates pressure against the intake port and vacuum at the outlet port located immediately forward of the propeller. Propeller action, then, plus the forward motion of the boat, provide water circulation.

At slow speeds, like those used for trolling, pressure of water from the propeller tips may not be enough to

force water through the cooling system. Sufficient cooling is still maintained, however, by the siphon (sucking) effect of the discharge channels as the boat moves through the water. When starting an engine with a pressure-vacuum system, the engine should be given a short burst of speed to fill the channels and water jacket with water, Fig. 7-11.

Another pressure-vacuum water flow system has the discharge ports located in the propeller blades. The centrifugal force created by the turning propeller aids in discharging the water. (Centrifugal force is the tendency of spinning matter to move away from the center of its path.) Since there are no moving parts, except the propeller, the system will function as long as the water channels and jackets remain unobstructed.

With this system, vacuum must be maintained, particularly at low speeds. Therefore, all water connections in the system are airtight. Air seepage into the cooling system would destroy the slow speed siphoning effect and cause overheating.

Worn propeller blades also can cause poor circulation. Their reduced diameter puts the tips further from the water scoop opening. This reduces water pressure. Salt water corrosion, marine growth and mud-clogged water channels are all causes for faulty water circulation in this type of system.

Other outboard water circulation systems are in use. Basic principles are the same. Some examples follow.

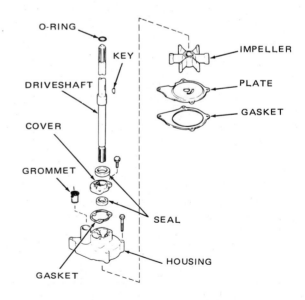

Fig. 7-12. Water pump for outboard engine is driven by vertical drive shaft. Some pumps are mounted on propeller shaft.

Fig. 7-12 shows a pump-driven cooling system. Water is drawn into the intake, pumped through the water jacket surrounding the cylinders and discharged. This is a relatively simple system. Notice that the pump is driven by the drive shaft.

In a similar system, the water pump is driven by the propeller shaft. When this pump location is used, the water is drawn in through ports in the propeller hub.

In some engines the temperature of the water is carefully controlled by a thermostat. This is simply a valve arrangement which stops circulation of coolant coming from the engine until it reaches operating temperature.

There are two types of thermostats, each opening and closing the valve at predetermined temperatures. Both work on the principle of the expansion of heated materials.

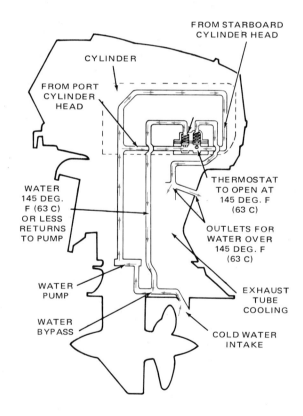

Fig. 7-13. Some outboard engines have a thermostatically controlled cooling system. Water recirculates until thermostat opens, discharging heated water and allowing more cold water to enter intake. This thermostatically controlled cooling system employs a double feed bypass.

In the old bellows type of thermostat, the valve is attached to a gas-filled, sealed bellows made of thin copper. Heat causes gas to expand, opening valve.

The more popular pellet thermostat operates in the same way but depends upon a small wax-filled copper cylinder. When heated, the wax pushes against a rubber diaphragm (disc). This moves a piston attached to the valve. In some cases, the piston remains stationary and the cylinder moves away to open the valve.

A thermostatically controlled engine, Fig. 7-13, will be kept at a constant temperature regardless of speed or outside temperature. Cold water enters the intake and passes through the pump to the water jacket. If the temperature of the water from the jacket is high enough to open the thermostat, the water is discharged from the engine. If the water is too cool to open the thermostat, it is recirculated through the pressure control valve and back to the pump. When water is recirculated, it retains some of its heat and eventually brings the cylinder head temperature up to thermostat temperature.

This type system can maintain a constant operating temperature and will automatically allow for even slight changes in cooling water temperature. A cold engine always has its thermostat closed until it reaches the thermostat opening temperature.

OBSERVING THERMOSTAT OPERATION

With the thermostat removed from the engine it is easy to see how it works. Place a beaker of water on a hot plate. Put the thermostat in the beaker of cool water. When the water has reached the proper temperature, the thermostat will open. By placing a thermometer in the beaker, Fig. 7-14, you can tell the temperature rating of the thermostat.

RADIATORS

Radiators are reservoirs for water made of many thin copper or aluminum tubes. These tubes are connected at top and bottom (or each side) to water tanks. The tubes are held in place by thin metal fins which increase cooling surface area of tubes. Tube and fin assembly shown in Fig. 7-15 is called "radiator core."

The water heated by the engine is pushed into the top or one side of the core by the water pump. The hot water travels downward or across through the tubes to the bottom tank. Cool air is forced between the fins in

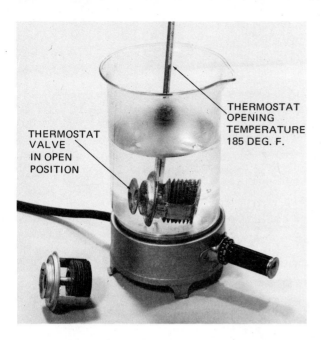

Fig. 7-14. A thermostat can be observed "in action" by heating it in a beaker of water. Thermometer registers temperature at which thermostat opens.

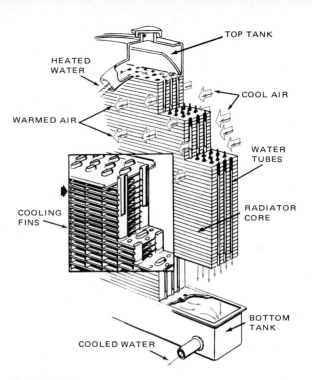

Fig. 7-15. Radiator carries water through tubes in core, where it is cooled by air forced across cooling fins.

the core by a fan driven by the engine. The air cools the water by conduction, radiation and convection. The cooled water passes from the bottom tank to the engine water jacket, ready to absorb more heat.

Engines cooled by radiators also have thermostats which open when the proper operating temperature is reached. When engine temperature is below normal, the thermostat remains closed, and the water is directed through a by-pass channel. The water by-passing the radiator retains its heat and returns directly to the engine. This action retards forced cooling until the engine reaches its normal operating temperature.

The fine balance of cooling maintained by the radiator and thermostat assures that the engine will not overheat under loads in the hot summer, nor run too cold in the winter.

OIL COOLING

Most of this chapter has been devoted to engine cooling by air or water. *Oil circulating through the engine also acts as a coolant.* The oil absorbs heat as it contacts the various metal parts. It then draws off the heat as it leaves the engine (two cycle engine) or transfers the heat through the crankcase to the outside air (four cycle engine).

REVIEW QUESTIONS – CHAPTER 7

1. The average temperature of the burned gases in the combustion chamber is:
   a. 2400 deg. F (1315.5 C).
   b. 3000 deg. F (1649 C).
   c. 6200 deg. F (3427 C).
   d. 3600 deg. F (1982 C).
2. The heat of combustion is carried away in three ways. One third is removed by the cooling system and one third by the exhaust system. That which remains is used to produce _____ _____.
3. Why is the temperature of the inside cylinder wall relatively cool compared to the combustion temperature?
4. The heat that reaches the cooling fins is carried away mainly by:
   a. Convection.
   b. Conduction.
   c. Radiation.
5. What are two ways to prevent overheating of air-cooled engines?
6. When used for engine cooling, water is about _____ times as effective as air.
7. Cylinders of water-cooled engines are surrounded by:
   a. Insulation material.
   b. A water reservoir.
   c. A water jacket.

d. Cooling water tubes.

8. Name three types of water pumps discussed in this chapter.

9. The _____ _____ water flow system in outboard engines utilizes the tips of the propeller to circulate water.

10. A heat sensing device that maintains a constant engine operating temperature is the:
   a. Check valve.
   b. Thermostat.
   c. Rheostat.
   d. Flexible impeller.

11. In cold weather, an _____ should be added to water-cooled engines.

12. What cooling system defect can cause "hot spots?"

13. What two fluids are used to cool engines?

SUGGESTED ACTIVITIES

1. Place a thermostat and thermometer into a beaker of water. Heat the water and check thermostat operation.

2. Operate an air-cooled engine and observe the flow of air through the shroud and around the cooling fins.

3. Inspect a cutaway of a radiator and explain how it cools water.

4. Disassemble the lower unit of an outboard engine and study the pumping and circulation system.

5. Place the outboard engine in a tank and operate it to observe the cooling system at work.

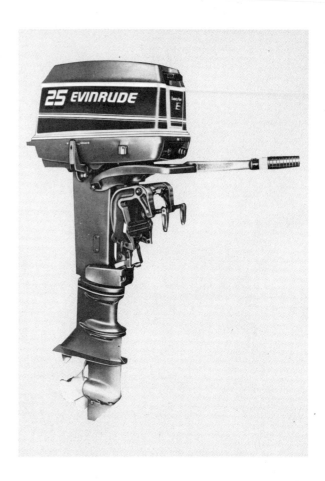

Cooling system of this 25 hp, two cylinder, two cycle outboard engine features a water pump, Fig. 7-10, and thermostatically controlled bypass system, Fig. 7-13.

A fully enclosed 8 hp small gas engine drives through a 3-speed tramission to power this medium size lawn tractor.

Chapter 8

MEASURING ENGINE PERFORMANCE

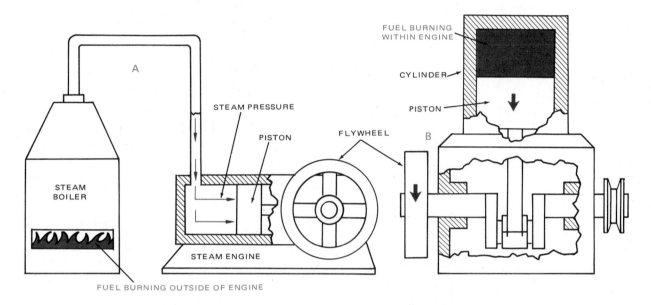

Fig. 8-1. External combustion engine burns fuel outside engine as in A. Internal combustion engine B burns fuel within engine.

The small gasoline engine belongs in the "heat engine" category. Other heat engines include the automotive reciprocating piston engine, the gas or steam turbine, steam engine, diesel engine, rotary combustion engine, rocket and jet engines. Only the small, one or two cylinder piston engines will be discussed here.

The small gasoline engine is called an internal combustion engine because the fuel (gasoline), combined with air, is ignited (fired) and burned inside the engine, Fig. 8-1. The heat of burning causes the gases to expand rapidly within the closed cylinder. The expanding gases apply strong force and push out in all directions within the cylinder, but only the piston can move.

The piston is pushed away from the center of combustion. If it were not fastened to the crankshaft, it would come out of the cylinder the way a bullet comes out of a gun. The crankshaft and connecting rod keep the piston under control and allow it to travel only a short distance, Fig. 8-2.

When the piston has moved downward as far as it can go, a port is opened. This allows burned gases to escape as the piston returns upward. The burned, escaping gases are called exhaust and come out through the exhaust pipe or manifold. This is a chamber which collects the exhaust and directs it to the exhaust pipe.

The piston does not pause long at the end of its stroke before the movement of the crankshaft and flywheel carry it back to the top of the cylinder. As you may have learned from your science class, bodies in motion tend to continue in motion. This tendency is called inertia. It is what keeps the crankshaft and flywheel moving in the engine.

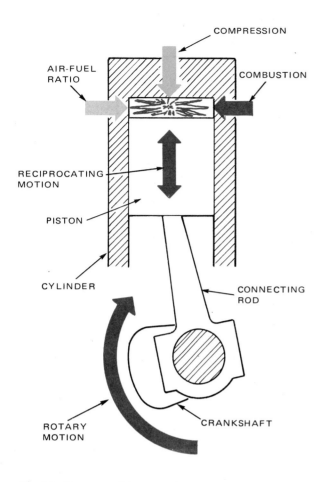

Fig. 8-2. Piston travel is controlled by connecting rod and crankshaft. Notice change from reciprocating (up and down) to rotary motion. (Deere & Co.)

When the piston reaches the top, it is ready to be forced back down again. Because the piston continues this back and forth or up and down motion, it is called a "reciprocating engine."

BASIC TERMINOLOGY

To better understand how a gasoline engine works, and to appreciate the power it provides, you must learn certain basic terms. These terms will be defined here only as far as necessary to provide a background for further discussion of measuring the work engines do and how well they do it. This is called "performance."

ENGINE BORE AND STROKE

Engine "bore" is the diameter or width across the top of the cylinder. "Stroke" is the up or down movement of the piston. Length of stroke is determined by the distance the piston moves from its uppermost position (top dead center or TDC) to its lowest (bottom dead center or BDC), or in reverse order.

The amount of crank offset (distance from center line of connecting rod journal to center line of crankshaft) determines the length of the stroke. A 2-in. offset would produce a 4-in. stroke, Fig. 8-3.

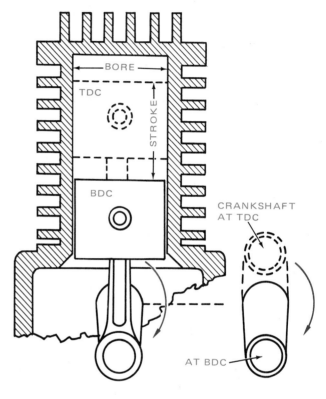

Fig. 8-3. Engine "bore" refers to diameter of cylinder. "Stroke" indicates length piston travels as it moves from TDC to BDC.

When the bore diameter is the same as the stroke, the engine is referred to as SQUARE. When the bore diameter is greater than the stroke, it is termed OVER SQUARE. Where bore diameter is less than the stroke, it is called UNDER SQUARE.

ENGINE DISPLACEMENT

Engine size or displacement, in a single cylinder engine, refers to the total volume of space increase in the cylinder as the piston moves from the top to the bottom of its stroke.

To work out a given engine's displacement, first determine the circular area of the cylinder (0.7854 by diameter^2). Then multiply that answer by the total length of the stroke (piston travel). The formula is:

ENGINE DISPLACEMENT =
0.7854 x D^2 x LENGTH OF STROKE

If the engine has more than one cylinder, multiply the answer to the above formula by the number of cylinders. For example, say that a two-cylinder engine has a bore of 3 1/4 in. and a stroke of 3 1/4 in. Using the formula, you would have:

0.7854 x D^2 (10.563) x LENGTH OF STROKE (3 1/4 in.) x NUMBER OF CYLINDERS (2) = 53.9 CU. IN.

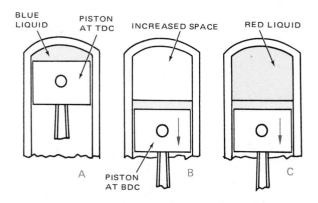

Fig. 8-4. Engine displacement is difference in volume of cylinder and combustion chamber above piston when it is at TDC and when it is at BDC. Red area shows displacement.

Fig. 8-4 illustrates piston displacement. In drawing A, the piston is at TDC (top dead center). Blue liquid has been added to fill up the space left between the piston and the cylinder head. The piston in B is at the bottom of its stroke. Note the increased space above the piston now. In drawing C, red liquid is added until the cylinder is once again filled. The amount of red liquid, in cubic inches, would represent the piston displacement for this cylinder.

COMPRESSION RATIO

The compression ratio of an engine is a measurement of the relationship between the total cylinder volume when the piston is at the bottom of its stroke as compared to the volume remaining when the piston is at TDC.

For example, if cylinder volume measures 6 cu. in. when the piston is at BDC (view A in Fig. 8-5) and 1 cu. in. when at TDC, (view B), the compression ratio of the

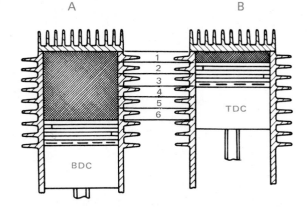

Fig. 8-5. Compression ratio is relationship between cylinder volume with piston A at BDC and piston B at TDC. Volume has been compressed to one-sixth of its original size, which indicates 6 to 1 compression ratio.   (Briggs and Stratton Corp.)

engine is 6 to 1. Many small gasoline engines have 5 or 6 to 1 compression ratios. Certain motorcycle engines have 9 or 10 to 1 compression ratios.

FORCE

Forces are being applied all around us. *Force is the pushing or pulling of one body on another.* Usually two bodies must be in contact for force to be transmitted.

For example, as you read this you are applying a force to a chair if you are sitting, or to the floor if you are standing. The force is equal to the weight of your body.

You can easily measure it with a scale, Fig. 8-6. This is known as gravitational force, and it acts upon all materials on and around the earth.

Some forces are stationary (motionless); others are moving. For example, if you push against a wall, force is applied but the wall does not move. The use of force may or may not cause motion. Force itself cannot be seen, but there are many ways of using it.

Centrifugal force acts upon a body whenever it follows a circular or curved path. The body tries to move outward from the center of its path. Modern examples are the man-made satellites that orbit the earth. The circular path and speed of the satellite produces a centrifugal force outward equal to the earth's gravitational force inward so that each is opposed and balanced. Therefore, the satellite neither goes up nor comes down. This is one case where a force

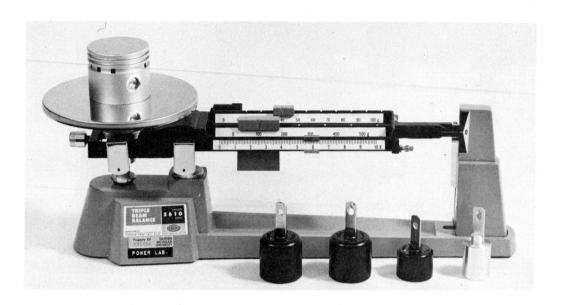

Fig. 8-6. A triple beam balance is used to measure force of gravity applied to an engine piston. Weight is read off three scales. Top scale is graduated in 10s; center in 100s; bottom in single grams.

is applied without one body touching another body. A ball swung on a string applies centrifugal force as shown in Fig. 8-7.

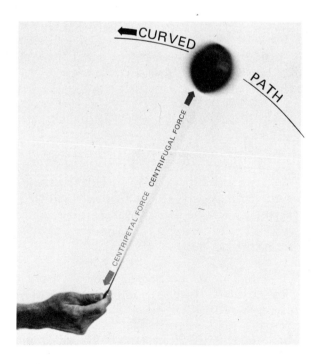

Fig. 8-7. An object in motion tends to travel in a straight line. Additional force required to cause deviation (or change) is called centripetal force. Reaction to centripetal force is called centrifugal force.

Fig. 8-8. A small gasoline engine being lifted with a force equal to, or slightly greater than, its own weight.

Many forces interact when a gasoline engine is operating. Rotational speed of the crankshaft and flywheel create centrifugal force which causes tensil (tension or pull) stress within the materials making up these parts. If the outward pulling force becomes greater than the strength of the material, the engine could fly apart. The rapid reciprocation (backward and forward or up and down motion) of the piston may put high forces on the connecting rod, crank journal and piston pin. Fig. 8-2 shows how these parts work together.

One of the forces used well in the gasoline engine is that applied to the top of the piston by the rapidly expanding gases. This is produced by burning gasoline mixed with air. *The greater the force applied to the piston, the greater the amount of power and work that can be done by the engine.*

Force is measured in units of some standard weight such as pounds, ounces or grams. For example, to support an engine weighing 16 lb., a person would have to apply a lifting force of 16 lb. Obviously, only half the lifting force would be needed to support an 8-lb. engine, Fig. 8-8.

Force and pressure are often confused. These terms should be understood in the way they are applied. *Pressure is a force per given unit of area.*

For example, a piston with a face area of 5 sq. in. may have a total force of 500 lb. applied to it by the expanding gases. However, the pressure being applied is 500 lb. divided by 5 sq. in., which equals 100 psi (pounds per square inch). This means that every square inch on the piston face has the same as a 100 lb. weight pushing on it.

When we speak of pressures in mathematical calculations we allow letters to replace several words. For example, psi means "pounds per square inch." As an example:

$$PSI = \frac{FORCE}{AREA}$$

$$or \quad FORCE = PSI \times AREA$$

$$or \quad AREA = \frac{FORCE}{PSI}$$

The area of a circle can be found by multiplying π (π = 3.1416) times the radius squared. (Written: AREA = πr^2.) Another method is to multiply the constant .7854 times the diameter squared. (Written: AREA = .7854 D^2.)

For example, we shall calculate a force applied to a 3-in. diameter piston, Fig. 8-9, if the cylinder pressure is 125 psi:

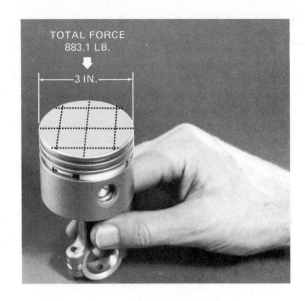

Fig. 8-9. Total force supplied to piston face is equal to its area in square inches times psi (pounds per square inch).

$$AREA = \pi r^2 \text{ or } 3.14 \times 1.5 \times 1.5$$
Then the piston area equals: 7.065 sq. in.
$$FORCE = PSI \times AREA \text{ or } 125 \times 7.065$$
This results in a total force of 883.1 lb.

WORK

Work is accomplished only when a force is applied through some distance. If a given weight is held so that it neither rises nor falls, no work is done even though the person holding the weight may become very tired. If the weight is raised some distance, then work is being done. The AMOUNT of work is the product or result of the FORCE and the DISTANCE through which it is moved.

If a weight of 20 lb. is lifted 3 ft., then 60 ft.lb. of work has been accomplished. *The distance must always be measured in the same direction as the applied force.* This results in the formula:

$$WORK = FORCE \times DISTANCE$$

Because the formula calls for multiplying feet times pounds, the answer is expressed in FOOT-POUNDS (ft.lb.)

The gasoline engine utilizes the principles of a number of simple machines. These machines include the lever, the inclined plane, the pulley, the wheel and axle, the screw and the wedge.

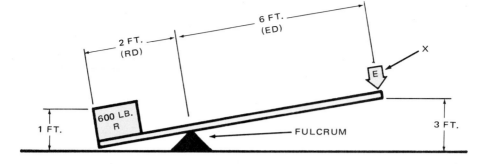

Fig. 8-10. Principle of lever is same as those applied to other forms of simple machines.

With each of the simple machines, you will find that to increase the output force with a given input force, the input distance will have to be increased in the same proportion or percentage. Therefore, without considering loss through friction, the foot-pounds of output are equal to the foot-pounds of input.

By using the principle of the lever, for example, a mechanical advantage is possible. Fig. 8-10 illustrates how a heavy load can be moved a short distance with a small force exerted through a relatively great distance.

The formula for computing leverage, as it applies to Fig. 8-10, is as follows:

$$\text{MA (mechanical advantage)} = \frac{\text{ED (effort distance)}}{\text{RD (resistance distance)}}$$

$$\text{MA} = \frac{\text{ED}}{\text{RD}} = \frac{6}{2} = 3$$

$$\text{E (effort)} = \frac{\text{R (resistance)}}{\text{MA}} = \frac{600}{3} = 200 \text{ lb.}$$

$$\text{or:} \quad X{:}6 = 2{:}600$$

$$X = \frac{2 \times 600}{6}$$

$$X = 200 \text{ lb.}$$

POWER

In studying the formula for work, note that it does not consider the time required to do the work. For example, if a small gasoline engine weighing 50 lb. is lifted from the floor to the workbench, a distance of three feet, 150 ft. lb. of work is done. The same amount of work would be performed whether it took 50 sec. to lift the engine or only five.

Because it is important to know the rate at which work is done, the word POWER enters the picture. *Power is the rate at which work is performed.* The rate (amount of time) is given in seconds. Power can then be considered as foot-pounds per second.

The formula for power is:

$$\text{POWER} = \frac{\text{WORK}}{\text{TIME}} \quad \text{or}$$

$$\text{POWER} = \frac{\text{FEET} \times \text{POUNDS}}{\text{SECONDS}} \quad \text{or}$$

$$\text{POWER} = \text{FT. LB. PER SECOND}$$

When the engine is lifted in 5 sec., 150 ft. lb. of work is performed. Using the power formula, it can be seen that:

$$\text{POWER} = \frac{\text{WORK}}{\text{TIME}} \quad \text{or}$$

$$\text{POWER} = \frac{150 \text{ LB.}}{5 \text{ SECONDS}} \quad \text{or}$$

$$\text{POWER} = 30 \text{ FT. LB. SECONDS}$$

When lifting the engine in 50 seconds, the formula shows:

$$\text{POWER} = \frac{\text{WORK}}{\text{TIME}} \quad \text{or}$$

$$\text{POWER} = \frac{150 \text{ LB.}}{50 \text{ SECONDS}} \quad \text{or}$$

$$\text{POWER} = 3 \text{ FT. LB. SECONDS}$$

Work is a force applied to an object causing the object to move, and power is the rate at which the work is done. The standard unit of power is termed HORSE-POWER.

ENERGY

Energy is the capacity to perform or do work. It is grouped into various types such as: POTENTIAL ENERGY (PE); KINETIC ENERGY (KE); MECHANICAL ENERGY (ME); CHEMICAL ENERGY (CE); THERMAL ENERGY (TE). These kinds of energy and some of their sources are shown in Fig. 8-11.

Many of these words just used will be unfamiliar to you. We will explain them as we go along. To begin with, energy is something we cannot define. It puts

	Type of Energy				
	PE	KE	ME	CE	TE
Coal, Oil, Gasoline				x	
Clock, Spring-Wound	x				
Moving Auto		x			
Water Behind A Dam	x				
Jacking Up An Auto	x				
A Cooking Fire					x
Dynamite, Stored				x	
Dynamite, Exploding			x		x
Crankshaft, Turning		x			
Man, Running		x			

Fig. 8-11. Potential, kinetic, mechanical, chemical and thermal energy and some of their sources. (Go-Power Corp.)

"life" into matter, giving it warmth, light and motion. Energy cannot be seen, weighed or measured. It does not take up space. However, we know it is there. The warmth and light of a bonfire, the electrical spark that jumps the gap of a spark plug, or the turning of a wheel are things we can sense. Thus, they must exist.

Matter and energy cannot be destroyed. Nothing is lost. Only the nature of matter and the forms of energy change. For example, when a piece of charcoal burns and disappears we may think it is completely gone. But the charcoal material has combined with air and formed a like quantity of ash, water and gases. The energy that did this existed as flame (light energy) and heat (heat energy). All of this will continue until another change takes place.

Whenever a form of matter can be separated from other forms of matter so that a part or all of its energy can be released, it is said to contain potential energy. Examples of such matter are crude oil and the gasoline taken from it. We have learned different ways of releasing a part of the energy stored in these substances.

Engines are designed to release and change the potential energy of gasoline into mechanical power. Mechanical power does the work at hand.

HORSEPOWER

Men used horses to perform work for hundreds of years. It was only natural that when machines were

made, their ability to perform work would be compared to the horse.

James Watt, in his work with early steam engines, wanted some simple way to measure their power output. In measuring the power or rate of work performed by a horse, he found that most work horses could lift 100 lb. 330 ft. in 1 min.

If 1 lb. is lifted 1 ft. in 1 min., 1 ft. lb. of work is done. The horse lifted 100 lb. 330 ft. in 1 min. Using the work formula,

$$(WORK = DISTANCE \times FORCE)$$

Watt found that the horse performed 33,000 ft. lb. of work.

In determining the RATE OF POWER developed by the horse, we use the formula:

$$POWER = \frac{WORK}{TIME} = \frac{33,000 \text{ FT. LB. (WORK)}}{1 \text{ MINUTE}} \quad or$$

$$\frac{550 \text{ FT. LB.}}{1 \text{ SECOND}} = 1 \text{ HORSEPOWER}$$

The 550 ft. lb. sec. (ability to lift 550 lb. 1 ft. in 1 sec.) was then established as 1 hp. This standard is still in use today.

HORSEPOWER FORMULA

Engine horsepower may be learned by dividing the total rate of work (ft. lb. per sec.) by 550 (ft. lb. per sec.)

If an engine lifted 330 lb. 100 ft. in 6 sec., its total rate of work would be:

$$5500 \frac{\text{ft. lb.}}{\text{sec.}}$$

Dividing this by:

$$550 \frac{\text{ft. lb.}}{\text{sec.}}$$

(1 horsepower), you find the engine rated at 10 hp. The horsepower formula would then be:

$$1 \text{ HORSEPOWER} = \frac{\text{RATE OF WORK IN } \frac{\text{FT. LB.}}{\text{SEC.}}}{550 \frac{\text{FT. LB.}}{\text{SEC.}}}$$

This formula may be used also to determine the exact horsepower needed for other tasks.

KINDS OF HORSEPOWER

The word horsepower is used in more than one way. Some of the common words are BRAKE horsepower,

INDICATED horsepower, FRICTIONAL horsepower and RATED horsepower.

BRAKE HORSEPOWER

Brake horsepower (bhp), indicates the actual usable horsepower delivered at the engine crankshaft. *Brake horsepower is not always the same. It increases with engine speed.* At very high and generally unusable engine speeds (depending on engine design), the horsepower output will drop off somewhat.

HORSEPOWER

| | ENGINE MODELS | |
RPM	ACN	BKN
1600	2.5	3.5
1800	2.9	4.0
2000	3.5	4.4
2200	3.7	4.9
2400	4.2	5.4
2600	4.5	5.8
2800	4.8	6.2
3000	5.2	6.5
3200	5.6	6.7
3400	5.8	6.9
3600	6.0	7.0

Fig. 8-12. Brake horsepower increases with engine speed. Note that hp at 3600 rpm is about twice that developed at 1600 rpm. (Wisconsin Motors Corp.)

Fig. 8-12 shows how the horsepower increases with speed for two different engine models. The top speeds in this chart do not run high enough to cause a drop in horsepower.

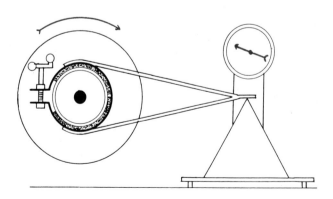

Fig. 8-13. When using Prony brake, one end of pressure arm surrounds a spinning flywheel driven by engine. By tightening friction device, torque is transmitted to and measured at scale.

MEASURING ENGINE BRAKE HORSEPOWER

Brake horsepower can be measured by using a PRONY BRAKE or an engine DYNAMOMETER.

The Prony brake is a friction device that grips an engine-driven flywheel and transfers the force to a measuring scale, Fig. 8-13. One end of the Prony brake pressure arm rests on the scale and the other wraps around a spinning flywheel driven by the engine under test. A clamp is used to change the frictional grip on the spinning flywheel.

To check brake horsepower, the engine under test is operated with the throttle wide open. Then engine speed is reduced to a specific number of revolutions per minute by tightening the pressure arm on the flywheel. At exactly the right speed, the arm pressure on the scale is read. By using the scale reading (W) from the prony brake, the flywheel rpm (R) and the distance in feet from the center of the flywheel to the arm support (L), bhp can be computed.

This formula is used to determine brake horsepower on the Prony brake:

$$BHP = \frac{2\pi \times R \times L \times W}{33,000} \quad \text{or: } BHP = \frac{R \times L \times W}{5252}$$

R = Engine rpm or speed.
L = Length from center of flywheel to the point where beam presses on scale in feet.
W = Weight as registered on scale in pounds.

NOTE: One hp is:

$$550 \; \frac{\text{ft. lb.}}{\text{sec.}}$$

Since engine rpm is on a per-minute basis, it is necessary to multiply the 550 by 60 (60 sec. per min.), giving the figure 33,000.

ENGINE DYNAMOMETER

As with the Prony brake, the DYNAMOMETER loads the engine and transfers the loading to a measuring device. Instead of using a dry friction loading technique (clamping pressure arm to a spinning wheel) the dynamometer utilizes either hydraulic or electric loading. Several different types are illustrated in Fig. 8-14.

In drawing A in Fig. 8-14, the engine drives an electric generator that is attached to a spring scale. When an electrical load is placed in the circuit, the

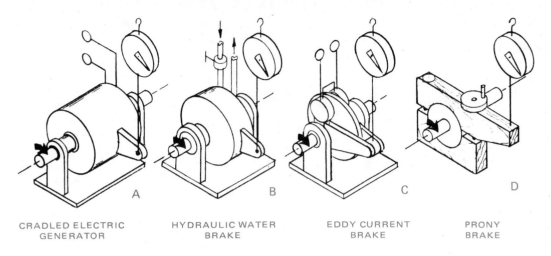

A B C D

CRADLED ELECTRIC
GENERATOR

HYDRAULIC WATER
BRAKE

EDDY CURRENT
BRAKE

PRONY
BRAKE

Fig. 8-14. Variety of dynamometers shown utilize different principles of construction.
(Go-Power Corp.)

generator housing (an enclosure holding the moving parts) attempts to spin, applying a force to the scale. In drawing B, a hydraulic water brake is attached to the scale. (Hydraulic means moved or operated by a liquid.)

The engine is loaded by admitting more and more water into the brake, causing the housing to try to rotate. This exerts a force on the scale. Drawing C shows an Eddy Current brake; D is the Prony brake.

Fig. 8-15. A small gas engine is readied for testing on a water brake
dynamometer. (Go-Power Corp.)

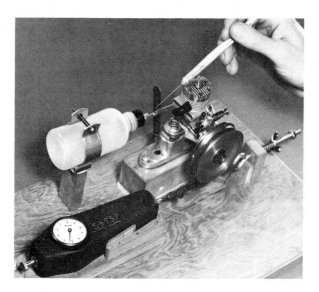

Fig. 8-16. A dynamometer built with a precision spring scale for testing a model airplane engine. Notice vibration tachometer being used to record rpm.

A hydraulic load cell is used with one type of water brake dynamometer. When the housing tries to rotate, it creates a force on the hydraulic load cell. This cell is a type of piston which pushes against the water. The "push" is transmitted through tubing to the gage, where it is read in pounds. Fig. 8-15 shows a small gas engine set up for testing on such a dynamometer. Note the load cell. See how it is attached to the housing. Fig.

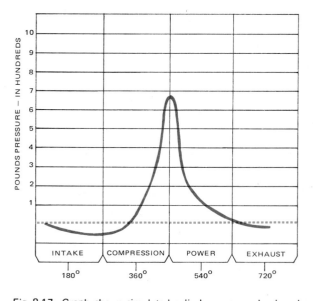

Fig. 8-17. Graph shows simulated cylinder pressure developed in cylinder of a specific four cycle engine. Atmospheric pressure is shown by dotted line. Graph, makes it possible to establish mean effective pressure (mep).

8-16 pictures a dynamometer which can be built to test model airplane engines.

INDICATED HORSEPOWER

Indicated horsepower (ihp) measures the power developed by the burning fuel mixture inside the cylinder. *To measure ihp, you must determine the pressure inside the cylinder during the intake, compression, power and exhaust strokes.* A special measuring tool is used to provide constant checking of cylinder pressure. This pressure information is placed on an indicator graph like the one in Fig. 8-17.

At this point, a mean effective pressure (mep) must be determined. To do this we subtract the average pressure during the intake, compression and exhaust strokes from the average pressure developed during the power stroke. Mep changes according to engine type and design. After finding the mep, the following formula determines indicated horsepower:

$$\text{INDICATED HORSEPOWER (IHP)} = \frac{\text{PLANK}}{33,000}$$

P = mep in pounds per square inch.
L = length of piston stroke in feet.
A = cylinder area in square inches.
N = power strokes per minute:
$\frac{\text{rpm}}{2}$ (four cycle engine).
K = number of cylinders.

FRICTIONAL HORSEPOWER

Frictional horsepower (fhp) represents that part of the potential or indicated horsepower lost because of the drag of engine parts rubbing together, Fig. 8-18.

Despite smooth contact surfaces and proper oiling of these parts, a certain amount of friction (resistance to movement between two objects that are rubbing together) is always present and represents a sizeable loss. Actual loss will vary with engine design and use, but will run about 10 percent. Friction loss is not always the same. It increases with engine speed.

Frictional horsepower is determined by subtracting brake horsepower from indicated horsepower or:

$$\text{FHP} = \text{IHP} - \text{BHP}$$

RATED HORSEPOWER

An engine used under a brake horsepower load that is as great as the engine's greatest brake horsepower will

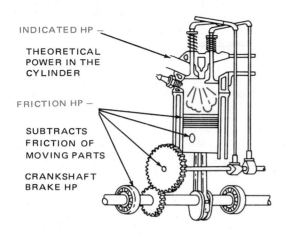

INDICATED HP —

THEORETICAL POWER IN THE CYLINDER

FRICTION HP —

SUBTRACTS FRICTION OF MOVING PARTS

CRANKSHAFT BRAKE HP

Fig. 8-18. Frictional horsepower is determined by subtracting crankshaft brake horsepower from indicated horsepower. (Deere & Co.)

overheat. Excessive pressure on bearings (loading) will then seriously shorten its service life. In some cases, complete engine failure occurs in a very short time.

As a general rule, never load an engine any more than about 80 percent of its highest brake horsepower

rating. For example, if a job requires a horsepower loading of 8 hp, you would use an engine with at least 10 hp. Then, the load would be no more than 80 percent of the engine's highest or maximum hp.

An engine's "rated" horsepower generally will be 80 percent of its maximum brake hp. Note in Fig. 8-19 how the "rated" horsepower (recommended maximum operating bhp) is less than engine maximum bhp.

CORRECTED HORSEPOWER

Standard brake horsepower ratings are based on engine test conditions with the air dry, temperature at 60 deg. F and with a barometric pressure of 29.92 (sea level). Horsepower, however, can be greatly affected by changes in atmospheric pressure, temperature and humidity (amount of moisture in the air).

Corrected horsepower is a "guess" at the horsepower of a given engine under specific operating conditions that are not the same as those present during actual dynamometer testing. Some facts to consider are:
1. For each 1000 ft. of elevation above sea level, horsepower will drop around 3 1/2 percent.

Fig. 8-19. Maximum operating brake hp loading is charted for a specific engine. At all speeds, "rated" hp is about 80 percent of maximum bhp. (Briggs and Stratton Corp.)

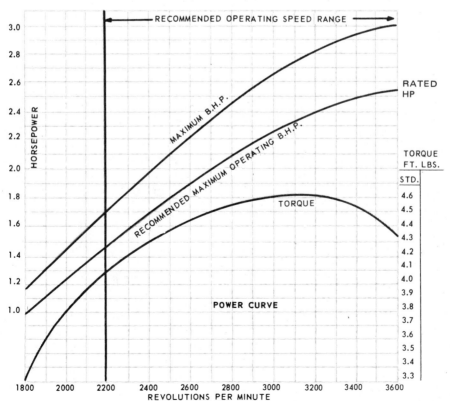

2. For each 1 in. drop in barometric pressure, horse-power will drop another 3 1/2 percent.
3. Each 10 deg. of temperature increase results in a horsepower loss of 1 percent.
4. New engines will develop somewhat less horsepower (due to increased friction) until they have been operated a number of hours.
5. An increase of 200-400 deg. F in head operating temperature can lower horsepower by 10 percent.
6. Quality of fuel, mechanical conditions and "state of tune" also affect horsepower.

When horsepower tests are conducted under conditions varying from standard, corrections must be applied to establish true horsepower.

CORRECTION FACTOR

The correction factor (a factor is a condition which would change an answer) is determined by using the following formula:

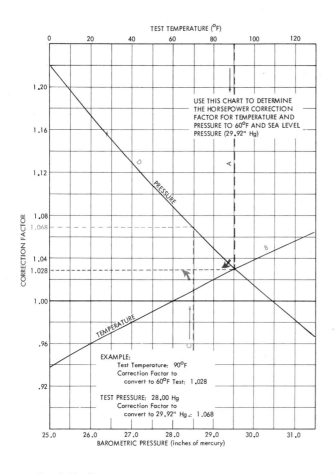

Fig. 8-20. Temperature and barometric pressure correction chart. (Go-Power Corp.)

CORRECTION FACTOR =
TEMPERATURE CORRECTION x PRESSURE
CORRECTION x HUMIDITY CORRECTION

For example, suppose that the dynamometer tests were carried on at a temperature of 90 deg. F with an atmospheric pressure of 28 in. Hg and a wet bulb temperature (determines humidity) of 73.5 deg. F. First, refer to the chart in Fig. 8-20. Follow dotted line A from top of chart (90 deg. F temp) down until it crosses the temperature line B. See that by now moving left along the chart, the dotted line shows a temperature correction factor of 1.028.

Now, follow dotted line C up from the 28 in. Hg marking at the bottom of the chart until it crosses pressure line D. By moving left at this point, a pressure correction factor of 1.068 is shown.

To find the humidity correction factor, use the chart in Fig. 8-21. Follow the dotted line up from the 90 deg. F dry bulb temperature mark until it meets the 73.5 deg. F dotted wet bulb temperature line. Move across to the right and note that the humidity correction factor is 1.0084.

Using the correction factor formula:
CORRECTION FACTOR =
TEMPERATURE CORRECTION x PRESSURE
CORRECTION x HUMIDITY CORRECTION
or
CORRECTION FACTOR = 1.028 x 1.068 x 1.009
or
CORRECTION FACTOR = 1.108

If the dynamometer test had shown 3.15 horse-power, this reading could be changed to standard test conditions by applying the correction factor of 1.108 as determined above. Thus:
CORRECTED HP =
CORRECTION FACTOR x TEST HP   or
CORRECTED HP = 1.108 x 3.15   or
CORRECTED HP = 3.49 HP

TORQUE

Torque is a twisting or turning force. Therefore, any reference to engine torque means the turning force developed by the rotating crankshaft.

In order to find torque, we must know the force (in pounds) and the radius (distance from the center of the turning shaft, in feet, to the exact point at which the force is measured). The formula could read:

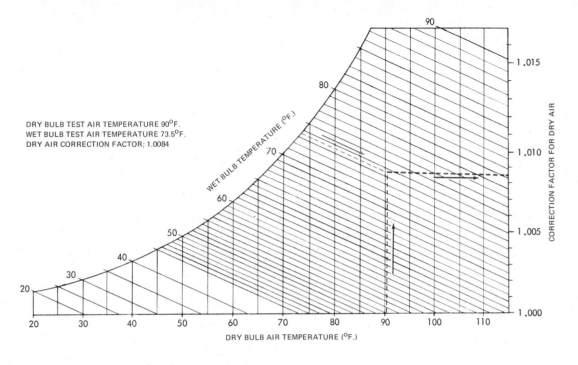

Fig. 8-21. Chart for determining humidity correction factor.
(Go-Power Corp.)

TORQUE = FORCE x DISTANCE (RADIUS) or
TORQUE = POUNDS x FEET or
TORQUE = LB. FT.

Fig. 8-22 shows a torque wrench attached to a rotating crankshaft. Imagine that it is attached with a friction device much like the Prony brake pressure arm shown in Fig. 8-13. This will allow the crankshaft to turn while still applying turning force to the stationary torque wrench. Suppose a scale placed exactly two feet from the center of the crankshaft indicates a force of 100 lb. By using the formula for determining torque (Torque = lb. ft. or Torque = 100 lb. x 2 ft.), we find that this engine is developing 200 lb. ft. of torque.

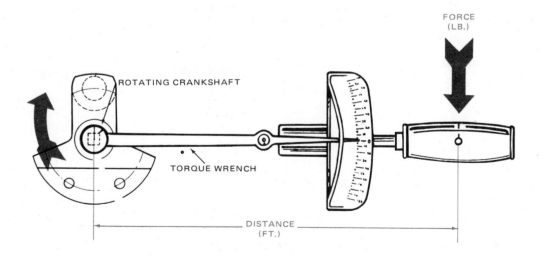

Fig. 8-22. Torque is determined by multiplying turning effort in pounds times distance from shaft center to point at which force is read. (Dresser Industries Inc.)

If the scale is 3 ft. from the shaft center and the force is 50 lb., the torque will be 150 lb. ft. *When measuring torque, the reading is given in lb. ft. When measuring work, the reading is given in ft. lb.*

TORQUE IS NOT CONSTANT

Engine torque, for any engine and set of test conditions, will change according to engine speed. The pressure of the burning air-fuel mixture against the piston is transferred to the crankshaft by the connecting rod. The greater the pressure, the greater the torque the crankshaft will develop.

The point where gas pressure will be highest is the speed at which the engine takes in the largest volume of air-fuel mixture. This point will vary according to engine design but will always be at a lower speed than that at which the greatest horsepower is reached. Horsepower generally increases to quite a high rpm before it finally begins to drop off. Torque, on the other hand, decreases at a much lower rpm.

As engine speed is increased beyond idle, its torque increases. As it continues to speed up, a point will be reached where the natural restriction to air flow through the carburetor, intake manifold and valve ports begins to limit the speed at which the air-fuel mixture can enter the cylinder. At this point, the highest torque is developed.

When engine speed goes higher than this point, the intake valve will open but the piston moves far down on the intake stroke before the mixture can get into the cylinder. This cuts down the amount of air-fuel mixture entering the cylinder. As a result, burning pressure is lowered as well as the torque. Beyond this point, as speed increases, torque will decrease.

VOLUMETRIC EFFICIENCY

How well an engine "breathes," or draws the air-fuel mixture into the cylinder, is referred to as its volumetric efficiency. It is measured by comparing the air-fuel mixture actually drawn in with that which could be drawn in if the cylinder were completely filled, Fig. 8-23.

Volumetric efficiency changes with speed. At high engine speeds, it can be very low. The reason for this is simple. As engine revolutions increase beyond a certain point, the piston moves down (intake stroke) so rapidly that it travels far down the cylinder before the air-fuel

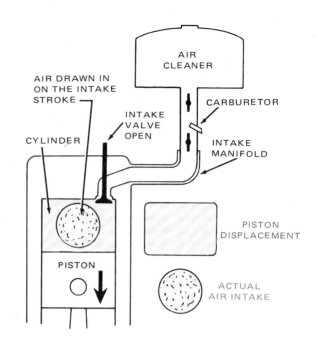

Fig. 8-23. Volumetric efficiency is measurement of an engine's "breathing" ability. It compares intake of air-fuel mixture with piston displacement. Note that actual air intake is considerably less than piston displacement. (Deere & Co.)

mixture begins to flow into the cylinder. The intake cycle can be complete (piston moving upward on compression stroke) before the cylinder is much more than half-filled.

Other factors can change the volumetric efficiency. For example: atmospheric pressure; air temperature; air cleaner, carburetor and intake manifold design; size of intake valve; engine temperature; throttle position; valve timing; camshaft design.

Efficiency can be increased by using a larger intake valve, altering cam profiles (shapes) or cam timing, increasing the size of the carburetor air horn, straightening and increasing the diameter of the intake manifold, improving exhaust flow or adding a supercharger.

Fig. 8-24 illustrates how volumetric efficiency varies with speed. There is a gradual buildup to a certain rpm, followed by a rapid decline as speed is increased. Remember that the speed at which top volumetric efficiency is reached will vary with engine design.

TORQUE AND HORSEPOWER

Unlike torque (which drops off when engine revolutions per minute exceed point of maximum volumetric

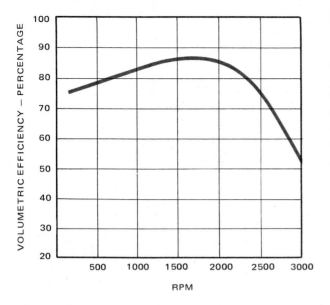

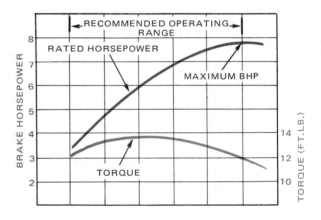

Fig. 8-24. Graph shows close relationship of volumetric efficiency and engine rpm. As engine speed reaches a certain point, efficiency declines rapidly. Torque is also greatest at point of highest volumetric efficiency.

efficiency), horsepower continues to increase until engine speed is very high. Beyond a certain speed, horsepower will actually decrease.

Keep in mind that torque measures the twisting force generated by the crankshaft while horsepower measures the engine's ability to perform work. Even though torque may decline at higher speeds, the shaft is turning much faster. Thus, it is able to perform work at a greater rate.

Fig. 8-25 shows the relationship between torque and horsepower curves for one specific engine. Note also the arrow indicating the rated horsepower. *This is the horsepower at which the engine can be operated continuously without damage.*

PRACTICAL EFFICIENCY

In theory, each gallon of gasoline contains enough energy to do a certain amount of work. This may be thought of as potential energy. Unfortunately, engines are not efficient enough to use all the potential energy in the fuel. Practical efficiency takes into consideration power losses caused by friction, incomplete burning of the air-fuel mixture, heat loss, etc. Practical efficiency is simply an overall measurement of how efficiently an engine uses the fuel supply.

MECHANICAL EFFICIENCY

Mechanical efficiency is the percentage of power developed in the cylinder (indicated horsepower) compared with that which is actually delivered at the crankshaft (brake horsepower). Brake horsepower is always less than indicated horsepower. The difference is due to friction losses within the engine. Mechanical efficiency runs about 90 percent, indicating an internal friction loss of about 10 percent.

The formula for mechanical efficiency is:

$$\text{MECHANICAL EFFICIENCY} = \frac{\text{BRAKE HORSEPOWER}}{\text{INDICATED HORSEPOWER}} = \frac{\text{BHP}}{\text{IHP}}$$

Fig. 8-25. An illustration of relationship between torque and horsepower for one particular engine. (Clinton Engine Corp.)

THERMAL EFFICIENCY

Thermal efficiency (heat efficiency) indicates how much of the power produced by the burning air-fuel mixture is actually used to drive the piston downward.

Much of the heat developed by the burning gas is lost to such areas as the cooling, exhaust and lubricating systems. Thermal efficiency will run about 20 to 25 percent. Keep in mind that the percentages are only "about right" and will vary depending upon engine design and operation, Fig. 8-26. The exhaust system syphons off about 35 percent of the heat. The cooling and lubricating systems combine to absorb a like amount. The rest is lost through radiation and incomplete combustion.

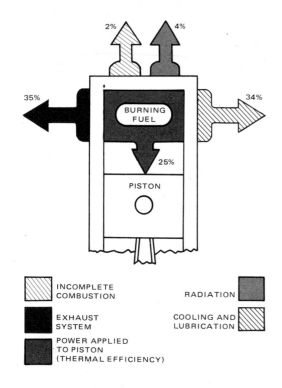

Fig. 8-26. Thermal (heat) losses make thermal efficiency rather poor. Note that it is about 25 percent.

Use this formula for computing brake thermal efficiency:

BRAKE THERMAL EFFICIENCY =
$$\frac{\text{BRAKE HORSEPOWER (BHP}\ \text{x}\ 33{,}000)}{778\ \text{FUEL HEAT VALUE}\ \text{x}\ \text{WEIGHT OF FUEL BURNED PER MINUTE}}$$

The 778 in the formula is Joule's equivalent (a number which is equal to another). The fuel heat (calorific) value is based on the Btu (British thermal unit) per pound.

REVIEW QUESTIONS – CHAPTER 8

1. What type of small gasoline engines are discussed in this text?
2. How do you compute engine displacement?
3. What is the force which acts opposite to the direction of centrifugal force?
4. The force applied opposite to the way a compressive force is applied is called what force?
5. If an object is raised 3 ft. and is then moved 5 ft. sideways, at the same level, work is being done over a distance of:
   a. 8 ft.
   b. 5 ft.
   c. 3 ft.
   d. 2 ft.
6. Give the formula for work.
7. If a force of 35 lb. is applied to an area of 5 sq. in., what is the pressure in psi?
8. The top area of a 2 1/2 in. diameter piston would be how many sq. in.?
9. If 60 lb. is lifted 5 ft. in 6 sec., what amount of power is exerted in ft. lb. per sec?
10. What is the definition of horsepower? How much is 1 hp?
11. A measure of the horsepower delivered at the engine crankshaft is called what?
12. A Prony brake with a 12-in. arm is applied to an engine flywheel. At 1200 rpm, the scale registers 20 lb. Calculate the horsepower to two decimal places.
13. What is indicated horsepower?
14. When testing an engine on the dynamometer, what are the standard test conditions?
15. What is the engine horsepower reduction for each 1000 ft. of elevation?
16. Explain the horsepower correction factor.
17. Volumetric efficiency is greatest at what speed?
18. Horsepower of an engine is greatest at maximum rpm. True or False?
19. On the average, what percentage of the energy from fuel is used to produce power?
20. An engine's "rated" horsepower is approximately what percent of its maximum brake horsepower?

SUGGESTED ACTIVITIES

1. Using the principles studied in this chapter, determine the horsepower required of several individuals to walk up one flight of stairs. The following items will be needed: a stopwatch, tape measure and bath scale.
2. Design and build a Prony brake for a small gasoline engine.
3. Use a dynamometer to develop a graph like the one in Fig. 8-25. Compare the graph with the manufacturer's graph for the same model engine.
4. On an engine with the head removed, measure the position of the top of the piston (in relation to the top of the block) at TDC and again at BDC. Determine the stroke of the engine.
5. On the same engine now, measure the diameter of the piston and determine if the engine is square, oversquare or undersquare.
6. Determine the cubic inch displacement of the engine from the facts learned in activities above.

Chapter 9

TROUBLESHOOTING, SERVICE AND MAINTENANCE

Most small gasoline engine service and repair jobs can be done without taking the whole engine apart. If the engine starts hard, runs rough, lacks power or does not run at all, troubleshooting (analysis of the problem) is necessary. Analysis is just a number of tests and steps you go through to find the problem.

Sometimes the cause of an engine problem is easy to find. Other times, checking out probable causes requires a certain amount of reasoning and use of the process of elimination. Test the simplest and the most probable cause first.

If the component or assembly tests good, test the next most probable cause, until you find the defect. *More than the one fault may exist at the same time, making it harder to locate the trouble.*

SERVICE INFORMATION

Before starting to work on an engine, look at the manufacturer's service manual or troubleshooting chart. Manuals generally include service "procedures." These tell the steps to take and the order in which the job must be done. Manuals also show exploded views of assemblies and systems. These detailed drawings help you disassemble and reassemble parts in the right order. A troubleshooting chart lists the most common and chronic troubles along with possible causes and suggested remedies. See Fig. 9-1.

Experienced mechanics have learned what the symptoms mean and go about solving the existing problem without having to use troubleshooting charts.

TOLERANCES AND CLEARANCES

Engine tolerances and clearances are given in chart form in service manuals and/or bulletins. See Fig. 9-2. When checking and adjusting spark plug gap, breaker point gap, ignition timing, etc., this chart gives the correct dimensions and piston location.

Note in Fig. 9-2, that the spark plug gap setting is .030 (thirty thousandths) in. Breaker point gap (magneto point gap) is .020 in. Ignition timing shows the

TROUBLESHOOTING GUIDE

PROBLEM	Fuel-Related Causes			Ignition Causes		Other Causes						
	No Fuel	Improper Fuel	Fuel Mix. Wrong	No Spark	Poor Ignition	Improper Cooling	Improper Lubrication	Poor Compression	Valve Problems	Carbon Build-Up	Governor Faulty	Engine Overloaded
Will Not Start	X			X				X	X			
Hard Starting		X	X		X	X		X	X			
Stops Suddenly	X			X			X		X			
Lacks Power		X	X		X	X		X	X	X		X
Operates Erratically		X	X		X						X	
Knocks or Pings		X	X			X				X		X
"Skips" or Misfires			X		X							
Backfires			X		X				X			
Overheats			X		X	X			X			X
Idles Poorly			X		X							

Fig. 9-1. Typical troubleshooting chart or guide for small gasoline engines.

TABLE 1. TOLERANCES AND CLEARANCES
FOR THE J-321 ENGINE

Cylinder Bore	2.1265 2.1260	Spark Plug Gap	.030
Piston Skirt Diameter	2.1227 2.1220	Magneto Point Gap	.020
Piston Ring Width	.0925 .0935	Ignition Timing in Degrees B.T.D.C.	$22°$(Full Retard)
Piston Pin Diameter	.5001 .4999	Piston Skirt to Cylinder Clearance	.0033 .0045

Fig. 9-2. Typical tolerance and clearance chart furnished by manufacturer of small gasoline engines.   (Jacobsen Mfg. Co.)

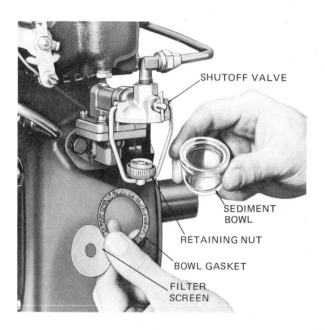

SHUTOFF VALVE

SEDIMENT BOWL

RETAINING NUT

BOWL GASKET

FILTER SCREEN

Fig. 9-3. Disassembled sediment bowl includes fuel bowl, strainer and gasket. If water was present, it would settle to bottom of bowl.

crankshaft throw at 22 deg. before top dead center (BTDC), which is the fully retarded position. Some engine charts show piston height in thousandths of an inch BTDC rather than degrees.

Where tolerance specifications show two values, the actual dimension must be within that range. In Fig. 9-2, for example, the cylinder bore (diameter) must measure somewhere between 2.1260 and 2.1265 in., a range of .0005 (5 ten thousandths) in.

TROUBLESHOOTING THE FUEL SYSTEM

If the symptoms of the engine's malfunction point to the fuel system, the problem could be located in any one of several parts. It could involve the fuel pump, the carburetor, reed valves (in two cycle engines), fuel lines, filters or air cleaner. Troubleshooting will involve checking and/or testing one part after another until the trouble is located and corrected.

FUEL PUMP

If no fuel is being delivered to the carburetor, and the fuel is gravity fed, check the following:
1. Is there fuel in the gas tank?
2. Is fuel flow blocked by clogged filter, Fig. 9-5, or obstructions in the line?

If engine is equipped with a fuel pump, and fuel is not being delivered to carburetor, check to see that:
1. There is fuel in the gas tank.
2. Fittings connecting the fuel line to tank and pump are tight; otherwise, the pump will draw air.
3. Pump filter is clean, and gasket on the filter bowl is in good condition. Method of inspection is shown in Fig. 9-3. Water in the fuel as well as other contaminants are easily seen. Water, being heavier than fuel separates and collects at the bottom of the bowl, along with other contaminants such as dirt and rust. Visual inspection can be made without dismantling the pump. If disassembly is required, consult the exploded view of the pump in Fig. 9-4.
4. Pump is actually working. Disconnect the fuel line between pump and carburetor. Turn engine with the

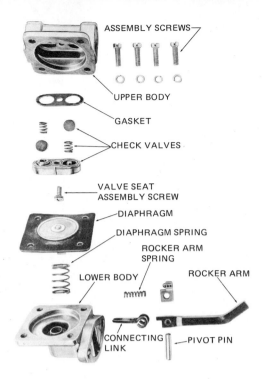

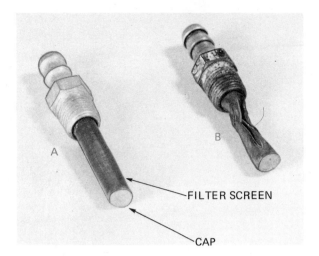

Fig. 9-5. Damaged gas tank filter can block fuel flow. A—New filter. B—Filter damaged when gasoline can nozzle was inserted too far into fuel tank.

Fig. 9-4. Exploded view of fuel pump is typical of those found in engine service manuals. These are a useful aid to repair and reassembly. (Wisconsin Motors Corp.)

starter. There should be a well-defined spurt of fuel at every stroke of the pump. This occurs every two revolutions of the engine.

Where fuel pump problems might be suspected, it is well to make sure that fuel flow from the supply tank is not interrupted before it gets to the fuel pump. Fig. 9-5 shows a filter that allowed the engine to start and idle but stalled whenever the throttle was opened. A replacement fitting (left) solved the problem.

CARBURETOR ADJUSTMENTS

If fuel pump problems are corrected but the engine still surges or lacks power, the cause may be poor carburetor adjustment or carburetor defects. Most carburetors have two needle valve adjustments. A third adjustment is the idle speed stop screw. Needle valves are not always found in the same location on the carburetor body. It is always a good practice to refer to the engine manual for needle positions and instructions on proper settings.

HIGH SPEED ADJUSTMENT

Each engine manufacturer will give a "rough" setting for the high speed and idle screws. This will permit the

engine to be started. It should then be run long enough to warm up before further adjustments are made.

General procedure for the first adjustment is to open throttle wide and turn the high speed adjusting needle forward then backward slowly until maximum speed is reached. See carburetor in Fig. 9-6 for general location of external parts. After reaching maximum speed, turn the needle counterclockwise very slightly so the engine is running a little rich.

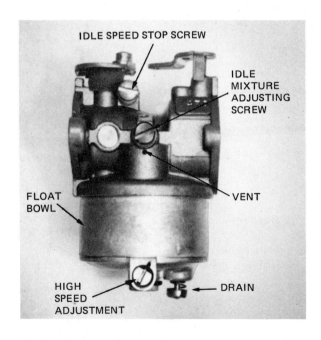

Fig. 9-6. External carburetor adjustments are high speed adjusting needle, idle fuel regulating screw and idle speed stop screw. (Jacobsen Mfg. Co.)

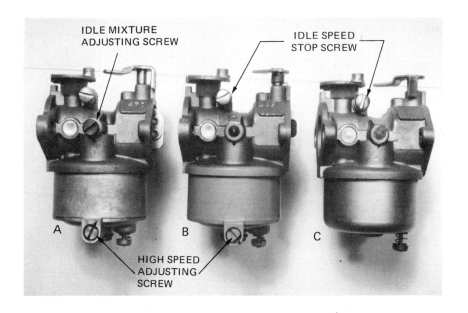

Fig. 9-7. Carburetors have one or several adjustments, depending on make or model. A—Carburetor with high speed adjustment, idle mixture adjustment and an idle speed stop screw. B—Carburetor with two points of adjustment, high speed and idle speed. C—Carburetor with idle speed stop screw only.

IDLE ADJUSTMENT

Now, move the throttle to the slow running position. Turn the idle adjusting screw slowly, first in one direction and then in the other, until idle is smooth. Now, adjust the idle speed stop screw to get the idle speed recommended in the manual. Check carefully, each make and model may be different.

Unlike Carburetor A in Fig. 9-7, not all carburetors have two needle settings. Carburetor B has only the high speed needle adjustment and C has no fuel adjustments. Each carburetor has an idle speed adjustment screw. Some of the needle adjustments are left off because the manufacturer has pre-set the adjustments and sealed them.

TESTING TWO CYCLE ENGINE REEDS

To determine whether the reeds are leaking on a two cycle engine, remove the air cleaner from the carburetor intake. Run the engine while holding a clean strip of paper about one inch from the carburetor throat, Fig. 9-8. If the paper becomes spotted with fuel, the reeds are not seating.

Operating the engine with leaking reeds will result in fuel starvation, poor lubrication, and overheating of the engine. The reeds should be replaced or repaired.

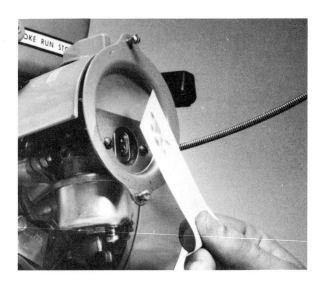

Fig. 9-8. Reeds can be tested for leakage by placing clean strip of paper one inch from carburetor intake. If spots of fuel show on paper, reeds are leaking.

AIR CLEANER

The carburetor air cleaner should be cleaned before each season of operation and at regular intervals thereafter. A plugged air filter like the one in Fig. 9-9 can cause hard starting, loss of power and spark plug fouling.

Fig. 9-9. Carburetor air filters must be kept clean. Their purpose is to trap dirt but, eventually, air flow is restricted. (Jacobsen Mfg. Co.)

Remove old oil from oil bath cleaners and wash them in gasoline or another solvent. Refill with engine oil to the correct level.

Polyurethane foam type air filters should be washed and rinsed in kerosene or similar solvent. About two tablespoons of clean SAE 30 oil should be evenly distributed in the filter material by compressing it in the hand.

Under severe dust conditions, air filters should be cleaned more often. If a dry canister filter is used, replace it with a new one.

DIAPHRAGM CARBURETOR REPAIR

If inspection and adjustment of the carburetor indicate need for repair, remove the carburetor from the engine. Wash the outside in a solvent. Disassemble and thoroughly wash each part. Lay the cleaned parts out on a clean white cloth so none become lost or damaged.

If the carburetor is very dirty, a commercial carburetor cleaner solution may be used. *Be careful not to get any solution on hands or clothing. Wear safety glasses. Put only the metallic carburetor parts in the solution and let them soak. Non-metallic parts can be damaged by harsh commercial cleaners.*

After the parts have soaked for several hours, rinse them with a milder cleaning solvent and dry with compressed air. Do not dry the parts with a rag or paper towel. Lint will get into passages. Never clean holes or passages with wire or similar objects. These will distort the openings and may prevent the engine from running properly.

After cleaning, inspect the parts for wear, material failure or damage. Check the carburetor body and crankcase for cracks or worn mating surfaces. See Fig. 9-10. Replace defective parts.

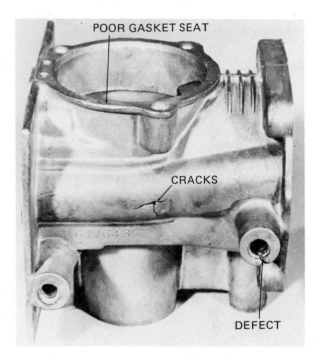

Fig. 9-10. This engine part has several defects. If repair is not possible, demaged parts should be replaced. (Lawn-Boy Power Equipment, Gale Products)

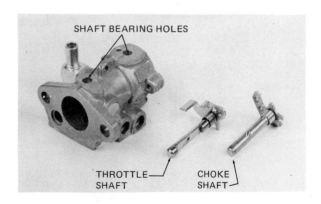

Fig. 9-11. Throttle and choke shafts must fit bores closely, but turn freely. (Tecumseh Products Co.)

Check throttle and choke shafts. They must fit closely but turn easily in their bearing holes, Fig. 9-11. If loose, they will cause poor engine performance.

The diaphragm should be checked for defects that would cause leakage. The diaphragm needle valve must be straight and fit the seat so that it seals when closed.

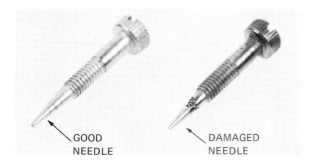

Fig. 9-12. Left. Needle valve has straight smooth taper. Right. Needle valve should be replaced.

The high speed and idle mixture needles should have straight, smooth tapers, Fig. 9-12. The O-rings or seals around the needle should be replaced if they are cut or deformed.

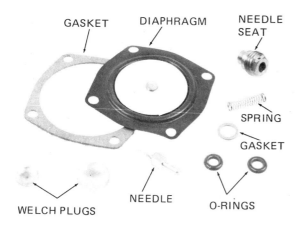

Fig. 9-13. Carburetor repair kit shown is for diaphragm type carburetor.

Some manufacturers supply carburetor repair kits, Fig. 9-13, which include those items that would most likely need replacing. Other parts can be bought as they are needed.

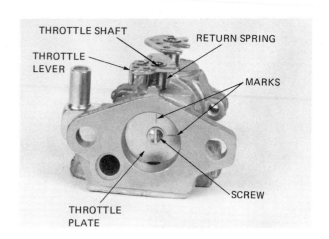

Fig. 9-14. Throttle shaft with return spring attached is placed in holes provided in throttle body. Throttle valve must be fastened to shaft in proper position. (Tecumseh Products Co.)

ASSEMBLING DIAPHRAGM CARBURETOR

Because of the many different carburetor designs, it is highly recommended that the manual be followed. The following procedure is only an example.

Place the throttle shaft in the bearing hole with the return spring on the shaft as illustrated in Fig. 9-14. The throttle plate must be placed properly in the carburetor bore. Usually identifying marks of some sort are put on the plate to assist the assembler. In the carburetor shown, Fig. 9-14, the marks have to be placed facing outward.

The choke plate is assembled in the same way as the throttle. It is vented, has openings to allow some air to

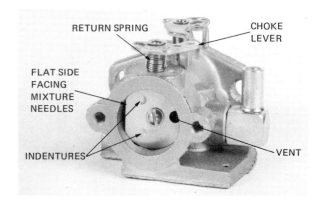

Fig. 9-15. Choke valve assembly is shown in place in air horn of carburetor.

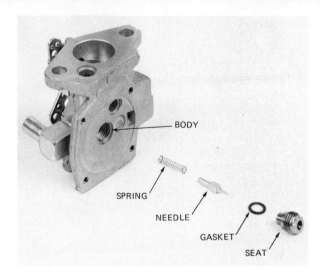

Fig. 9-16. Diaphragm needle valve is installed in carburetor body in sequence shown.

enter even when it is closed. Fig. 9-15 shows the flat side facing the mixture needles and the indentations facing outward.

In Fig. 9-16, the spring and needle valve are installed in the order shown. The spring and needle valve are inserted into the body. The gasket and seat are then screwed into the threaded hole in the carburetor body.

Assemble the diaphragm as shown in Fig. 9-17. Four screws placed through holes in diaphragm cover draw the assembly tight to the body. The rivet head in center of diaphragm is turned toward the needle valve.

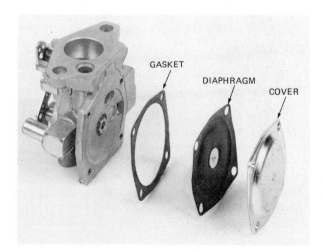

Fig. 9-17. In this application, gasket, diaphragm and cover are fastened to carburetor body with four screws. (Tecumseh Products Co.)

The high speed and idle mixture needles are installed in the order shown in Fig. 9-18. Adjust the needles as prescribed by the manual. This will get the engine started. Finer adjustments can be made when the engine is warmed up.

The idle stop screw, Fig. 9-18 is turned one full turn after it first makes contact with the throttle lever.

The carburetor can now be attached to the engine. Install the linkage and attach the air cleaner with new gaskets.

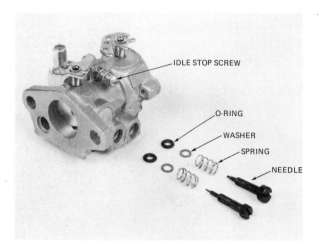

Fig. 9-18. High speed needle, idle mixture needle and idle stop screw are installed and preset.

FLOAT-TYPE CARBURETOR REPAIR

Except for the float and float chamber, the procedure for repairing float-type carburetors is like diaphragm-type carburetors.

Inspect the float for dents, leaks and worn hinge bearings, Fig. 9-19. The float bowl gasket should be replaced if it shows any signs of damage.

ASSEMBLING FLOAT SYSTEM

The float valve seat is installed in the carburetor body as shown in Fig. 9-20.

After assembling the float valve and valve clip as shown in Fig. 9-19, install the assembly in the carburetor body with the hinge pin. Turn the carburetor body over and check the height of the float using the

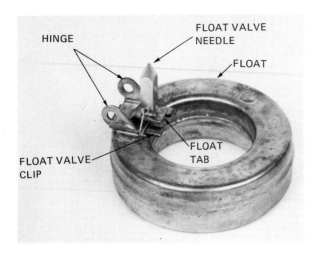

Fig. 9-19. Hollow metal float is typical of type used in small gasoline engine carburetor. Note inlet needle.

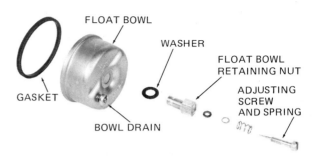

Fig. 9-22. Float bowl is assembled in order shown. Gasket must be in good condition with no leaks. (Tecumseh Products Co.)

specified drill size as shown in Fig. 9-21. Bend the float tab if adjustment is necessary until the setting is correct. See the manual.

Assemble the float bowl in the order shown in Fig. 9-22. On this carburetor, the bowl assembly must be installed as shown in Fig. 9-23. The high speed mixture needle, idle mixture needle and idle stop screw are then assembled, Fig. 9-24. The carburetor can be re-installed. Needle valves should be adjusted according to the manufacturer's recommended settings.

Fig. 9-20. Float valve seat is placed in carburetor body before installing float. (Tecumseh Products Co.)

Fig. 9-21. Float must rest on needle at a specified height. If float is high, too much fuel will be used. If float level is low, lean mixture may cause overheating.

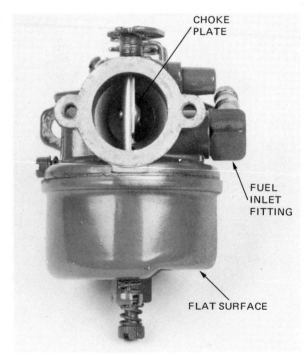

Fig. 9-23. Float bowl may be revolved on installation on this carburetor. Specifications give proper position.

Fig. 9-24. On this float type carburetor, high speed mixture needle fits into float bowl retaining nut. (Tecumseh Products Co.)

Labels in figure:
IDLE STOP SCREW
HIGH SPEED MIXTURE NEEDLE
IDLE MIXTURE NEEDLE

ENGINE GOVERNOR ADJUSTMENTS

Small engine governor designs are of many types and so are the methods of adjustment. The best advice is to refer to the engine manual for the specific make and model of the engine being serviced.

Only two types of governors are used on small engines — air-vane and centrifugal flyweights. The centrifugal flyweight governor is by far the most popular. Often, the method of adjusting the governor can be determined by good judgment and reasoning if it is kept in mind that centrifugal force and spring pressure are opposed to each other and work to open or close the throttle.

TESTING COMPRESSION

Sometimes it is necessary to check further into the condition of an engine. A cylinder compression test can be a first step toward determining the condition of the upper major mechanical parts of the engine. This test is especially valuable if an engine is losing power, running poorly and shows little or no improvement after fuel system and ignition adjustments. Use the following procedure for a compression test:

1. Run the engine until it is warm.
2. Disconnect all drives to the engine.
3. Open choke and throttle valves wide.
4. Remove the air cleaner.
5. Remove the spark plug and insert compression gage, Fig. 9-25.
6. Crank the engine as fast as possible to obtain an accurate test. Repeat to insure accuracy.

Engines equipped with compression release camshafts may have to be cranked in reverse rotation to obtain an accurate test. Most can be cranked forward.

An engine producing a compression less than the minimum suggested by the manufacturer usually has one or more of the following problems:

1. Leaking cylinder head gasket.
2. Warped cylinder head.
3. Worn piston rings.
4. Worn cylinder bore.
5. Damaged piston.

Fig. 9-25. Compression test can indicate condition of various mechanical components of engine.

6. Burned or warped valves.
7. Improper valve clearance.
8. Broken valve springs.

To determine whether the valves or rings are at fault pour a tablespoonful of SAE 30 oil into the spark plug hole. Crank the engine several times to spread the oil and repeat the compression test. The heavy oil will temporarily seal leakage at the rings. If the compression does not improve, the rings are satisfactory and leakage is due to valves, cylinder head, or a damaged piston. If the compression is much higher than original test, the leakage is due to defective piston rings.

GENERAL PREVENTIVE MAINTENANCE

There are certain maintenance tasks that must be performed regularly to keep an engine working properly and less apt to break down under use. These tasks come under the heading of preventive maintenance. They are so called because they help prevent premature wearing of the machine.

KEEPING ENGINE CLEAN

Cleaning a small air-cooled engine periodically can prevent overheating. For proper cooling action, air must pass across the extended metal surfaces (cooling fins) of the cylinder block and cylinder head. If the fins are insulated by dirt, leaves and grass clippings, engine parts will retain most of the heat of combustion. Parts will expand, probably distort and possibly seize. Therefore, all finned surfaces should be cleaned to the bare metal.

Methods of cleaning a small air-cooled engine vary. You can scrape with a piece of wood, then wipe with a clean cloth. Or, you can blow debris from the fins with compressed air, then use a cleaning solvent.

Various aerosol spray cleaners are suitable for use in small engine work. *When using compressed air, be extremely careful where you direct the blast of air. Wear safety goggles. Never direct the blast toward someone else.*

CHECKING OIL LEVEL AND CONDITION

Crankcase oil in four cycle engines should be checked periodically. Preferably, it should be checked each time fuel is added. The engine manufacturer provides a means of visually inspecting the level and condition of the oil. Use the type and viscosity grade of oil recommended and maintain it at the proper operating level.

To check the oil level, withdraw the dipstick, wipe it dry. Re-insert the dipstick as far as it will go. Withdraw it a second time and observe the oil level, Fig. 9-26. Add oil if the level is below the ADD or LOW mark.

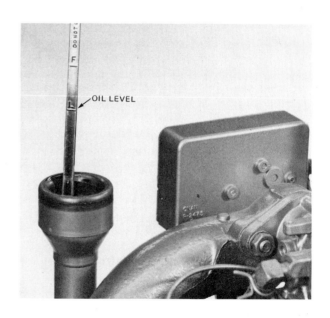

Fig. 9-26. Oil level should be maintained between full and low marks. Never overfill.   (Onan Corp.)

Do not run the engine with oil showing above the FULL mark on the dipstick. If the crankcase oil level is high, drain some oil. Over filling can foul plugs and cause engines to use too much oil.

Some small gasoline engines do not have dipsticks. Instead, they have a filler plug that seals out dirt and seals in the oil. Fig. 9-27 shows the proper method of

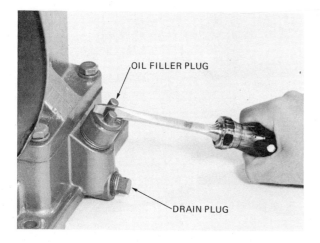

Fig. 9-27. Removing oil filler plug from engine crankcase. (Briggs and Stratton Corp.)

loosening one type of filler plug. When the plug is removed, the oil level should be to the top of the filler hole, or to a mark just inside the filler hole. Drain plug removal is shown in Fig. 9-28.

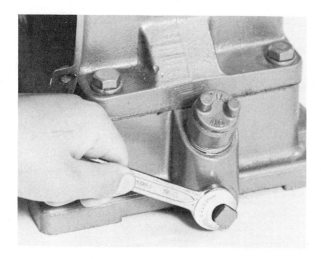

Fig. 9-28. Oil drain plug is located at a low point in crankcase to permit complete drainage of used oil.

If the engine oil level drops at an excessive rate (requires addition of oil frequently), look for the cause. Refer to the troubleshooting chart for that particular engine. Typical causes are: external leaks, worn oil seals around the crankshaft, worn valve guides, worn piston rings, or a hot running engine.

The color of used oil is not always an accurate indication of its condition. Additives in the oil may cause it to change color while not decreasing its lubricating qualities.

WHEN TO CHANGE OIL

The small engine manufacturer will recommend oil changes at intervals based on hours of running time. A new engine should have the first oil drained after only a few hours of running to remove any metallic particles from the crankcase. After that, the time specified may vary from 10 to 50 hours.

Engine oil does not "wear out." It always remains slippery. However, oil used for many hours of engine operation becomes contaminated with dirt particles, soot, sludge, varnish-forming materials, metal particles, water, corrosive acids and gasoline. These contaminants finally render the oil "useless." The harm they cause outweighs the lubricating quality of the oil.

The time interval for oil changes is selected so that the oil never reaches a "loaded" level of contamination. Loaded oil is that condition when the oil cannot absorb any more contaminants and still be an effective lubricant. When oil reaches a loaded condition, varnish deposits begin to form on the piston and rings, and sludge collects in the crankcase.

CHANGING OIL

Draining the oil is not difficult. First, run the engine until it is thoroughly warmed up. Warm oil will drain more completely, and more contaminants will be removed if the oil is agitated.

Stop the engine and disconnect the spark plug. The oil drain plug is located at a low point on the crankcase, usually along the outside edge of the base, Fig. 9-28. Some engines are drained through the filler cap and have no drain plug.

Clean the dirt from the drain plug area, then remove the plug with a proper fitting wrench. Drain the oil for approximately five minutes to remove as much contam-

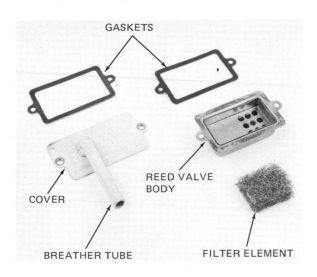

GASKETS

COVER

REED VALVE
BODY

BREATHER TUBE

FILTER ELEMENT

Fig. 9-29 Crankcase breather elements need periodic cleaning and inspection. Reed valve function is to only permit air to leave crankcase. (Deere & Co.)

inated oil as possible. Tilt the engine toward the drain hole if it is located on the side or top. Draining completed, replace the drain plug.

Before putting fresh oil in the engine, clean the filler opening, funnel (or other container) and the top of the can of oil. Be sure to use the correct type, viscosity grade and quantity of oil recommended by the manufacturer. Pour it in the crankcase of the engine and check the level. Then replace the filler cap and connect the spark plug lead to the spark plug.

Start and run the engine for a few minutes. Stop the engine and recheck oil level. Check for oil leaks.

Finally, wash or destroy oily rags. Storing them may cause spontaneous combustion.

If any oil is left in the can, keep it clean during storage by installing a plastic top from a vacuum packed food container. If kept in screw top cans, keep the cap tightly sealed to prevent condensation.

MIXING OIL AND FUEL

A two-cycle engine can be seriously damaged by improper mixing of the gasoline and oil, even though the recommended proportions are used. The proper way to mix gasoline and oil is to pour some of the gasoline into a clean metal container first. Add the oil to the gasoline and agitate (shake) this partial mixture. Add rest of the gasoline and agitate thoroughly again.

Many service station attendants will pour the oil into the container first, then allow pump pressure to mix the gasoline and oil. However, this method does not always thoroughly mix the two elements.

Once the gasoline and oil are thoroughly mixed, the oil will remain in solution indefinitely. If fuel is to be stored for several weeks or more, add a gasoline stabilizer to prevent oxidation and formation of varnish and corrosive acids that can ruin an engine.

For safety, always store fuel in metal containers clearly marked "gasoline." If possible, keep the storage container full to prevent moisture condensation.

Condensation forms in partially filled containers because of temperature changes. During the day, the air above the fuel may become very warm and hold considerable moisture. At night, the warm air cools, contracts and loses the water vapor.

As water droplets gather on the inside walls of container, they run together and flow into the fuel. Each night, more water is added to fuel in this manner. Partially filled fuel tanks react to temperature changes in this way. For this reason, fuel tanks should be kept "full" of gasoline or the gasoline/oil fuel mixture.

CRANKCASE BREATHER SERVICE

If the small gasoline engine has a crankcase breather, it should be removed and cleaned periodically. The breather assembly is located over the valve stem chamber. It is held in place with two or more screws.

Remove the screws. Under the cover is a filter element and a reed valve unit similar to the one in Fig. 9-29. The breather allows outward air flow only. Inspect the reed valve to make sure it is not damaged or distorted. Wash the parts in a cleaning solvent and replace damaged gaskets with new ones. See that the drain hole in the reed valve body is open. It permits accumulated oil to return to the engine. Then, replace the assembly and tighten the screws.

MUFFLER SERVICE

An engine takes in large quantities of air mixed with fuel, then burns the mixture. Unless it readily rids itself of the by-products of combustion, its efficiency will be greatly reduced. This is the task of the exhaust system which, in small gasoline engines, mainly consists of exhaust port(s) and a small muffler.

The muffler is designed to reduce noise and allow gases to escape. When it becomes clogged with carbon soot, gases cannot get out quickly enough to allow fresh air and fuel to enter. Then, a power loss occurs along with a tendency toward overheating.

If a muffler is designed to be taken apart, as shown in Fig. 9-30, it should be disassembled and cleaned in a solvent. If it is a sealed muffler, and clogging is suspected, replace it with a new one and check for improved engine efficiency.

Fig. 9-30. Accumulations of carbon in the exhaust muffler can seriously retard scavenging of gases from cylinder. This muffler can be disassembled for cleaning. (McCulloch)

MAINTAINING WATER COOLING SYSTEMS

Water cooling systems, where they are used in small engines, require some maintenance similar to that employed in the automobile engine. Because the combination of water and metal sometimes produces harmful chemical reactions that attack the water jacket, a chemical rust inhibitor should be added to the cooling system.

Whenever a system is drained and refilled, it is advisable to add an inhibitor. If rust and scale are allowed to form and accumulate, the walls of the water jacket will become insulated. This will cause engine heat to be retained rather than removed.

Scale settling to the bottom may plug water passages in the cylinder block and clog the water tubes in the radiator. Without free circulation of water, the engine will run hot even when the thermostat is open. Local "hot spots" can occur in the engine when the passages in the block are obstructed.

In severe cases, the water may boil inside the block, and steam will prevent water from contacting and cooling the inner walls. Then, serious overheating and damage to parts of the engine are bound to occur.

The cooling fins surrounding the tubes of the radiator should be kept clean for good heat transfer. Compressed air or a pressure water hose will remove any accumulations that might prevent air from passing through the fins and across the tubes. To remove the debris, direct the flow of air or water in the opposite direction of normal air flow.

Engine blocks and radiators may be cleaned periodically by reverse flushing the system with water under pressure. Disconnect the hoses from the radiator and block. Force clean water in the opposite direction of normal circulation. This will push loose sediment out. Continue flushing until the water runs clear. Flushing should be done with the engine stopped and cool.

To remove additional rust clinging to inner surfaces, use a commercial cooling system rust remover. Follow the instructions given by the manufacturer.

SYSTEMS COOLED WITH SALT WATER

Outboard engines operated in salt water are exposed to extremely corrosive conditions. Exposed engine parts require very careful maintenance.

Outboards used in salt water should be removed from the water immediately after operation. If the engine cannot be removed, tilt the gearcase out of the water and rinse it with fresh water. (The gear case must not remain in the water while not in use.)

Flushing the internal cooling system of an outboard engine is extremely important. Flushing is done by attaching a fresh water hose to the water scoop or by operating the engine in a barrel of fresh water for several minutes.

Rinse the engine with fresh water and wipe all lower unit parts with a clean, oily cloth. Ignition leads and spark plug insulators should be wiped frequently to prevent an accumulation of salt residue.

STORING WATER-COOLED ENGINES

Storing water-cooled engines for lengthy periods, particularly during winter, calls for certain special maintenance measures. If the engine is "radiator cooled," antifreeze must be added to the water to protect against the lowest possible freezing temperatures. If rust inhibitor is not supplied in the antifreeze, it should be added.

If the engine will not be started at any time during storage, drain the cooling system completely. Then, tag the engine to indicate its drained condition.

When storing outboard engines, remove all plugs from the gearcase and driveshaft housing. This allows accumulated water in the gearcase and cooling system to drain off.

Failure to take this precaution when winterizing may result in cracked cylinder blocks and/or gearcase, plus possible damage to water channels and tubes.

Rock the engine from side to side to make certain all water has drained. Refill the gearcase with type of grease specified by the engine manufacturer. Attend to all other lubrication recommendations made by the manufacturer for care of engines being stored.

TORQUE SPECIFICATIONS

Before attempting troubleshooting or maintenance of any kind, you must be familiar with torque specifications for fasteners used in assembling different parts of the engine. Likewise, you must learn how to use the torque wrench. Torque refers to the effort extended to turn something. Bolts and nuts then, must be "torqued" to a given tightness to hold mating parts together under specified tension.

Fig. 9-31. Head bolts should be tightened evenly and in sequence recommended by manufacturer. (Deere & Co.)

Obviously, if a bolt or nut is too loose, vibration may loosen it more. If the same bolt is turned too tight, the threads may be stripped or the bolt broken off in the hole.

If the installed bolt does not break or strip the threads, expansion from heat will weaken the metal in the bolt. This expansion may go beyond the elastic limit of the bolt (a point where the bolt stretches but does not return to its original length upon cooling). When this happens, the bolt or part may fail.

There is still another reason why bolts and nuts should not be overtightened. Excessive internal stresses within a part may cause warpage or failure of the part.

Where a number of bolts are required to fasten mating parts together (such as cylinder heads to cylinder block), it is necessary to tighten them evenly and according to a certain sequence (order) to prevent warping and leaks. Where gaskets are used, overtightening of fasteners will crush the gasket under the bolt heads and distort the metal between the bolts.

Fig. 9-31 shows correct cylinder head bolt tightening sequence in a typical small gasoline engine. To illustrate the procedure, the bolts are numbered in the particular order or sequence in which they are to be tightened. After the first bolt is tightened to 100 in. lb., the one directly opposite is tightened to the same torque reading. Next, the bolts that lie perpendicular to a line between the first two are similarly "torqued." Then, tighten the bolts 90 deg. to the left and right, etc.

Generally, bolts are tightened in a sequence from one side to another in a kind of extending rotation, first tightening each bolt to 100 in. lb., then going back to tighten each an additional 20 in. lb. until specified torque is reached.

USING A TORQUE WRENCH

Torque wrenches vary in design and method of operation. Fig. 9-32 shows how to use a "preset" torque wrench. The desired amount of torque tightness is set, then the wrench socket is placed on the bolt head and the handle is drawn until a click can be felt and heard. The socket should be held down firmly while pulling on the handle.

The torque wrench shown in Fig. 9-33 also uses a pointer that moves upscale as torque is applied. With this wrench, the mechanic pulls the wrench handle until

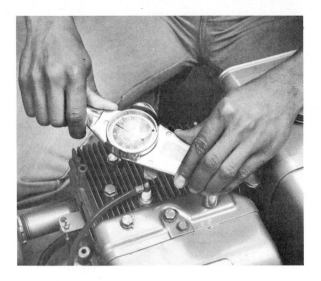

Fig. 9-32. Proper procedure for using a torque wrench is to support socket with one hand and apply turning effort at right angles to handle.

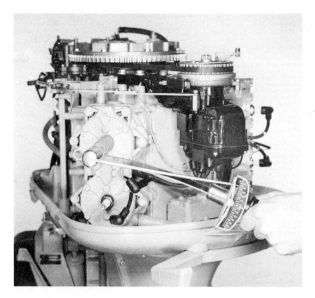

Fig. 9-33. A variety of styles of torque wrenches are available. This particular wrench uses a pointer and a flexible arm.

the pointer reaches the correct torque tightness. Torque data charts like the one in Fig. 9-34 supply the necessary data to correctly tighten critical parts.

Most small engine torque charts specify both inch-pounds (in. lb.) and foot-pounds (ft. lb.), both torque wrenches are calibrated in either of these increments (units). The scale on the wrench is clearly marked whether the reading taken is in in. lb. or ft. lb.

The torque reading in any case is the product of the length of the wrench handle in feet or inches and the applied force. For example, applying one pound of force through a handle one foot long would produce 1 ft. lb. or 12 in. lb. *In order to convert ft. lb. to in. lb., multiply by 12.*

REVIEW QUESTIONS – CHAPTER 9

1. If the gasoline supply and fuel pump operation are satisfactory, but an engine still idles and accelerates poorly, where would you check for the cause of the trouble?

FOUR CYCLE TORQUE SPECIFICATIONS

		INCH POUNDS	FT. POUNDS
Cylinder Head Bolts		140 - 200	12 - 16
Connecting Rod Lock Nuts	1.5 - 3.5 H.P. 4 6 H.P.	65 - 75 86 - 110	5.5 - 6 7 - 9
Cylinder Cover or Flange to Cylinder		65 - 110	5.5 - 9
Flywheel Nut		360 - 400	30 - 33

Fig. 9-34. Torque specifications are provided in engine manuals for all critical bolts and nuts.   (Tecumseh Products Co.)

2. Name the three basic carburetor adjustments in the order in which they are performed.

3. Which of the following procedures would not be recommended in carburetor maintenance?
   a. Clean non-metallic parts in a commercial carburetor cleaner.
   b. Wash metallic parts in commercial carburetor cleaner.
   c. Dry parts with compressed air.

4. Make sure all passages and holes are open by pushing a stiff wire of the proper size through them. True or False?

5. Use paper toweling to dry parts. Yes or No?

6. In checking out an engine you find it has low compression. How do you determine if rings are the cause?

7. How can you check the condition of a reed valve without removing the carburetor and the valve?

8. If the head bolt torque data specifies 14 ft. lb., and your torque wrench is calibrated in in. lb., what torque would you apply in in. lb.?

9. List five possible causes of low compression.

SUGGESTED ACTIVITIES

1. Make a complete carburetor adjustment, following the procedure described in the text. Set with a tachometer.

2. Look up governor adjustments in a manual for a specific make and model engine. Explain how the governor works and demonstrate the correct adjustments.

3. Make a complete compression test on an engine and use the results to determine engine condition.

4. Disassemble an engine that has had a great deal of use. Compare the dimensions of its parts with the dimensions specified in the service manual.

5. Rebuild a float type carburetor.

6. Rebuild a diaphragm type carburetor.

Tune-up procedure starts with spark plug removal and compression test. Next, shroud must be removed to gain access to flywheel and ignition parts.

Chapter 10

ENGINE TUNE-UP

While the electrical systems of small gasoline engines are durable, they do need periodic inspection and maintenance. When these services are performed, it is called an engine tune-up.

A tune-up involves the entire ignition system from coil windings to spark plugs. The small engine mechanic will check or test part after part until the entire system is working well. In the process, worn or defective parts must be replaced.

IGNITION SYSTEM AND SPARK PLUGS

Often, when a small engine is difficult to start, a new spark plug may seem to solve the problem. However, the mechanic cannot assume that this is the only fault. Often a less obvious problem has caused the plug to fail.

Although a magneto may be able to supply 20,000 volts, it produces only enough voltage to jump the spark plug gap. So the condition of the spark plug determines the amount of voltage the other ignition parts must produce.

Spark plugs used in normal operation will wear out from erosion caused by combustion. A new plug may need only 5000 volts to fire. But after many hours of running, the same plug may require 10,000 volts. If a pull on the starter cord produces less than 10,000 volts, the engine equipped with such a plug will not start. Certainly not all cases of hard starting are caused by a bad spark plug. Therefore, replacing plug without further checking is not a good practice. The problem may lie in the malfunction of other ignition components.

Changing the plug merely means less voltage is needed to fire it. Carbon deposits will again build up in the cylinder and exhaust ports due to poor combustion. More carbon will form on the spark plug electrodes and cause further hard starting.

You can analyze the quality of combustion that has been taking place in the cylinder by examining carbon deposits on the spark plug. Deposits showing beige to gray-tan color indicate normal combustion of the air-fuel mixture at proper operating temperature. Fig. 10-1 shows how a "normal" used spark plug looks. Fig. 10-2 compares a normal plug with one that is extremely carbon fouled.

Fig. 10-1. A normal spark plug, taken from an engine, will be clean and dry, showing a beige to gray-tan color on porcelain shell insulator. (Champion Spark Plug Co.)

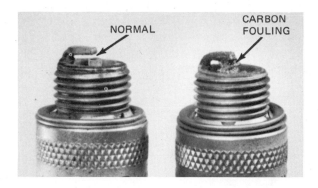

Fig. 10-2. Normal spark plug on left is compared to a carbon fouled plug on right. (Jacobsen Mfg. Co.)

Fig. 10-3. An oil-fouled spark plug can be indication of some mechanical malfunction. (Champion Spark Plug Co.)

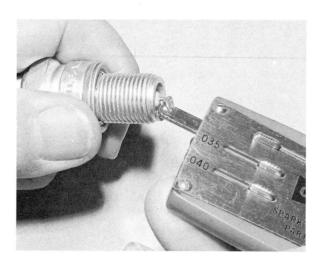

Fig. 10-5. Use spark plug gapping tool to bend outer electrode toward or away from center electrode. (Champion Spark Plug Co.)

An oil-fouled plug like the one shown in Fig. 10-3 is saturated with wet oil. The eroded plug pictured in Fig. 10-4 got that way from many hours of use.

Fig. 10-4. Spark plug electrodes tend to erode from normal combustion after many hours of use. (Jacobsen Mfg. Co.)

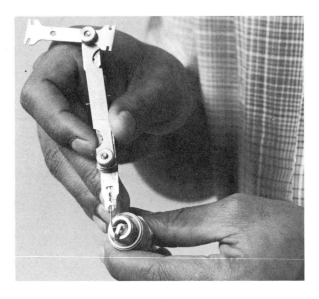

Fig. 10-6. Gap should be carefully adjusted and measured with wire gages.

Study the situation carefully before changing to a "hotter" spark plug than the one specified. The "hot" plug may stop carbon buildup, but other, more serious problems develop. Spark plug deposits usually are caused by ·weak magneto voltage, incorrect carburetor adjustment, poor air cleaner maintenance, incorrect gasoline or oil, or incorrectly mixed gasoline and oil.

When gapping the spark plug, bend the outer electrode toward or away from the center electrode. For best results, use a gapping tool, Fig. 10-5. Note the built-in wire gages (.035, .040).

Standard leaf-type feeler gages may be used if the plug is new or if the electrodes are in good condition. Otherwise, wire gages should be used, Fig. 10-6.

The reason why wire type thickness gages are recommended for use on worn spark plugs is shown in Fig. 10-7. The flat leaf-type thickness gage would leave an additional gap between the electrodes.

SPARK PLUG REMOVAL

The following four steps for spark plug removal are simple and should become a regular habit. They can save troublesome problems from occurring later.
1. Gently rotate and then pull the spark plug boot

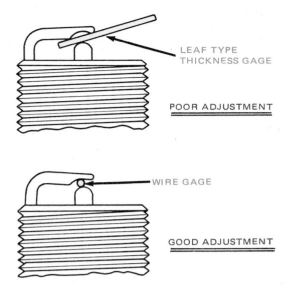

LEAF TYPE THICKNESS GAGE

POOR ADJUSTMENT

WIRE GAGE

GOOD ADJUSTMENT

Fig. 10-7. Leaf type thickness gages may not give an accurate measurement of gap if electrodes are not flat and parallel.

Fig. 10-7A. File oxide from electrodes with a spark plug file.

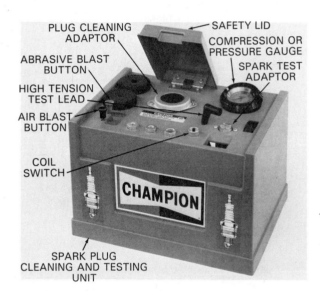

Fig. 10-7B. Clean threads with a power wire brush.

and cable from the spark plug. DO NOT GRAB OR VIOLENTLY PULL ON THE SPARK PLUG CABLE. Pull on the boot only.

2. Use a correct size deep spark plug socket with a rubber sleeve installed to protect the spark plug insulator from breakage. NOTE: To remove spark plugs from engines with aluminum heads, allow the engine to cool. The heat of the engine, in combination with a spark plug that has run for many hours, may cause the spark plug to seize.

3. Before removing the loosened spark plug, air blast dirt away from the area around the plug.

4. Remove spark plug and check its appearance. Refer to Spark Plug Analysis Table in the back of text.

CLEANING SPARK PLUGS

1. Wipe all spark plug surfaces clean. Remove oil, water, dirt, and moist residues.

2. Check the firing end or tip of the spark plug for oily or wet deposits. If the spark plug has deposits, brush the spark plug in a non-flammable and non-toxic solvent. Then dry the plug with compressed air to prevent caking of cleaning compound deep within the spark plug shell.

3. Clean the spark plug in a spark plug cleaning machine. See USING THE SPARK PLUG CLEANING MACHINE.

4. File spark plug electrodes after cleaning to square them and to remove any oxide or scale from the surfaces. Open the gap so that a spark plug file can be put between the electrodes, Fig. 10-7A.

PLUG CLEANING ADAPTOR

SAFETY LID

ABRASIVE BLAST BUTTON

COMPRESSION OR PRESSURE GAUGE

SPARK TEST ADAPTOR

HIGH TENSION TEST LEAD

AIR BLAST BUTTON

COIL SWITCH

CHAMPION

SPARK PLUG CLEANING AND TESTING UNIT

Fig. 10-7C. This is a special spark plug cleaning unit which removes carbon from electrodes and insulator.

5. Clean spark plug threads with a hand or powered wire brush, Fig. 10-7B. Take care not to damage the electrodes or insulator. If threads are nicked or damaged, discard the plug.

USING THE SPARK PLUG CLEANING MACHINE

1. Select the proper size adaptor and position it in the plug cleaning recess. Lock the adaptor to cleaning recess with retaining ring.

2. Press the firing end of dirty spark plug into hole in adaptor firmly, Fig. 10-7C.

3. Wobble or swivel the terminal end of the spark plug in a circular motion (one inch circle). With the other hand, depress the "Abrasive Blast" lever for three or four seconds.
4. Release the "Abrasive Blast" lever. Depress the "Air Blast" lever and wobble the plug for three or four more seconds to remove loose particles of compound that may be lodged in spark plug.
5. Inspect the spark plug with an inspection light. If the insulator is not white, clean the plug again.

CAUTION: Prolonged application of the cleaning blast will wear down the insulator and electrodes and damage the spark plug.

The spark plug shown in Fig. 10-8 is being compression tested. A spark coil provides the voltage while a valve wheel increases compression. The intensity of the spark is observed through the window of the tester. If the spark is intermittent, or stops before the compression reading reaches the area on the dial marked good, the plug should be discarded.

When installing plugs, clean the spark plug seat on the cylinder head. Also clean the threads and make sure the gasket is in good condition.

Not all plugs require gaskets. Fig. 10-9 shows one type that does not.

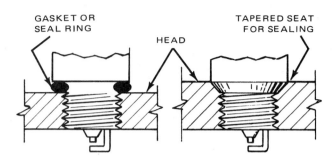

Fig. 10-9. Two types of plugs. Tapered seat at right requires no gasket. Clean area around plug hole for good seal. When reinstalling plugs, inspect gaskets carefully. A damaged gasket will not seal properly.

ANALYSIS OF USED SPARK PLUGS

Spark plugs that are cleaned and gapped at regular intervals can provide many additional hours of useful life in an engine using leaded gasoline. Spark plugs in high energy ignition systems (TCI or CDI), burning unleaded fuel, can provide more than twice as many hours of useful service.

Spark plugs in two different engines of the same make and model may show a wide variation in appearance. Engine condition, carburetor settings, and operating conditions such as sustained high speeds, or continual low speed stop-and-start operation are all variables that affect spark plug life.

Spark plugs are sometimes incorrectly blamed for poor engine performance. Replacing an old plug may temporarily improve engine performance because of the lessened demand a new spark plug makes on the ignition system. But this is not the cure-all for poor performance due to worn rings or cylinders, carburetion troubles, worn ignition system parts, or other engine problems.

SPARK PLUG INSTALLATION

Spark plugs must be installed properly. Heat dispersing properties of spark plugs depend upon correct seating. If the spark plug is tightened excessively, the gasket will be crushed. Internal leakage may also result. Attempting to remove an overtightened spark plug can possibly strip cylinder head threads. Seating the spark plug too loosely can result in pre-ignition and possible engine damage due to overheating of the spark plug.

To install spark plugs:
1. Make sure cylinder head and spark plug threads

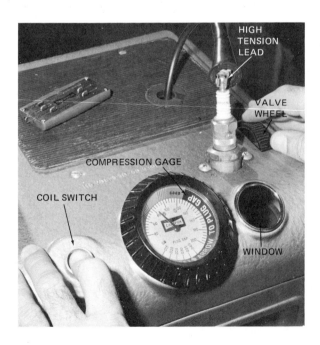

Fig. 10-8. After cleaning and gapping, spark plug can be spark tested under compression and compared to a new one. (Champion Spark Plug Co.)

are clean. If necessary use a thread chaser and seat cleaning tool.

2. Make sure that the spark plug gasket seat is clean. Thread the gasket to fit flush against the gasket seat on the spark plug.
3. Make sure the spark plug has the correct gap.
4. Screw the spark plug finger-tight into the cylinder head. Use a torque wrench to tighten the spark plug. Refer to the Installation Torque Specification Chart in the rear of the text.

IGNITION TESTING PROCEDURE

Certain basic inspections and tests are necessary to determine whether the ignition system is working properly. These tests include:

1. Make certain thin ignition ground wire is not grounding out entire system. Examine high tension spark plug lead for voltage leaks. Look for worn insulation or cuts where metallic contact is made.
2. Remove spark plug and examine condition of electrodes and porcelain insulator. If plug is carbon-grounded, replace it. Adjust plug electrodes to specified gap. Then, install plug and tighten it to specified torque, or turn plug finger-tight, then tighten it a half turn with a wrench. This allows proper heat transfer from plug to cylinder head.
3. Hold plug wire by insulation (well away from metal connector) so that connector is 3/16 in. from tip of plug. See Fig. 10-10.
4. Pull starter cord. An orange blue spark should jump the gap between connector and plug. If it does, ignition is good.
5. If no spark occurs, hold the wire 3/16 in. from base of plug, Fig. 10-11. Pull starter cord once more. If spark occurs at base of plug and not at tip, plug is failing under compression. Replace it.
6. If a spark does not occur at either position, problem is in the magneto.

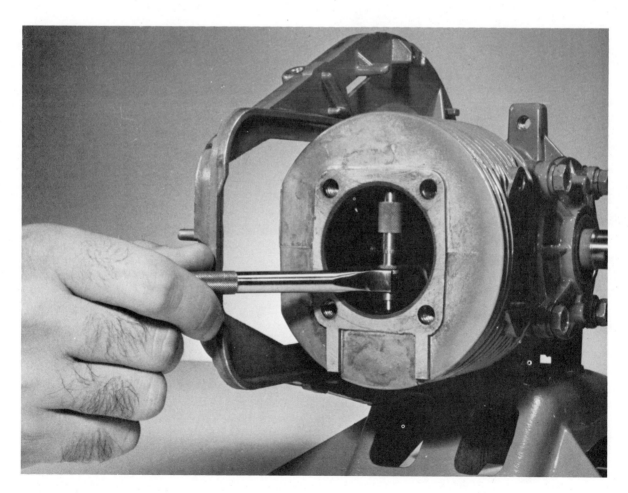

Internal engine inspection begins with use of an inside micrometer. Measure cylinder diameter at several points to check for excessive wear and/or taper.

Fig. 10-10. If spark does not jump a 3/16 in. gap at plug tip but jumps same gap at base, plug is defective.

Fig. 10-11. If no spark occurs at plug tip nor at plug base, magneto is faulty.

Fig. 10-12. A neon tester can be used to determine intensity of secondary current. A light will flash in window each time spark ignition occurs.

A spark plug tester, Fig. 10-12 also can be used to test the ignition circuit. If current is flowing in the spark plug lead, it will induce a voltage in the tester and light the neon tube. The greater the voltage, the brighter the flash will be in the window of the tester.

MAGNETO SERVICE

On some magnetos, the coil is mounted outside of the flywheel and the gap between the laminated core and the flywheel magnets is adjustable. *This adjustment must be made carefully or magnetic strength and voltage will be reduced.* Feeler gages or non-magnetic shim stock of the correct thickness can be placed in the gap, Fig. 10-13. Then the adjustment screws are tightened.

Fig. 10-13. Magneto air gap is checked with a non-magnetic thickness gage on some engines.
(Lawn Boy Power Equipment, Gale Products)

Many magnetos are completely contained under the flywheel. The flywheel must be removed from the crankshaft to service breaker points, condenser and, on some engines, the coil. The flywheel is mounted on the tapered end of the crankshaft and keyed for alignment, Fig. 10-14.

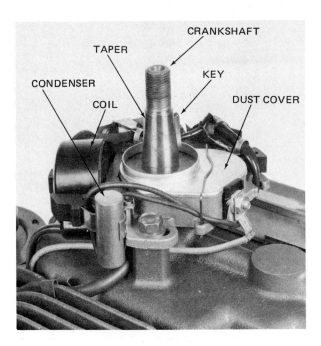

Fig. 10-14. Crankshaft end is tapered and keyed to hold flywheel in exact position. Flywheel often needs a special tool for removal from taper. (Deere & Co.)

Fig. 10-15. A knock-off tool is used to remove a flywheel. Tool must be tight on shaft before being hit. Pry bar provides a valuable assist. (Jacobsen Mfg. Co.)

Fig. 10-16. A wheel puller can be used to remove a flywheel. Always tap wheel puller after tension is applied with a wrench. Take care not to bend flywheel. Use light pressure only.

REMOVING THE FLYWHEEL

Since the flywheel is fastened to the crankshaft with a nut, it may be hard to remove. One way is to remove the flywheel nut and other accessory parts and place a "knock-off tool" on the crankshaft. Turn it on all the way and tap the tool with a hammer. See Fig. 10-15. The sudden jolt will loosen the flywheel. *Never use the flywheel nut as a knock-off tool or the crankshaft threads will be damaged.*

Another method of flywheel removal is to use a wheel puller like the one shown in Fig. 10-16. Tighten the puller with a wrench, then tap the puller with a mallet. If the flywheel does not come loose, tighten further with the wrench and tap again. If the puller is not tapped, the flywheel can be bent, and damaged. Do not use this tool on aluminum flywheels.

INSPECTING THE FLYWHEEL

When the flywheel is removed, inspect it. Make sure there are no cracks and that the mounting hole and keyway are not damaged. Inspect the key. If it has begun to shear, as in view A in Fig. 10-17, or if the key is too narrow as in view B, the engine will be out of time. In either case, the key must be replaced.

To check the strength of the magnet, place a 1/2 in. socket on the magnet and shake the flywheel. The socket should remain in place on the magnet, Fig. 10-18. The magnet may lose its magnetism.

ADJUSTING BREAKER POINTS

To adjust ignition timing, first adjust the breaker points to the correct gap. Look in the engine manual

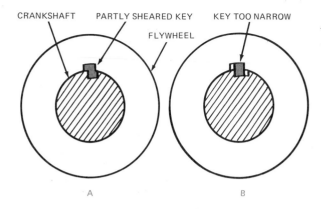

Fig. 10-17. Partly sheared flywheel keys or keys that do not fit well can put ignition system out of time.

Fig. 10-18. Magnetism is heart of magneto. Test magnets by placing a socket on them, then shake flywheel. Socket should remain in place.

Fig. 10-19. To adjust breaker point gap, loosen stationary point set screw first.

for the proper setting. Then, remove the dust cover from the stator plate to expose the breaker points. Turn the crankshaft slowly until the high point of the cam lobe is directly under the wear block or cam follower. Then slightly loosen the point adjustment screw as demonstrated in Fig. 10-19. NOTE: On the magneto shown in these illustrations, a screwdriver can be used to move the stationary part of the breaker point. A fulcrum point of leverage is provided, Fig. 10-20.

Fig. 10-20. Using screwdriver as a lever resting on fulcrum provided, move stationary breaker point to vary size of gap. Wear block should be on high point of cam.

To adjust, use a feeler gage of the thickness called for. First, move the stationary point until it lightly touches the feeler gage, Fig. 10-21. Then, tighten the adjusting screw that holds down the stationary breaker point. Finally, recheck the point gap. The stationary point may move when the screw is tightened. Repeat the adjusting procedure until the gap setting is right.

Fig. 10-21. A feeler gage is used to set correct gap of breaker points, then set screw is tightened to lock in adjustment.

Model	A	B	C	D	E V & H40	F V & H60	G V & H50	H	I
Displacement	6.207	7.35	7.61	8.9	11.04	13.53	12.176	7.75	9.06
Stroke	1-3/4"	1-3/4"	1-13/16"	1-13/16"	2-1/4"	2-1/2"	2-1/4"	1-27/32"	1-27/32"
Bore	2.125 2.127	2.3125 2.3135	2.3125 2.3135	2.5000 2.5010	2.5000 2.5010	2.6250 2.6260	2.6250 2.6260	2.3125 2.3135	2.5000 2.5010
Timing Dimension Before Top Dead Center for Horizontal Engines	H$\frac{.060}{.070}$	H$\frac{.060}{.070}$	H$\frac{.060}{.070}$	H$\frac{.030}{.040}$	H$\frac{.090}{.100}$	H$\frac{.090}{.100}$	H$\frac{.090}{.100}$	H$\frac{.060}{.070}$	H$\frac{.030}{.040}$

Fig. 10-22. A table of specifications in service manual will provide data concerning piston height (timing dimension before top dead center) for use in timing opening of breaker points. (Tecumseh Products Co.)

ADJUSTING PISTON HEIGHT

Ignition spark occurs at the instant the breaker points open. When this happens, the piston must be at the proper position (near top dead center) in the cylinder. Fig. 10-22 shows piston heights for a full line of Tecumseh engines. Model A, for example, requires the piston to be between .060 and .070 in. BTDC.

One method of setting the piston is to use a timing tool, Fig. 10-23. Install the cylinder head over the

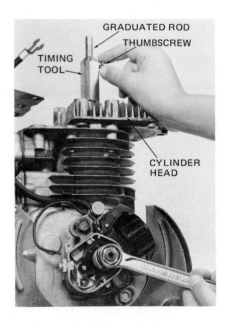

Fig. 10-23. A special timing tool can be used to adjust piston to specified location BTDC.

cylinder and fasten it by one or two head bolts. Do not use the gasket. Locate the head so that the spark plug hole is over the piston and both valves can move freely.

To use the timing tool, screw it into the spark plug hole with its graduated rod riding on the piston top. Note that each marking on the rod is 1/32 in. (.031 in.). Place the nut on the crankshaft so it can be turned with a wrench. Then, turn the crankshaft in the direction the engine runs until the piston is at TDC. Since proper piston position is between .060 and .070 in. BTDC, two graduations or marks (.062 in.) will be required. However, to take care of any backlash, back the piston down six graduations and bring it back up four marks. This places the piston exactly two graduations or .062 BTDC. Lock the thumbscrew and remove the wrench.

There is still another method of adjusting piston height. With this technique, Fig. 10-24, depth micrometer is set for correct piston height BTDC, then the piston is brought up to it.

TIMING IGNITION SPARK

With the correct breaker point gap set, and the piston positioned at the point where the spark should occur, the next step is to time the "breaking" of the points. You can do this with a simple, battery-operated continuity tester.

First loosen the two stator adjustment bolts so the stator can be turned or rotated. Disconnect the coil lead to the points and connect one end of the continuity tester to the breaker point terminal. Touch the other end of the continuity tester to the stationary breaker

Fig. 10-24. A depth micrometer also can be used to measure piston height BTDC. Piston can be raised or lowered by turning wrench on crankshaft nut.

point. Then, rotate the stator plate until the test light goes out, indicating that the points have opened, Fig. 10-25. Next, carefully tighten the two stator adjustment bolts. Reconnect the coil lead to the breaker point terminal.

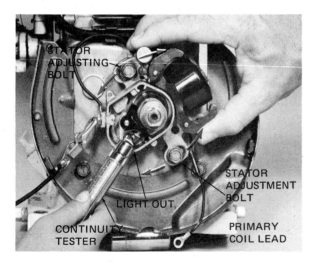

Fig. 10-25. A continuity tester is attached. Then stator plate is loosened and turned until continuity light goes out, indicating points have opened. Stator plate is locked in place. Coil primary lead is disconnected during this test.

Fig. 10-26. Felt oiler should have several drops of oil to lubricate cam lobe. Some felt oilers are made to be reversible. (Jacobsen Mfg. Co.)

Before replacing the dust cover over the stator plate, place one or two drops of oil on the felt cam oiler, Fig. 10-26. Lubrication reduces wear on the point wear block, but DO NOT OVER OIL. Excess oil will foul the breaker points.

REINSTALLING THE FLYWHEEL

If you have carefully followed the recommended procedure, the engine ignition system will be "in time." The next step is flywheel installation.

Turn crankshaft to place keyway in 12 o'clock position, then insert key. If shaft uses a Woodruff key, position it as shown in Fig. 10-27. *Make sure key seats properly in keyway before starting flywheel on shaft.*

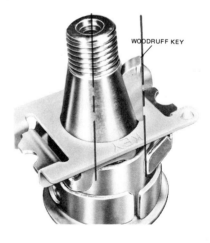

Fig. 10-27. If a Woodruff type key is used on crankshaft, it should be placed so top is parallel to center line of shaft. (Lawn Boy Power Equipment, Gale Products)

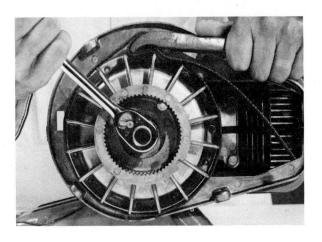

Fig.10-28. Flywheel can be held with a strap wrench while tightening flywheel nut. (Jacobsen Mfg. Co.)

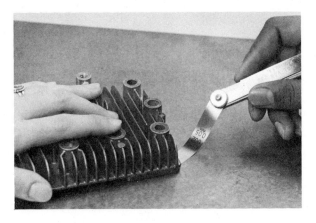

Fig. 10-30. Cylinder head should be placed on a surface plate and tested for flatness with a thickness gage.

Next, align flywheel keyway with crankshaft key and install flywheel. Then, tighten crankshaft nut to correct torque. NOTE: Use a strap wrench, Fig. 10-28, or a spanner wrench, Fig. 10-29, to hold flywheel during torqueing operation.

Fig. 10-29. A spanner wrench also can be used to hold flywheel when torqueing flywheel nut.

REINSTALLING THE CYLINDER HEAD

First remove carbon deposits from the cylinder head. Then, check the head for flatness by laying it on a surface plate. Use an .002 in. feeler gage between the mating surfaces to detect any warpage, Fig. 10-30.

If there are any broken cooling fins, cracks, nicks or burrs on the machined surface, the cylinder head should be resurfaced or replaced. Always install a new head gasket and follow the bolt tightening sequence path recommended by the manufacturer.

TESTING ELECTRICAL SYSTEM

Condensers, like coils, do not cause much trouble. A condenser should not be replaced unless you know it is

Fig. 10-31. Condenser leakage and short test. Test equipment usually has instructions for use. Note placement of red (positive) lead on breaker point terminal.

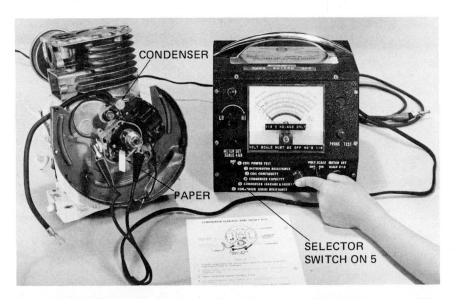

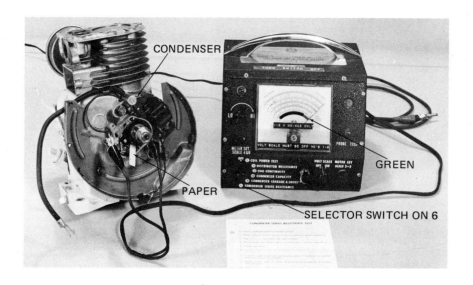

Fig. 10-32. Condenser series and resistance test.

bad. The only way to be sure is to check it for continuity (complete circuit) and capacity with ignition test equipment.

A condenser is a container for storing electricity. Generally, it is made of two foil strips with paper insulation between them. The foil strips and insulator are wound around each other in a tight, cylinder-shaped package. This package is encased in a metal container. One foil is grounded to the container, which is bolted to the engine. The other foil is connected by an insulated wire to the breaker point terminal. The condenser acts as a storage reservoir or "surge tank" for electricity when the breaker points are open.

CONDENSER LEAKAGE AND SHORT TEST

Certain tests will reveal the condition of a condenser. One is the leakage and short test. If the paper insulation deteriorates and current leaks from one foil strip to the other, or if the strips contact each other, these defects will show up on a test meter, Fig. 10-31.

The test equipment operating instructions describe each test and how to carry it out. The instruction manual also includes specifications of electrical values of the components tested.

To make the leakage and short test with the tester shown in Fig. 10-31, connect the black lead to the stator plate. Attach the red lead to the breaker point terminal, then push in the red button and take a reading.

The meter pointer will move to the right end of scale No. 5, then return to the left. If not, the condenser is leaking or shorted and should be replaced.

Notice in Fig. 10-31 that the breaker points have a strip of paper inserted between them. This keeps them from making contact and allows the current to flow through the condenser. The same test can be made with the condenser removed from the engine.

CONDENSER SERIES RESISTANCE TEST

The series resistance test determines whether the condenser will charge and discharge fast enough. Fig. 10-32 shows the proper hookup for this test. Connect the test leads as shown, then turn the selector switch to the No. 6 to test for condenser series resistance. Clip the red and black leads together and adjust the meter set scale so that the pointer is at the right end of scale No. 6. Then connect the black lead to the stator plate and connect the red lead to the breaker point terminal. Install the paper strip to keep the points separate. Under test, the pointer should remain in the green arc of scale No. 6, Fig. 10-32.

If the needle moves into the red arc on this scale, the condenser is defective. *It is good practice while conducting this test to wiggle the condenser lead and look for a needle deflection. (a sudden movement of the needle).* If the needle does deflect, replace the condenser. It may fail from engine vibration after testing good on the meter.

Condenser No.	Mfg.	Mfg. No.	Capacity Reading in Microfarads Min. Max.
25809	Fairbanks-Morse	R-2433	.17—.23
29559	Wico	X-11672	.16—.20
30548			.12—.14
30548-A			.16—.18
610253	Phelon	FG-159	.15—.19

Fig. 10-33. Typical condenser specifications are provided.   (Merc-O-Tronic Instrument Corp.)

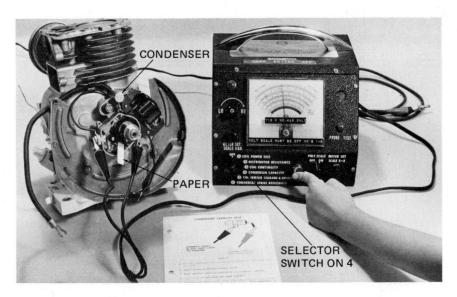

Fig. 10-34. Condenser capacity test can prevent burning out breaker points prematurely.

CONDENSER CAPACITY TEST

Condenser capacity is measured in electrical terms (microfarads, meaning millionth of a farad). Fig. 10-33 shows a list of condensers and the range of capacities for each. To use the list, first find the condenser number. It is usually stamped on the condenser case. The particular condenser under test in Fig. 10-34 is No. 30548-A. If good, it will test in the .16—.18 microfarad range. If condenser capacity is too high or too low, the breaker points will arc and pit more quickly than normal, Fig. 10-35. If capacitance is too high, metal transfers from the stationary point to the movable point. If too low, metal transfers from the movable to the stationary point.

In the test shown in Fig. 10-34, the meter needle shows a reading of .17 microfarads. Based on the .16—.18 specification, this indicates a good condenser. Always follow the instructions of the meter manufacturer when making the tests.

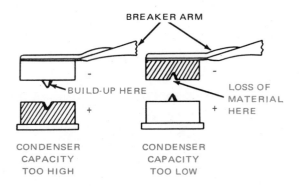

Fig. 10-35. Results of wrong capacitance in condenser.

COIL TESTS

Coils generally are trouble free. However, it is good practice to test the coil when other parts fail to show a defect. The primary resistance test shown in Fig. 10-36, for example, measures resistance to electron flow in the

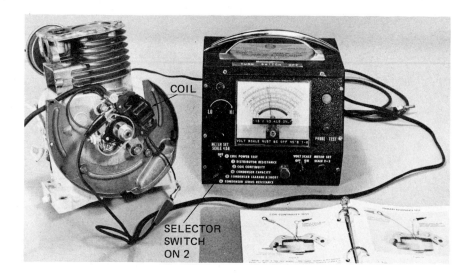

Fig. 10-36. A primary circuit resistance test being performed on a coil.

SMALL ENGINE COIL SPECIFICATIONS

Coil No.	Mfg. Model No.	Mfg. No.	Operating Amperage	Primary Resistance Min. Max.	Secondary Continuity Min. Max.
26787	Fairbanks-Morse	HX-2477	1.90		50—60
28259	Wico	X-11654	2.1		40—55
29176	Phelon	FG-4081	2.8		40—60
29632	Lauson	5022 (Syncro)	2.3	.5—1.5	40—55
30560	Lauson VH60	5160 (Syncro)	2.9	.35—.45	40—55
	John Deere (Replaces 30546)				
32014	Lauson HH80	8	1.4	.37—.45	55—65

Fig. 10-37. Coil specifications are provided for testing by manufacturer of test equipment.   (Merc-O-Tronic Instrument Corp.)

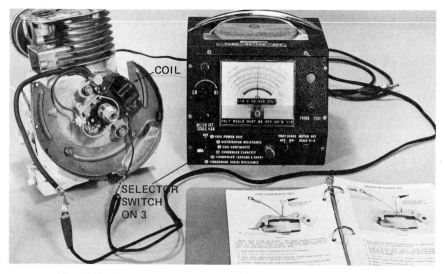

Fig. 10-38. Electrical continuity in secondary circuit of coil being tested.

Fig. 10-39. Connections are shown for making coil ground test.

primary winding of the coil. Resistance is measured in units called ohms.

A sample coil specification chart, Fig. 10-37, shows that the resistance range for a No. 30560 coil is .35—.45 ohm. With coil under test, the ohmmeter reading must be between the two values shown in chart.

COIL CONTINUITY TEST

The coil continuity test determines whether or not there is a break or a high resistance somewhere in the secondary circuit of the coil. Fig. 10-37 shows a value range of 40-55 for the 30560 coil.

The test hookup for a 30560 coil is shown in Fig. 10-38. Connect the red test clip to the spark plug "high tension" lead. Read the resistance value on scale No. 3 of the meter.

A test result that is lower than the low end of the specified range indicates that the secondary winding is shorted. A reading higher than the highest specified value indicates that the secondary winding has a break in it and is "open." In either case, the coil should be replaced.

COIL GROUND TEST

Check coil specifications carefully. COILS THAT ARE NOT permanently grounded to the coil laminations *should not indicate any meter pointer movement when tested.*

Coils that ARE permanently grounded to the coil laminations must show a full deflection of the meter needle to the right of the scale. See Fig. 10-39. Ground the red test clip to the stator plate. Connect the black test clip to the primary coil lead.

COIL POWER TEST

The coil power test determines the coil's ability to provide an adequate spark. With the test leads properly connected, Fig. 10-40, advance the current control knob until specified operating amperage is reached on scale No. 1. At 2.9 amps., a spark should be firing steadily between the 5 mm spark gap behind the window. If there is no spark, or it is faint or intermittent, the coil is bad.

If the spark is good at 2.9 amps, turn the current control knob to the far right for the high speed test. The coil should continue to give out a strong steady spark.

If the coil fails any of the tests, replace it with a new coil, or with one known by test to be good. It has been said, "One test is worth a thousand expert opinions."

CHECKING RPM

Often in small engine service work, it is necessary to test or set maximum idle or governor rpm. One way of doing this is by means of a device that converts engine vibration from power pulses to rpm. See Fig. 10-41.

Fig. 10-40. Coil power test shows whether coil will function at specified operating amperage and at high speeds.

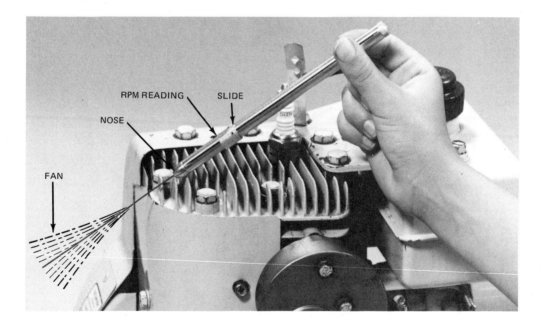

Fig. 10-41. Test speed of engine through vibrations caused by power pulses of piston. Note adjustable slide. (Vibra-Tach)

Place the nose of the instrument against the running engine. A thin wire is moved in or out of the instrument barrel until it vibrates into a fan pattern. When the fan shape is at its widest point, take the rpm reading from the scale on the barrel of the instrument.

Many mechanics use an electronic unit called a stroboscope (strō'-bō-scōpe), Fig. 10-42. This method is very accurate. To prepare for the test, make a chalk mark on the crankshaft of the engine.

The stroboscope produces a high intensity light, which flashes on and off at a controlled rate. Aim the light at the chalk mark on the rotating shaft. Then adjust the flashing frequency until the chalk mark appears to stand still. At this point, the time interval between each crankshaft rotation is equal to the time interval between each flash of the light. The scale on the stroboscope shows the number of light flashes per minute. This, in turn, is equal to the rpm of the crankshaft.

Fig. 10-42. Flashing light of stroboscope unit can be adjusted to match revolutions of crankshaft, which gives engine rpm.

When a stroboscope is used, parts which are actually moving at a high rate of speed appear to be stopped. *Never attempt to touch the parts or the chalk mark.*

SERVICING BATTERY IGNITION SYSTEMS

Battery operated ignition systems, are much like magneto systems. However, they have additional components which need maintenance. While more common to automobiles, trucks and tractors, they are sometimes used on small engines.

Presence of a battery and a starter does not mean that the ignition is battery operated. Some magneto ignition systems also use these parts. You can identify the battery ignition system by the can-shaped coil in addition to the battery and generator.

BATTERIES

Storage batteries need regular maintenance to keep them in good operating condition. Remove the caps about twice a month and add distilled water, as needed, to bring the electrolyte above the plates. This prevents sulphation (forming of salt-like deposits when air combines with the electrolyte), which will ruin the battery. These deposits are the same as the greenish deposits that collect around the posts and cable clamps of an automobile battery.

Keep the battery case and terminals clean. Remove the battery cables occasionally and clean all parts thoroughly. Reinstall the cables and coat them with petroleum jelly to retard further corrosion. Keep these connections clean and tight to avoid short circuits and/or voltage loss.

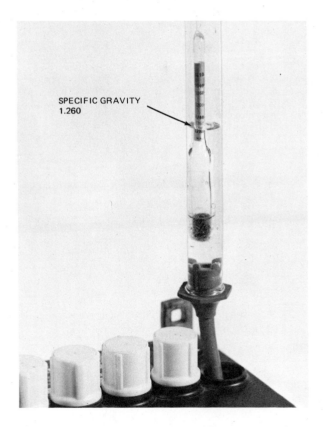

Fig. 10-43. Hydrometer reading shows fully charged condition of a battery. Specific gravity reading is in 1.260 to 1.280 range.

HYDROMETER TEST

If battery condition is doubtful, test it. Use a hydrometer to determine if the battery is fully charged. Remove one of the battery vent caps and draw enough electrolyte into the hydrometer to make the float move freely, Fig. 10-43. No further testing should be attempted if battery is less than 1.250. The battery must be recharged, Fig. 10-44.

Fig. 10-44. Battery charger cables must be attached to battery posts, positive to positive and negative to negative. Otherwise the battery could be ruined. If battery markings are unreadable, dip leads from battery terminals in a weak solution of sulphuric acid. More gas bubbles will collect around negative lead.

CHECKING VOLTAGE

Another way of testing a battery is to measure cell voltage. This can be done with an accurate voltmeter, Fig. 10-45, or a high-rate discharge tester. The tester puts a heavy load on the battery for three seconds.

Fig. 10-45. An accurate voltmeter can be used to measure battery voltage and cell voltages.

Check each cell in turn. If voltage checks out from 1.95 to 2.08V, the battery is good. If there is a difference of 0.05V between any cells, replace the battery.

Some batteries with one-piece hard covers cannot be checked by testing individual cells. In this case, the hydrometer test is made or an instrument equipped with a carbon pile rheostat is used. There is also a third tester with probes which are dipped into the electrolyte. The probes carry an electrical impulse to the voltmeter.

BATTERY RECHARGING

Only direct current can be used to recharge a battery. The battery charger automatically rectifies (converts) alternating current (ac) available from electrical receptacles to direct current (dc). Add water to bring the electrolyte in the battery cells to the right level. Make sure the outside of the battery is clean, then connect the charger leads positive to positive and negative to negative. Charging times will vary, depending upon battery condition and charging rate. Follow makers' instructions.

Badly sulphated batteries can sometimes be reclaimed by recharging them very slowly. This reconverts the sulphate to electrolyte.

Most new batteries are dry-charged. To place this type of battery in service, add the electrolyte solution according to instructions given by the maker of the product. Some new activated batteries require a short period of charging; others can be placed in service immediately.

CHECKING THE DC STARTER CIRCUIT

There are certain steps to follow in checking out or troubleshooting the direct current (dc) starter-generator circuit found on some small engines.
1. Disconnect all equipment, place transmission in neutral and turn on ignition switch.
2. If generator warning light comes on, or if ammeter shows a discharge, these units are working and the battery is supplying current.
3. Start engine. If generator warning light stays on, or if ammeter continues to show discharge, look for trouble in generator circuit.
4. If system is charging, check out starting circuit. There are four separate checks:
   a. Test ground connection at starter-generator or at battery, using a voltmeter. If negative battery

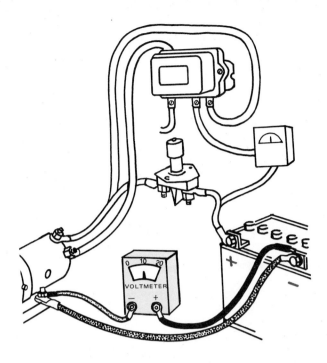

Fig. 10-46. Voltmeter can be used to check battery and starter-generator system ground connections.

cable goes to ground, attach negative voltmeter lead to starter-generator mounting frame, and clip positive voltmeter lead to negative post of battery, Fig. 10-46. Reverse this procedure if system is positive ground.

If voltmeter shows about 10V, move on to next step. A lower reading indicates poor ground. Clean and tighten all ground connections. Replace ground cable if worn or corroded. Operate starter and check voltage.

b. If voltage is still low, check circuit between battery and starter switch. To do this, leave negative cable grounded to mounting frame. Move positive lead to switch terminal nearest battery. Then turn on starter switch.

A low reading on the voltmeter indicates a poor connection between battery and starter switch. If engine will not crank, clean and tighten cable connections. Then turn on switch and recheck voltage.

c. Check starter switch. Leave negative lead of voltmeter attached to starter-generator frame. Move positive lead to switch terminal nearest starter-generator. Turn on switch. If there is little or no voltage, starter switch is not closing circuit. Repair or replace. If voltage is still low, go to next step.

d. Move positive lead to armature post of starter-generator. Leave negative lead grounded. Starter should operate and voltmeter should read about 11V. If engine does not start, and voltage is normal, starter is faulty. No start and little or no voltage indicates a loose or broken connection between starter switch and starter-generator connection. Clean, tighten and inspect or replace wiring. Recheck voltage.

COMMUTATORS AND BRUSHES

All dc starters and generators have commutators and brushes that occasionally need service. Start by cleaning the metal housing. Avoid getting cleaning solvent on insulated wiring.

Check for a worn bearing at either end of the armature shaft. (You can feel play with your hands and, usually, worn bearings are noisy when operating.)

If there is a cover band, remove it. A ring of solder along the inside of the band indicates the unit has overheated. Further repair must be done by an experienced mechanic.

If the generator has no cover band, remove the long bolts running through the housing and pull off the end plate nearest the commutator. The commutator is the group of bars arranged in cylinder-like fashion inside

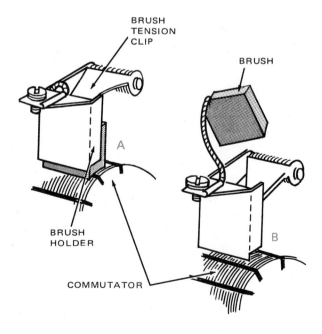

Fig. 10-47. Brushes are held against commutator by brush clip. A—Brush is worn down so clip rides on holder. B—Brush is removed for replacement. Never pull on brush leads.

the generator or starter. Spring loaded brushes rub on it, Fig. 10-47. Check the brushes for wear. They should move freely and press firmly against the commutator.

A generator will have either two or three brushes. A starter-generator will have two. If brushes are worn to half-length, or if clips are resting on the brush holders, replace the brushes. If brushes are binding in the holders, wipe holders with a clean, dry rag.

Check and tighten all electrical connections. Inspect the commutator for damage and/or wear. If the bars are rough and no longer round, the armature will have to be chucked in a lathe. Then, the commutator can be turned to a smooth finish, and the mica (insulating material between bars) undercut.

If the commutator is only dirty and glazed, clean it up with fine sandpaper while the engine is running slowly. *Do not leave any wires disconnected during this operation. You will burn out the generator!*

You can clean the starter commutator while the starter is turning over the engine. Disconnect the spark plug wire so the engine does not accidentally start. Use No. 00 sandpaper wrapped over the squared end of a stick. Move back and forth on the commutator until all dirt and glaze is removed. Never use emery cloth or a solvent. Emery will cause arcing. Solvents will soften the insulation.

Install new brushes, if needed. If they do not seat squarely, pull sandpaper back and forth between brush and commutator letting the sandpaper work the brush down to the shape of the commutator. Blow out the dust and replace the band.

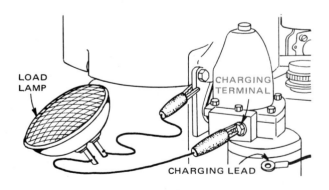

Fig. 10-48. Load lamp hooked up to charging terminal and ground will show if alternator is producing current. Load lamp is an automobile sealed beam unit with leads and clips attached. (Briggs and Stratton Corp.)

Polarize the generator at the regulator by placing one end of a jumper wire on the battery terminal. Touch the other end, momentarily, to the generator terminal. This is necessary if you have disconnected any of the wire leads.

If you do not do this, and polarity has reversed, you may burn out the generator, the cutout relay points in the regulator or run down the battery.

MAINTAINING THE ALTERNATOR

Small engine alternators come in different sizes rated from 1.5 amp. output up. Maintenance involves lubrication (on some units, but not on others). Periodically, see that brushes, slip rings and bearings are in good condition. If the battery is run down, test alternator output and, possibly test separate parts of the alternator. But first check battery polarity. Is the proper terminal going to ground? If not, reverse them.

CAUTION: Never reverse polarity when servicing the battery. It could burn out alternator diodes and damage wiring by overheating. Do not try to polarize an alternator. It is NOT necessary. Polarity of an alternator cannot be lost or reversed.

OUTPUT TEST
Unhook charging lead from the charging terminal, Fig. 10-48. *Do not allow lead to touch engine or equipment.* This will ground or short the system and cause possible circuit damage.

Attach a load lamp between the charging terminal and ground. If the lamp lights when the engine is started, the alternator is working.

If the lamp does not light, test the stator. This is the heavy core of iron wrapped with many coils of wire. Disconnect the charging lead from the battery to the rectifier.

Examine the rectifier until you find two terminals with red and black leads connected to them. Again, do not allow the charging lead to touch the engine.

With the engine running, touch the load lamp leads to the terminals. If lamp does not light, the stator is defective and must be replaced.

TESTING RECTIFIERS
As mentioned earlier, a rectifier, or diode, changes alternating current (ac) to direct current (dc). It has a

high resistance to current flowing in one direction and a low resistance to current flowing in the opposite direction.

You will have to test it to see if it is doing its job. Do not run the engine. Touch one lead of an ammeter to the charging terminal and the other to ground. Then reverse the leads. Current should flow in only one direction. Replace the diode if the ammeter needle registers both ways.

Larger alternators will have additional parts to check out. For example, there may be more than one diode. Each diode of the assembly is tested by the method just described. Some charging systems may have a regulator to control the charging rate. Consult the engine service manual for the proper test procedure. This test will be similar to the one described for the stator and rectifier.

VOLTAGE REGULATOR SERVICE

Do not attempt to service voltage regulators until the engine has run for about 20 min. This will allow the temperature of the regulator to stabilize. Then check:
1. Condition of ground strap running to engine. Connections must be clean and tight.
2. Appearance of breaker points. They tend to wear concave and should be filed as necessary.
3. Wipe filed points with a clean, lintless cloth soaked in carbon tetrachloride. *Never use emery cloth or sandpaper.*

ADJUSTING CURRENT VOLTAGE
There are two points of adjustment in the current voltage regulator Fig. 10-49. Disconnect battery, then:
1. Check armature air gap. Push down on armature until contact points just touch. Measure air gap and check against manufacturer's specifications (usually .075 in.). Adjust by loosening contact mounting screws and raising or lowering contact bracket. Align points and retighten screws before making voltage setting.
2. Adjust voltage setting by turning adjusting screw. Clockwise adjustment will increase voltage setting; counterclockwise movement decreases it. Replace cover. Run engine to stabilize voltage and recheck.

CUTOUT RELAY SERVICE
The generator regulator usually has a cutout relay unit with three adjustments, Fig. 10-50. Disconnect battery, then:
1. Measure and adjust air gap. Place finger pressure on armature, directly above core until points close.

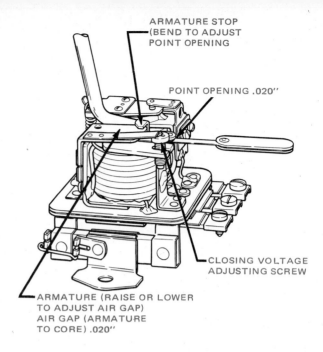

Fig. 10-49. Current-voltage adjustments on voltage regulator are called out. (Kohler Co.)

Measure gap and adjust to manufacturer's specifications. Generally, this is about .020 in. Retighten screws.
2. To adjust point opening, bend armature stop. See

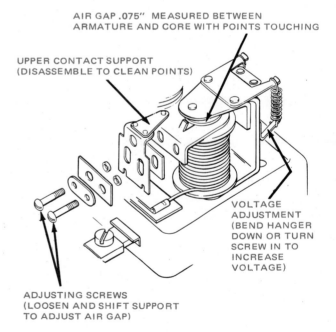

Fig. 10-50. Points of adjustment are shown on cutout relay. To clean points, loosen upper contact support.

service manual for correct gap.

3. Adjust closing voltage by turning screw clockwise to increase tension and voltage; counterclockwise to decrease. Closing voltage must be at least .5V less than current voltage regulator unit setting.

DISTRIBUTORS

Small engines having more than one cylinder and a battery ignition system must use a distributor to bring spark to the right cylinder at right time. During tune-ups, distributor cap should be removed and inspected for cracks, carbon tracking and pitted contacts.

The rotor also requires inspection. This is the small plastic "arm" mounted on top of the distributor shaft that revolves during engine operation. The metal tab on top of the rotor must make good contact with the metal inset in the center tower of the distributor cap. Therefore, the firing end of the rotor should not be worn or irregular.

The breaker points should be filed flat or replaced as described for magneto points. Spark timing can be set by rotating the distributor. Usually, the distributor is clamped in place to lock in the ignition timing adjustment.

IGNITION TIMING

To time the engine shown in Fig. 10-51, the No. 1 spark plug must be removed. Next, while turning the

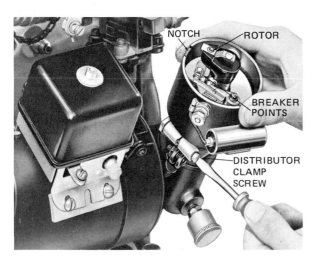

Fig. 10-51. With No. 1 piston at TDC, rotor should point to notch in distributor housing. To time ignition, loosen clamp screw, rotate distributor housing until breaker points close. Then turn distributor in opposite direction until points begin to open. Tighten clamp screw.   (Wisconsin Motor Corp.)

engine over slowly by hand, hold your finger over the spark plug hole to determine when the piston is on the compression stroke. Then, with the piston at TDC, the rotor should be approximately in line with a timing notch in the distributor housing. Loosen the clamp and rotate the distributor until the points just begin to open. Hold the distributor in place and retighten the clamp. This is accurate enough to start and operate the engine.

More accurate timing can be done with a neon timing light. In this method, a mark on the flywheel is aligned with a fixed mark on the engine block or shroud while the engine is idling.

LUBRICATION

Distributors should be lubricated at several points. Put a small amount of oil or grease in the reservoir or cup provided for the shaft, a film of grease on the breaker cam and a drop or two of oil on the pivot for the breaker points. *Use care, too much lubricant could cause the points to burn.*

MAINTAINING SOLID STATE IGNITION

Solid state ignition systems are offered in some small engines. They offer good service with no moving parts to oil and few adjustments. Certain parts come as an assembly and cannot be serviced.

If the system fails to produce a spark between the plug cable terminal and the cylinder head, check the condition of the cable. If it has cracked insulation or shows the aftereffects of arcing (black powdery discoloration or deposits), replace the pulse transformer and high tension lead assembly. See Fig. 10-52.

In some solid state systems, you can adjust the air gap between the trigger module and the flywheel trigger projection. Rotate the flywheel until the projection is next to the trigger module. Loosen the trigger retaining screw and move the trigger until the air gap is right. Use a feeler gage, and make certain that the flat surfaces on the trigger and projection are parallel before tightening the retaining screws.

Examine the low-tension lead. If the insulation is cracked or in poor condition, replace it. The ignition charging coil, electronic triggering system and mounting plate are available as an assembly only. If you must replace this assembly, carefully follow the manufacturer's instructions.

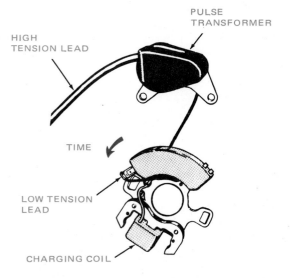

HIGH TENSION LEAD
PULSE TRANSFORMER
TIME
LOW TENSION LEAD
CHARGING COIL

Fig. 10-52. Solid state ignition assembly shown is type used on a two cycle engine. A spark will indicate that unit is satisfactory. Only spark plug need be replaced in such a case.

Before replacing the pulse transformer, attach leads from a new transformer, ground the unit and test for spark. If you have a spark where none existed before, replace the unit with the new one.

REVIEW QUESTIONS – CHAPTER 10

1. Which of the following determines the amount of voltage induced by the magnets when the engine is running?
   a. Breaker points.
   b. Spark plug.
   c. Condenser.
   d. Engine speed.
2. What color should the internal porcelain insulator be on a used but normal spark plug?
3. What five conditions cause spark plug deposits?
4. If no torque specification is given, what is the procedure for tightening a new spark plug?
5. If no spark occurs at the plug tip but it does occur when held near the engine cylinder head while cranking, the problem lies with the:
   a. Magneto.
   b. High tension lead.
   c. Coil.
   d. Plug.
6. Name two tools used to remove flywheels from small engine crankshafts.
7. If the key that positions the flywheel is deformed or partly sheared, the engine will most likely:
   a. Lose the flywheel.
   b. Be out of time.
   c. Run exceptionally fast.
   d. Burn excessive fuel.
8. The three major steps in timing a small engine are to adjust the piston height BTDC; rotate the stator plate until the points open; set the breaker point gap. List these three steps in this correct order of procedure for timing the engine.
9. Why would the stator on a magneto be rotated?
10. Explain the cause of condenser leakage.
11. The coil continuity test determines whether there is a break or high resistance in the:
    a. Secondary coil circuit.
    b. Primary coil circuit.
    c. Both secondary and primary circuits.
    d. Core of the coil.
12. Name the single component which indicates for certain that an engine has a battery ignition system:
    a. Battery.
    b. Distributor.
    c. Canister type coil.
    d. Starter.
    e. Generator.
3. Explain why emery cloth should not be used to clean a commutator of a generator or starter.
4. Describe under what conditions you would polarize an alternator.

SUGGESTED ACTIVITIES

1. Perform the complete timing procedure on an engine.
2. Demonstrate the proper technique for removing, inspecting and replacing a flywheel.
3. Obtain some old condensers and coils. Test them to determine their condition.
4. Examine some old spark plugs. Attempt to analyze the engine condition by their appearance. Clean, gap and test them.
5. Make a cutaway from a defective coil. Open up a condenser and examine its construction.
6. Place a bad plug in an engine and test the spark first at the plug tip, then at the base. Explain the difference.

This is a photo of a one-cylinder, four-stroke cycle, gasoline engine. Note mount on bottom so that crankshaft output is horizontal.

Chapter 11

CYLINDER RECONDITIONING

When you repair an engine, it is best to work in a clean and well-lighted work area. Tools should be clean and close at hand.

When using tools, safety rules should be observed. For example, safety glasses should be worn to protect the eyes. Oily and gasoline soaked rags should be placed in flame-proof cans, and heat or flame should never be allowed near solvents or other materials that will burn. A safe work place is not safe unless you use good judgment while working.

Some engines are easier to work on if they are mounted on an engine stand. Other engines in need of major repairs can be removed from the implement and torn down right on the workbench, Fig. 11-1.

You will need a service manual for the engine you are working on. Every engine make and model is built to certain dimensions and specifications that are different from any other engine. The kind of job you do depends a great deal upon the care and accuracy used in following the engine manual.

ENGINE INSPECTION

It is good practice to look for causes of engine problems. Do this even before removing it from the machine or vehicle. Such things as loose or broken engine mounts, misaligned pulleys or unevenly worn drive belts can cause excessive vibration. Wet oil on the outside of the engine may mean there are loose parts, leaking gaskets, leaking oil seals or a cracked casting.

Wires may need to be disconnected before the engine can be removed. If so, make flags of masking tape for the wire ends. Identify them with matching numbers, Fig. 11-2. This will avoid damage from wrong connections and will save reassembly time.

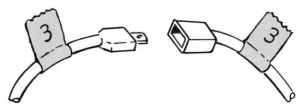

Fig. 11-2. If wires are coded with masking tape, they can be properly identified and reconnected.

DISASSEMBLING THE ENGINE

Before removing the engine, take out the spark plug. It could fire accidentally or break during disassembly of

Fig. 11-1. Small gasoline engines are easier to work on when removed from implement.

other parts. *When working on lawn mowers, snow throwers and other implements with exposed moving parts, never touch the blades or driven parts until the spark plug has been removed from the engine.*

After the engine has been removed, the starter unit can be unbolted. It may be a simple pulley-rope starter, a retractable starter rope or it may operate electrically. Usually only a few bolts fasten the starter unit to the engine. See Fig. 11-3.

Fig. 11-3. With engine out of implement, starter unit should be removed first.

The exhaust manifold pipe and muffler can be taken off next. Set them aside and out of the way. The carburetor and intake manifold pipe also can be removed, Fig. 11-4. It sometimes helps to sketch the carburetor and linkage locations. This can save time in reassembly. Check all gasket surfaces for defects. On some engines, the fuel tank is fastened to the carburetor. In any case, the tank can be removed and fuel lines disconnected.

The air shroud, blower housing and baffles can be removed to uncover the flywheel. Flywheel removal is explained in Chapter 10. Always use the right puller to prevent damage to the crankshaft or flywheel. After the flywheel is removed, magneto components can be disassembled and set aside.

Fig. 11-4. Carburetor and fuel tank should be removed next. Sketch carburetor linkage hookup for later reference.

ORGANIZE YOUR WORK

Organizing the job saves time and effort. At this point, the outer parts have been taken off the engine. They should be set aside in their own groups. Magneto and flywheel shroud and starter parts should be in one group; carburetor, fuel tank and exhaust manifold in another. See Fig. 11-5.

Keeping the work clean is part of good organization. Outside surfaces may be cleaned at this point of tear-down, Fig. 11-6. Inside surfaces will be cleaned later. Grass clippings and other debris should be removed by scraping and brushing fins and housings before using cleaning fluid.

A safe engine cleaning solvent should be used next to remove grease, oil and grit. Some parts, like the coil and condenser, may be cleaned by wiping with a clean cloth moistened with solvent. This is better than total immersion. *Never use solvents that burn easily nor those which may be harmful to humans.*

As each part is washed, wipe it with a clean cloth and set it aside to dry. If the workbench is oily and greasy, clean it before doing disassembly work on it. Hands should be washed often to keep dirt off the cleaned parts.

Fig. 11-5. A clean workbench and organized work procedure can speed the job.

Fig. 11-6. Grass and other debris should be removed from cooling fins before using a solvent.

CYLINDER INSPECTION

Several types of cylinders may be found on small engines. One is the separate cylinder block as shown in Fig. 11-7. Another is an integral (one-piece) cylinder block and crankcase, Fig. 11-8.

If the engine is of the separate cylinder type, remove the bolts and lightly tap the cylinder block with a soft leather hammer, Fig. 11-7. A slow, smooth pull will remove the cylinder from the piston assembly. An alternate procedure is to remove the connecting rod cap and pull the cylinder and piston from the crankcase as a unit. Then, pull the piston from the cylinder.

The piston must be removed from the top of the cylinder in the integral cylinder block and crankcase

type engine. First, remove the cylinder head. Next, take the crankcase cover from the side of the block and remove the connecting rod cap. Then, push the piston

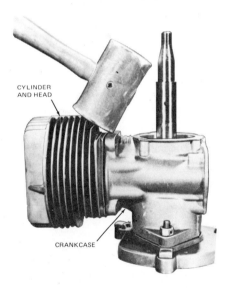

CYLINDER AND HEAD

CRANKCASE

Fig. 11-7. After bolts are removed, a light tap with a soft hammer will loosen cylinder block from crankcase. (Lawn-Boy Power Equipment, Gale Products)

out the top of the cylinder as shown in Fig. 11-8. When there is a heavy ridge around the inside top of the cylinder, the piston rings will not pass the ridge without

Fig. 11-8. Piston may be removed from cylinder if there is no noticeable ridge at top of cylinder above ring wear area. (Wisconsin Motors Corp.)

damage. Special ridge reaming tools, Fig. 11-9, are available for cutting the ridge to make this part of the cylinder the same diameter as the worn portion below.

Fig. 11-9. Maximum cylinder wear takes place near the top of the cylinder. Unworn portion at top of cylinder called the "ring ridge" must be removed. Cut it flush to cylinder wall with a ridge reamer. (Tecumseh Products Co.)

With the piston out, inspect the cylinder block. Look for areas of scuffing or scoring on the walls. Check for nicks or grooves in gasket surfaces. Inspect for chipped or broken fins. Examine head bolt holes and spark plug holes for damaged or stripped threads.

A worn cylinder has a narrow, unworn part at the top. The bottom of the ridge indicates the extent of the top piston ring travel. Right below the ridge is the area of most cylinder wear. The wear will be the greatest on two opposite sides 90 deg. to the crankshaft center line. The cylinder, therefore, wears into an oval shape. This increased wear is due to several things:
1. Less lubrication at this portion of the cylinder wall.
2. The diluting effect of raw gas on the engine oil.
3. The pressure that builds up behind the rings at their highest position.

Below the point where cylinder diameter is the greatest, cylinder wear lessens rapidly. Because of this wear pattern, there is a gradual taper toward the bottom of the ring travel. Below ring travel there is almost no wear. This area is well lubricated, and there is light wall pressure from the piston skirt.

CYLINDER MEASUREMENT

When the cylinder wall looks smooth and free of scuff and score marks, you are ready to measure the cylinder for wear and out-of-roundness.

Finding the amount of cylinder taper is the first important measurement in determining cylinder condition. First, measure the cylinder diameter below the ring travel, then just below the ring ridge. The difference is an accurate indication of the amount of cylinder taper. The taper measurements should be taken both parallel and at right angles to the crankshaft to determine the greatest amount of wear and out-of-roundness.

How much taper is allowed before the need for reboring depends on engine design, its general condition and the type of service in which it is used. There are no rules that apply to all engines regarding cylinder taper. The manufacturer, however, may set a limit. Beyond a certain point, he will advise reboring or cylinder replacement.

Out-of-roundness for small engines is generally limited to .005 in. or .006 in. Beyond this limit, engine performance is greatly reduced.

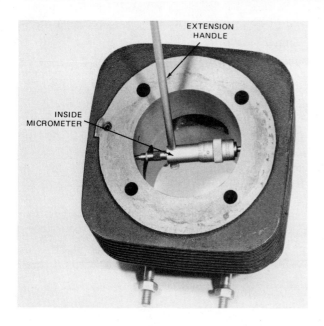

Fig. 11-10. An inside micrometer can be used to measure cylinder diameter. (Kickhaefer Mercury)

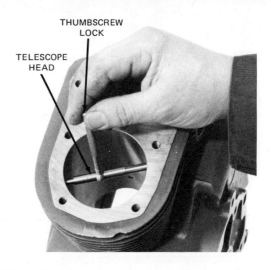

Fig. 11-11. A telescoping gage also can be used to measure cylinder diameter. Gage is adjusted to cylinder size, then is measured with an outside micrometer. (Deere & Co.)

Several methods can be used to measure cylinders. Fig. 11-10 shows an inside micrometer equipped with an extension handle. This precision instrument must be carefully adjusted to cylinder size. It must give the exact diameter of the cylinder.

Fig. 11-11 shows a telescoping gage being used to measure cylinder size. The gage head is spring loaded to expand when the thumbscrew is released. Once located in the cylinder, the gage is located in place by tightening the thumbscrew. Then it is removed from the cylinder, and an outside micrometer is used to measure the length of the telescope head. This, then, is cylinder size, or cylinder diameter.

There is a convenient tool designed for measuring small engine cylinders. Fig. 11-12 shows the various sections of the set gage part of this tool.

To use the tool, look up the standard cylinder size in the engine manual. Then place the right spacers on the shaft. Place the stop bar on the shaft with the remaining spacers and nut.

The cylinder gage attaches to the set gage as shown in Fig. 11-13. The needle can be set to zero by turning the knurled ring to the left. Lock the ring with the thumbscrew and remove the gage. The extension sleeves are used only when necessary.

When the cylinder gage is adjusted properly (set zero at specified diameter), put it into the cylinder, Fig.

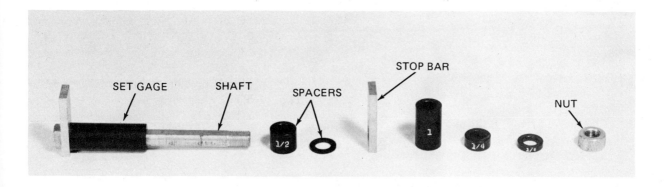

Fig. 11-12. Parts of disassembled set gage are shown. Gage is preset to a specified size.

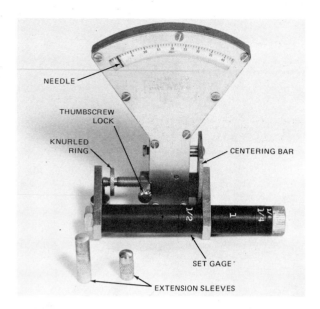

Fig. 11-13. Cylinder gage is adjusted to read zero when located in set gage by turning knurled ring.

11-14. The centering bar automatically centers the gage in the cylinder. Since the gage is spring loaded, it is free of the measuring spindle. The amount indicated by the needle is cylinder diameter oversize in thousandths of an inch. Take readings in three or four directions across

Fig. 11-14. Cylinder gage is self centering and will provide a direct reading. Needle deflection from zero indicates amount of wear in thousandths of an inch.

the diameter. Check these measurements against manufacturer's specifications.

Cylinder service needs are determined by cylinder condition. If the bore is not damaged, and if taper and out-of-round readings are within specified limits, only a light deglazing with a fine emery cloth may be needed. The engine block must be thoroughly washed afterwards.

REBORING THE CYLINDER

There are several ways to repair a cylinder that shows too much wear. What will be done depends on engine type. Some engines have chrome plated aluminum cylinders, Fig. 11-15. Worn or damaged cylinders

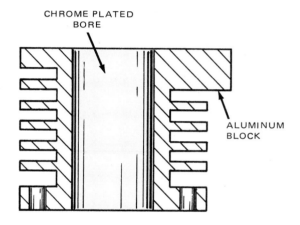

Fig. 11-15. Special cylinder construction is shown with a chrome plated surface.

of this type should be thrown away and replaced with a whole new cylinder. Other engines have a flanged cylinder sleeve, Fig. 11-16, which can be removed and replaced with a new sleeve. The cast-in sleeve, Fig. 11-17, or the solid cast iron cylinder, Fig. 11-18, can be rebored to a larger size.

Two problems must be solved when reboring a cylinder. The first is to resize and maintain original alignment while producing a round, straight bore. The second is to produce the correct cylinder wall finish.

Cylinders are usually rebored in steps of .010 in. If the cylinder is being rebored for the first time, its diameter will be increased .010 in. over the standard

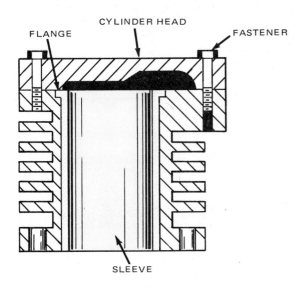

Fig. 11-16. Pressed-in sleeve can be secured by forming a flange on upper end. Cylinder head will hold sleeve in place.

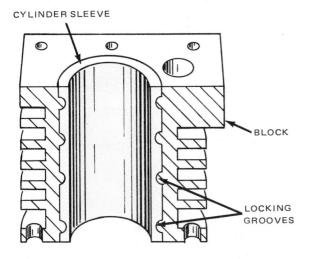

Fig. 11-17. Some cast iron cylinders are cast in place in block.

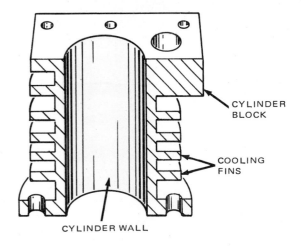

Fig. 11-18. Most cast iron cylinders are part of the engine block.

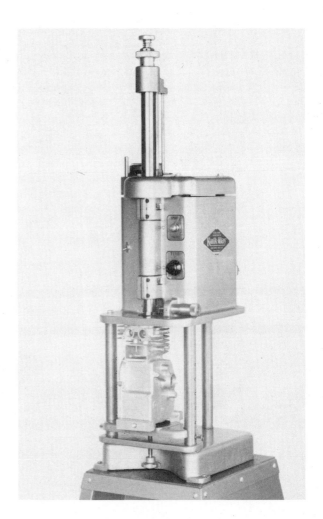

Fig. 11-19. A small engine cylinder boring machine is prepared for operation. Engine block is clamped in place. (Cedar Rapids Engineering Co.)

size. If this does not "clean up" the cylinder (remove imperfections), the next step is .020 in. over standard. When replacing pistons and rings, order .010 or .020 in. over standard size to match the new cylinder bore.

The boring machine shown in Fig. 11-19 is designed for use on small engines. In setting up the machine, the engine block or cylinder block can be clamped in place below the boring head. Fig. 11-20 shows the cutter being adjusted to the correct diameter for the new bore. A built-in micrometer provides accurate setup to .0001 in. (one ten-thousandth).

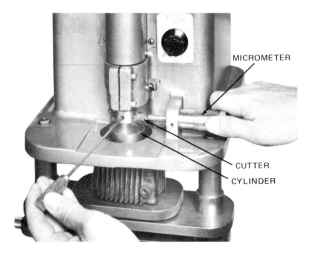

Fig. 11-20. Cutter on boring machine is set to give a precise bore diameter, using a built-in micrometer gage.
(Cedar Rapids Engineering Co.)

An electric motor drives the spindle to rotate the cutter. The feed rate controls the distance the cutter advances into the bore during each cutter revolution and can be changed by moving the feed dial. See Fig. 11-20 for example.

Boring the cylinder will produce a straight, round bore. However, the boring operation does not produce a satisfactory surface finish. The boring tool leaves microscopic furrows and surface fractures, so it is recommended that an overall minimum of .0025 in. of stock should be left for finish honing.

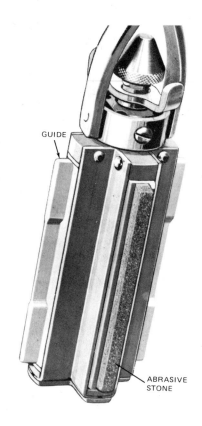

Fig. 11-21. A typical cylinder honing tool has two abrasive stones and two guides. (Sunnen Products Co.)

HONING THE CYLINDER

Honing is an abrasive (sandpaper-like) finishing process which removes the boring tool marks and

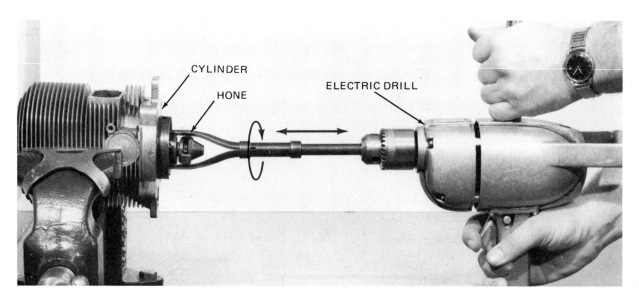

Fig. 11-22. Cylinder boring tool leaves microscopic furrows and surface fractures. These are removed and the cylinder is brought down to correct size by honing. (Kickhaefer Mercury)

surface fractures. Honing also produces the desired cylinder wall finish. Smoothness of finish depends upon the grit size of the stones used. See Fig. 11-21.

Ideally, a newly honed cylinder will wear smooth at just about the same rate the new piston rings need to wear in. When this "break-in" process is completed, both cylinder and rings will be smooth and last for hundreds of hours of engine operation.

Fig. 11-21 illustrates a typical cylinder honing tool for small engine work. It has two abrasive stones and two guides to keep the tool aligned with the cylinder. The stones can be removed and replaced with new ones as necessary.

An electric drill is used to rotate cylinder hone. In operation, assembly is slowly and steadily moved in and out of cylinder, Fig. 11-22. *Stones should not be permitted to extend far out of the cylinder end as uneven wearing of the stones may occur.* For best results, the honing process should produce a fine surface pattern like that shown in Fig. 11-23.

After reboring and honing the cylinder, use a piece of fine emery cloth to remove any burrs that may have developed around ports. Clean the cylinder walls and block thoroughly with kerosene and clean rags. Apply a light coat of SAE 10 oil to the cylinder to prevent rust.

Fig. 11-23. Honing process produces a fine cross-hatch surface pattern created by in-and-out motion of revolving hone.

REVIEW QUESTIONS — CHAPTER 11

1. When should the small engine mechanic begin looking for engine defects or problems?
2. Excessive vibration could be caused by:
   a. Loose engine mounts.
   b. Pulleys out of line.
   c. Worn drive belts.
   d. All of the above.
   e. Only a and c.
3. Before removing an engine from an implement, it is safe practice to:
   a. Drain the oil.
   b. Turn off the ignition switch.
   c. Remove the spark plug.
   d. Crank the engine slowly to remove all the fuel from the cylinder(s).
4. When you remove a carburetor and tear it down for repair, what will help you remember how it goes back together?
5. In what ways can you save time and effort when reconditioning an engine?

a. Consult the operator's manual.
   b. Keep bench and engine parts clean.
   c. Organize your work by grouping certain engine parts.
   d. All of above.
6. Explain how the piston is removed if there is a heavy ridge at the top of the cylinder in an integral cylinder block and crankcase type engine.
7. The location of cylinder diameter measurements should be:
   a. At the very top and bottom of the cylinder.
   b. Taken 90 deg. to each other at the center of the cylinder.
   c. Below the ridge and at the bottom of the cylinder to determine taper.
   d. Only below the ridge, but at 90 deg. to each other.
8. What do you call the special tool designed to remove cylinder ridges?
9. After a cylinder is rebored, it must be:
   a. Lapped.
   b. Reamed.
   c. Deglazed.
   d. Honed.

SUGGESTED ACTIVITIES

1. Design a well-organized workbench and tool panel for repairing small engines.
2. Completely disassemble an engine that has been run for many hours and needs reconditioning.
3. Inspect disassembled parts for wear or damage. Discuss possible causes.
4. Measure a worn cylinder with an inside micrometer or other measuring tool. Record readings for taper and out-of-round.
5. Rebore and hone or deglaze a cylinder.
6. Ream a cylinder ridge.

Measuring piston height is key step in relating position of piston in cylinder to spark timing. Engine manufacturer specifies thousandths of an inch before top dead center when breaker points should open.

Chapter 12

PISTONS AND PISTON RINGS

In reconditioning small gasoline engines, pistons and piston rings are critical service items. Generally, reboring and/or honing of the cylinders is necessary, followed by thorough inspection and repair of the parts closely fitted to them.

To do his job well, the small engine mechanic must understand the stresses to which a piston and its rings are subjected. He also must know the kinds of materials they are made from. Lastly, he must know what to do to put these parts back into top shape, and how to reassemble them for efficient engine operation.

The condition of rings and pistons can be learned by observing and inspecting the parts during disassembly of the engine. The need for service is evidenced by low compression, blow-by, oil pumping and fouled plugs.

The piston slides up and down in the cylinder. It sucks in the air-fuel mixture and compresses it, then carries the force of the burning fuel to the crankshaft through the connecting rod. On the final stroke of the engine cycle, the piston pushes burned gases from the cylinder.

In normal operation, the piston travels up and down in the cylinder more than a thousand times per minute. Subjected to high heat, pressure and friction, the piston must be lightweight, strong and properly fitted. It has a lot of work to do.

PISTON CONSTRUCTION

Pistons can be made of aluminum or steel. Aluminum is by far the most popular metal for this

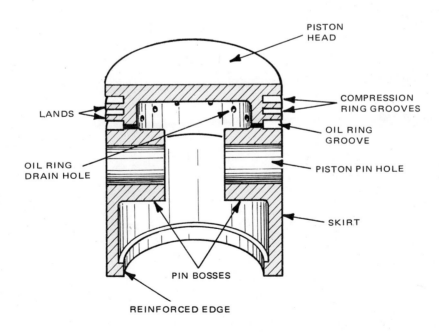

Fig. 12-1. A typical four cycle engine piston is cut away to show construction features.

application. The surface may be coated with a special break-in finish (tin or other coating). Sometimes pistons are chrome plated for installations where they operate directly on aluminum alloy cylinder walls.

The type of piston often used in a four cycle engine is shown in Fig. 12-1. The head is quite thick, giving this hardworking part strength and resistance to over-heating. The area below the head has grooves for the piston rings. Full-diameter ridges between the grooves are called "lands." The wall or bottom, of the oil ring groove is either slotted or pierced with holes. Oil wiped from the cylinder wall by the oil ring flows through these holes back into the oil sump. The sump is the low area of the engine block where the oil collects.

Pistons may have grooves for from one to four rings. Generally, the two cycle engine piston has one or two grooves. Both are compression ring grooves. Four cycle engine pistons will generally have three grooves, two for compression rings and one for an oil control ring.

The section of the piston surrounding the piston pin hole is called the pin boss. It is thick and often reinforced with cast-in webs. The piston skirt is the part from the bottom of the lower ring groove on down. The skirt is designed to be as light as possible to hold down the weight of the assembly.

The skirt actually guides the piston and keeps it from tipping from side to side. Portions of it may be cut away for lightness. Also, in some two cycle engines, portions may be cut away to allow the air-fuel mixture to pass through the piston skirt into other parts of the cylinder.

PISTON FIT

The piston is subjected to high temperatures, causing it to expand during operation. To allow for this increase in size, there must be a specific amount of clearance between the piston skirt and cylinder wall.

The cylinder also expands, but not as much as the piston. Normal clearance must be great enough to allow for lubrication and piston expansion. Different engines have different clearances. The amount depends upon engine design and use. Most small engines call for .003 to .005 in. piston-to-cylinder wall clearance, Fig. 12-2.

CAM GROUND PISTONS

When the designer wants the smallest possible clearance between the piston skirt and cylinder wall,

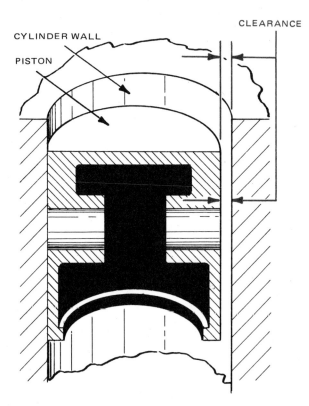

Fig. 12-2. Sufficient clearance must be allowed between piston skirt and cylinder wall to permit adequate lubrication and to allow for expansion of parts due to high temperatures.

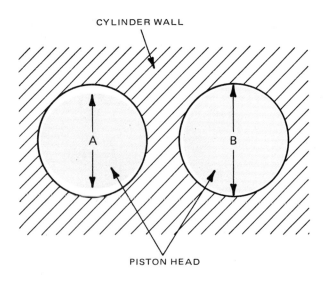

Fig. 12-3. Exaggerated top views of a cam ground piston as it would fit in cylinder. A—Cold. B—Hot. Arrows indicate piston pin position.

skirts are often cam ground to an elliptical (oval) shape shown at A in Fig. 12-3. The oval shape allows the thrust surfaces (sides of skirt forced against cylinder during compression and firing) to fit more closely, even when cold. As the piston heats up, the diameter across the thrust surfaces remains constant and the piston enlarges parallel to (in the same direction as) the piston pin. See B in Fig. 12-3. These exaggerated views illustrate how a cam ground piston expands to a round shape as it becomes hot.

PISTON THRUST SURFACES

During the compression stroke, the pressure of the confined air-fuel mixture forces the piston toward one side of cylinder. See view A in Fig. 12-4. When the crankshaft throw passes TDC, burning and rapidly expanding gases push hard on the piston, forcing it against opposite side of cylinder. See B in Fig. 12-4.

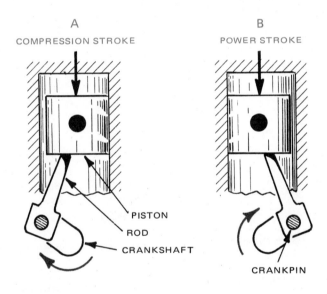

Fig. 12-4. Thrust surfaces of piston must resist heavy side pressure against cylinder walls. A—Upstroke. B—Downstroke.

In each instance, the sides of the piston forced against the cylinder wall are called thrust surfaces. These surfaces are at right angles (90 deg.) to the center line of the crankshaft and piston pin.

If the piston has too much clearance in the cylinder, side thrust during the compression and firing strokes will make it move, or "slap," from one side of the cylinder to the other. As it moves sideways, the piston will tend to tip or cock in the cylinder. This "loose fit" can be very harmful to the piston and rings. The piston must fit the cylinder properly to avoid slapping.

PISTON HEAD SIZE

The piston head receives the brunt of combustion heat, so it runs hotter than the skirt and expands more. Because of this, the head of the piston often is made with a smaller diameter than the skirt. Fig. 12-5 shows an exaggerated view of a piston with the head smaller than the skirt. The actual difference is only a few thousandths of an inch.

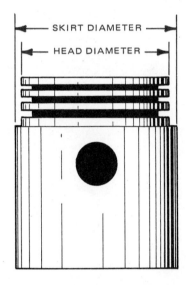

Fig. 12-5. Piston head receives greatest heat and is sometimes made smaller to compensate for expansion.

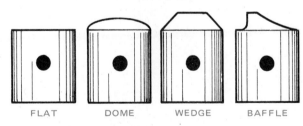

Fig. 12-6. Small gasoline engine piston heads are manufactured in a wide variety of shapes.

PISTON HEAD SHAPE

Piston heads are manufactured in many different shapes depending upon type of small engine and its use. On four cycle engines, the piston head can be flat, domed or wedge-shaped. Pistons used in two cycle engines generally are flat when used with loop-scavenging design. Cross-scavenging types use a raised baffle or deflector head piston, Fig. 12-6.

PISTON RING DESIGN

All small engine pistons must have clearance for lubrication and expansion. At the same time, they must have "rings" to help do the job of sealing the cylinder(s). Without piston rings, the piston could not compress the fuel charge properly. Also, burning gases would leak out between the sides of the piston and cylinder wall.

In performing their job, the piston rings ride against the cylinder wall, separated from it only by a thin film of oil. The rings rub freely against the sides of the ring

grooves, which hold the rings square to the bore and force them to slide up and down the cylinder with the piston, Fig. 12-7. Since the ring face is in steady contact with the cylinder walls, Fig. 12-7, an effective seal is formed.

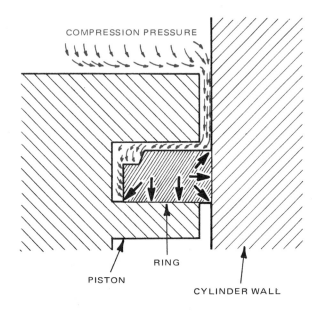

Fig. 12-8. Combustion chamber pressure forces ring against cylinder wall and bottom side of groove.

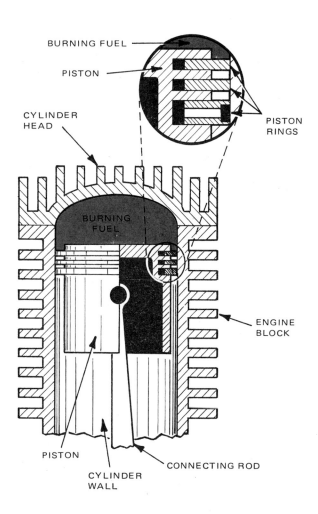

Fig. 12-7. Piston rings form a seal between the piston and cylinder wall.

Fig. 12-8 shows a compression ring in its groove. The sides of the ring groove are flat, parallel, and smooth. The ring has proper side clearance. In operation, expanding gases force the ring down against lower side of groove. At the same time, gases behind the ring force it against cylinder wall. A good seal is formed.

PISTON RING CONSTRUCTION

Piston rings are made of cast iron or steel. Both types may be plated with chrome or other long-wearing materials, Fig. 12-9. Most pistons use cast iron compression rings. Steel, when used, generally goes into the construction of the oil control ring. In some installations, a cast iron center spacer-scraper may be combined with steel side rails.

RING TENSION

To permit the piston rings to expand and contract under varied temperatures and operating conditions, the rings are cut through at one place at the time of manufacture. See Fig. 12-10. The size of this opening

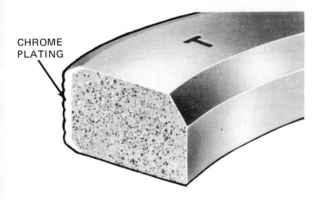

Fig. 12-9. Piston rings are often plated with chrome or other materials to reduce wear.

Fig. 12-10. End gap is cut through piston ring to permit ring to enter cylinder, then exert tension on cylinder wall.

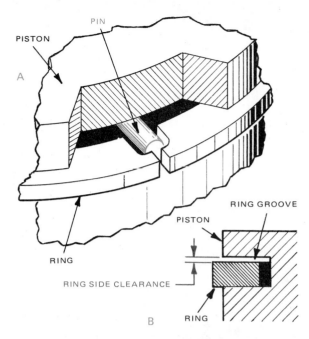

Fig. 12-11. A—Many two cycle engines have pinned rings to prevent ring rotation. B—Ring side clearance allows movement, admits lubricating oil and permits expansion of parts due to heat.

between the ends of the ring (with piston and rings in the cylinder) is called the ring end gap. Although a great number of gap joint designs have been used in an effort to seal against gas leakage, the plain butt joint is in common use.

In another design feature, the outside diameter of a piston ring is made slightly larger than cylinder bore diameter. *This permits the ring to exert some tension on cylinder wall when installed.* It is the reason why the ends of each ring in a single-piston set must be squeezed together to get the piston assembly into cylinder.

RING MOVEMENT

Piston rings are free to move inward and outward in their respective grooves in the piston. See B in Fig. 12-11. In addition, the rings will gradually work their way around (float) in the grooves, unless each ring is pinned in place, as shown at A in Fig. 12-11. In most four cycle engines, the rings float. Some two cycle engines have the rings pinned in position. This is to prevent ring ends from catching on edge of intake or exhaust port and cutting into the cylinder wall.

Floating rings must be installed with the ring end gaps staggered to avoid gap alignment and possible oil flow through the series of gaps to the combustion chamber. Pinned rings are held in position by a short pin manufactured into the ring groove in the piston. The ring ends are cut out to straddle the pin, Fig. 12-11. Obviously, the pin prevents rotation of the ring around the groove.

Fig. 12-12. A special tool is being used to clean carbon from ring grooves of piston. Do not widen nor deepen any part of grooves.

RING SIDE CLEARANCE

Piston rings must have the right amount of side clearance, which permits them to move in and out in the groove while exerting tension on the cylinder wall. Side clearance also provides for adequate lubrication and heat expansion. Fig. 12-12 shows how the groove is cleaned with a special tool having different size scrapers that are pulled around the grooves. Fig. 12-13 illustrates how side clearance is checked with a feeler gage.

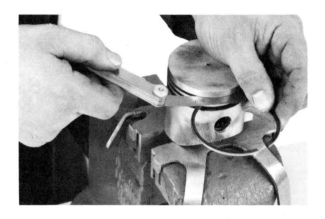

Fig. 12-13. Side clearance of groove is checked with a piston ring and a thickness gage.

COMPRESSION RINGS

Generally, the first and second rings from the top of the piston are compression rings. Compression rings are designed to provide a strong seal, keeping the compressed fuel mixture and the burning gases above the piston by preventing passage between the piston and cylinder wall. Compression ring shapes vary somewhat with scraper grooves, beveled faces and grooves or bevels on the inner side of the ring. See Fig. 12-14.

The various bevels and grooves are designed to create an internal stress in each compression ring. The stress causes the ring to twist slightly in its groove during the intake stroke of the piston. The twisting action places the lower edge of the ring, rather than the face, in contact with the cylinder wall. This allows compression ring to act as a mild scraper to aid in oil control. See view A in Fig. 12-15.

On the compression and exhaust strokes (four cycle engine), with the rings in a tipped position, they tend to slip lightly over the oil film on the cylinder, as in view B. On the power stroke, the pressure of the gases forces the ring flat so that the entire edge bears firmly against the cylinder wall. Maximum sealing is provided during this critical time, as in view C.

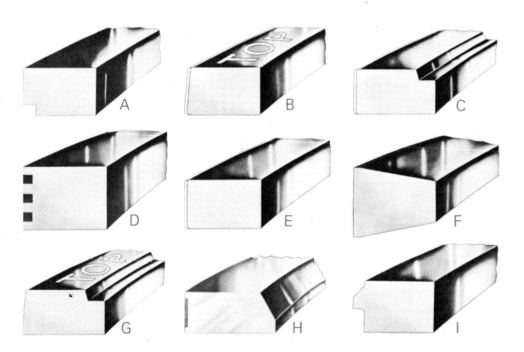

Fig. 12-14. Compression ring shapes. A—Outer groove. B—Chrome plated, taper face. C—Inner groove, chrome face. D—Groove faced-ferrox filled. E—Plain chrome face. F—Keystone. G—Inner groove, taper face. H—Inner chamfer, molybdenum filled, grooved face. I—Scraper face.   (Perfect Circle Products)

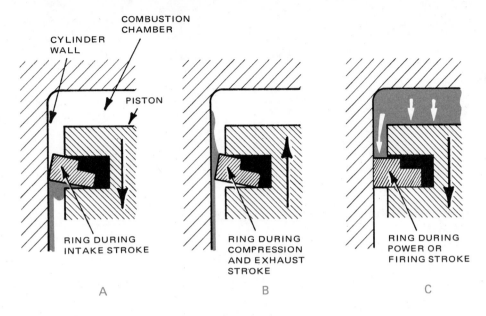

Fig. 12-15. Inner groove causes ring to twist slightly, aiding oil control and compression, plus reducing wear.

OIL CONTROL RINGS

The oil control rings are designed to remove surplus oil from the cylinder walls. They do this through a light scraping action against the walls. The ring and groove are both slotted and perforated (having holes). Oil trapped by the ring passes through the slots or holes of both the ring and groove, Fig. 12-16. It then flows down inside the piston where it drops into the oil pan or crankcase. A three-piece oil control ring with a hump type, spring steel expander spring is shown in Fig. 12-17.

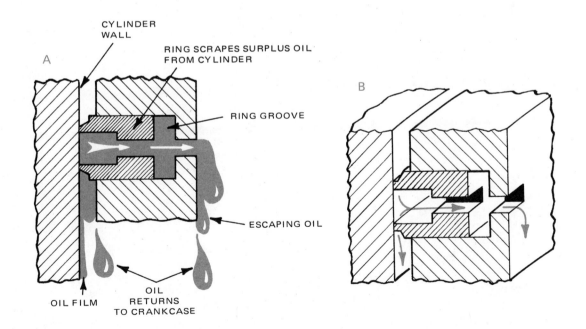

Fig. 12-16. An oil control ring removes surplus oil from cylinder walls. A—Oil being removed. B—Path of oil during removal.

Fig. 12-17. Three-piece oil control ring with flat steel (hump type), expander in place. (Perfect Circle Products)

RING END GAP

The inside diameter of a piston ring is always made smaller than piston diameter. This being the case, each ring must be expanded to get it over the piston head and into the ring groove. The amount of end gap is critical (vital to success of rebuilding operation), so manufacturer's specifications should be followed. *As a rule of thumb, however, allow .004 in. of end gap for every inch of cylinder diameter.* For example, minimum end gap for a 2 1/2 in. cylinder is .010 in.

Too much end gap will allow the gases to leak between the ring ends. Too little gap is even more serious. When the rings heat up in service, they will expand and close up. If the rings continue to heat and expand, they will break and score the cylinder wall.

To measure ring end gap, place the ring in the cylinder. Then turn a piston upside down and push the ring to the lower end of the cylinder. When the ring reaches the proper depth, remove the piston and try various size feeler gages in the gap until one fits, Fig. 12-18. If the ring gap is too small, dress the butt ends of the ring with a file, Fig. 12-19.

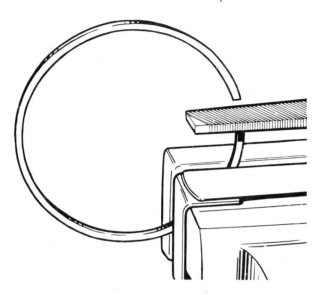

Fig. 12-19. Ring ends can be dressed with a file if end gap is too small. Use copper vise jaws to protect ring.

When piston-ring-to-groove side clearance and end gap are satisfactory, install the rings on the piston, Fig. 12-20. Next, use a piston ring compressor to compress

Fig. 12-18. Ring end gap is measured by pushing ring into cylinder with an inverted piston. Then, piston is removed and a thickness gage is used to measure gap.

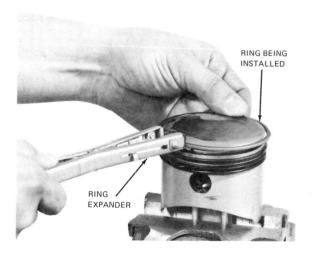

Fig. 12-20. Rings must be expanded with a special tool. Otherwise, danger or ring breakage is increased. (Tecumseh Products Co.)

Fig. 12-21. A ring compressor is used to squeeze ring ends together while piston is pushed into cylinder. (Kohler Co.)

the rings flush with the grooves, Fig. 12-21. Hold the compressor firmly against the top of the block, then use a wooden dowel or hammer handle to tap the piston out of the compressor and into the cylinder. Once free of the compressor, the rings will maintain firm contact with the cylinder wall.

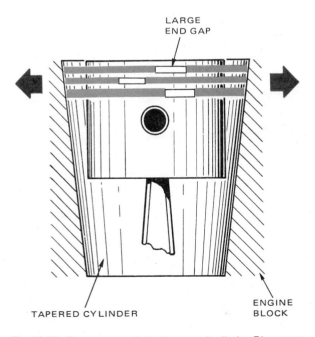

Fig. 12-22. Piston at top of a badly tapered cylinder. Rings must expand a great distance to stay in touch with cylinder walls.

RINGS AND CYLINDER WALLS

When cylinders wear, they become larger at the top. This is due to the wearing action of burning gases and from dust and grit brought in with the air-fuel charge. Lack of lubrication in the upper cylinder also increases wear in the upper area of ring travel.

Even though the cylinder walls have a tapered wear pattern, the rings must stay tight against them. When taper becomes greater than .008 or .010 in., ring tension on the cylinder wall is lost and the end gaps open wide to allow oil to pass through to the combustion chamber, Fig. 12-22.

When the piston is at the bottom of its stroke, the rings are forced inward by the walls, Fig. 12-23. When the piston travels to the top of the cylinder, the rings must constantly expand to maintain contact with the ever-widening cylinder, Fig. 12-22.

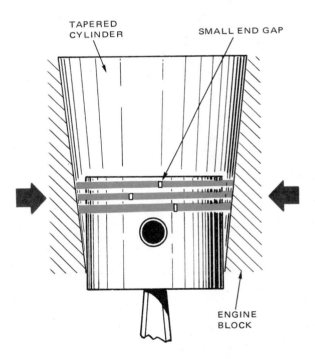

Fig. 12-23. Piston at bottom of a badly tapered (exaggerated) cylinder. Rings are forced into grooves.

When the engine is running at high speed, the piston can move to the top of a badly tapered cylinder and back down again so quickly that the rings do not have a chance to expand. This leaves the rings completely free of the walls at the top of the stroke. An engine with cylinder walls in this condition would burn large amounts of oil. It is possible, too, that the piston would tip and slap, causing ring wear and damage.

Once the tapered condition exceeds .010 in., it can only be corrected by boring the cylinder oversize, or by unit replacement.

PISTON DAMAGE

Many damaging things can happen to pistons. They should be cleaned, examined and measured after removal. The accompanying illustrations are good examples of conditions to look for.

Most piston and cylinder damage can be traced to one or more of the following causes: lack of oil, use of the wrong oil or oil-fuel mixture, use of other than specified gasoline, foreign particles in the cylinder, overheating caused by clogged cooling fins, excess carbon buildup in the cylinder exhaust ports, badly fitted rings and pistons.

Fig. 12-24 shows scoring and light scuffing of both piston and rings due to overheating. This occurs when high friction and combustion temperatures approach the melting temperature of the piston materials.

Fig. 12-24. Scored and scuffed piston due to overheating. (Tecumseh Products Co.)

Check and correct:
1. Dirty cooling shroud and cylinder head fins.
2. Lack of cylinder lubrication.
3. Poor combustion.

4. Improper bearing or piston clearance.
5. Overfilled crankcase causing fluid friction.

Fig. 12-25. Piston has rings stuck due to lacquer, varnish and carbon buildup resulting from high temperatures.

Fig. 12-25 illustrates a piston with rings stuck and broken from lacquer, varnish and carbon buildup due to abnormally high operating temperatures.

Check and correct:
1. Overloading the engine.
2. Ignition timing.
3. Lean fuel mixture.
4. Clogged cooling fins.
5. Wrong oil.
6. Low oil.
7. Stale fuel.

Fig. 12-26 reveals vertical scratches on ring faces and piston caused by the presence of abrasive particles.

Check and correct:
1. Damaged or improperly installed air cleaner.
2. Air leaks between air filter and carburetor.
3. Leak in gasket between carburetor and block.
4. Air leak around throttle shaft.
5. Improperly cleaned cylinder bore after reconditioning engine.

Compare the oil ring (bottom ring) in Fig. 12-27 with the one in Fig. 12-26. The rails of the oil ring in Fig. 12-27 are worn down to the drain holes, and the ring surface is flat. This type of wear results from extended use and possibly from abrasives in the engine.

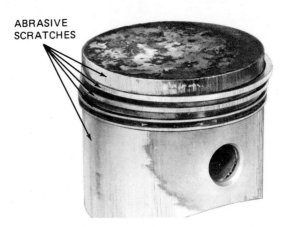

Fig. 12-26. Vertical scratches on ring faces and piston are caused by abrasive materials entering engine.

Fig. 12-27. Piston has extremely worn rings because of long use and possible abrasives. (Tecumseh Products Co.)

Fig. 12-28. Burned top land results from detonation.

Check and correct:
1. Rails worn down.
2. Low ring tension.

The piston in Fig. 12-28 has a burned top land, resulting from detonation. Detonation is abnormal combustion which causes too much pressure and temperature in the combustion chamber. Detonation is sometimes referred to as carbon knock, spark knock or timing knock. It occurs when the air-fuel mixture ignites spontaneously and interferes with the normal combustion flame front.

Check and correct:
1. Lean fuel mixture.
2. Low octane fuel.
3. Over-advanced ignition.
4. Engine lugging.
5. Excessive carbon deposits on piston and cylinder head (increasing compression).
6. Milled cylinder head (increasing compression ratio).

Preignition is the burning of the air-fuel mixture before normal ignition occurs. Preignition creates a pinging sound, resulting from severe internal shock. It is accompanied by vibration, detonation and power loss. If allowed to continue, it could cause severe damage to the piston, rings and valves. See Fig. 12-29.

Check and correct:
1. Internal carbon deposits remaining incandescent (red hot).
2. Spark plug heat range too high.

Fig. 12-29. A hole burned through piston head was caused by preignition.

3. Spark plug ceramic shell broken.
4. Thin edges on valves or elsewhere in combustion chamber.

If the connecting rod and piston are not aligned, a diagonal wear pattern will show on the piston skirts, Fig. 12-30. This condition can occur, along with poor ring contact, if the cylinder is bored at an angle to the crankshaft.

Fig. 12-31. Objects left inside engine can cause serious damage. A needle bearing is shown imbedded across ring. (Jacobsen Mfg. Co.)

Fig. 12-30. A diagonal wear pattern indicates improper alignment of connecting rod and piston.   (Tecumseh Products Co.)

Check and correct:
1. Rapid piston wear.
2. Uneven piston wear.
3. Excessive oil consumption.

Piston damage can be caused by foreign objects carelessly left inside an engine during reconditioning. Fig. 12-31 shows the results of a needle bearing that became imbedded in the piston. *Don't be careless.* Do the job with painstaking care from the start to final assembly.

PISTON RING WEAR-IN

After the small engine is reconditioned, a short wear-in period occurs. Wear-in is the process in which the face of each ring wears off until it is a perfect fit against the cylinder wall. To help the rings seat quickly, the face is covered with microscopic grooves. During the first few hours of operation, these grooves rub against the cylinder wall and all high spots are worn off.

As grooves wear away, both face of ring and cylinder wall become very smooth. Under normal operating conditions, wear beyond this point is very small.

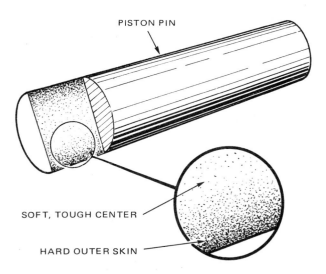

Fig. 12-32. Solid piston pin has hardened and ground surface.

PISTON PINS

A piston pin is used to secure the connecting rod to the piston. These pins are made of case-hardened steel and ground to exact size. They may be hollow or solid.

A typical solid piston pin is shown in Fig. 12-32. Note the finer grain size of the metal at the surface. This is due to a heat treatment that provides a hard durable case with a soft, tough center.

Many different piston pin assemblies have been used. The full floating pin arrangement shown in Fig. 12-33, is free to turn in the rod as well as in the piston bosses. When both the connecting rod and piston are of aluminum alloy, the pin can operate directly against this material. If the rod is steel, either a bronze bushing or a needle roller bearing is used in the rod. The piston bosses also may have bronze bushing inserts for the pin. The snap rings in each boss prevent the pin from rubbing on the cylinder surface. Fig. 12-33 shows a full floating pin with a steel connecting rod. This setup requires the use of a bushing, or bearing.

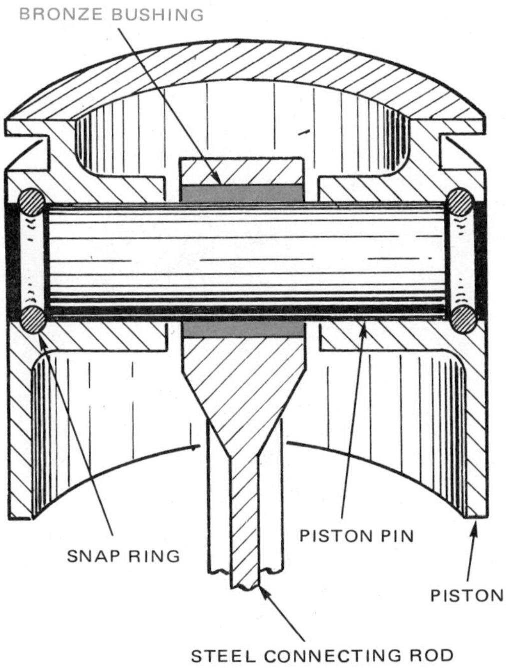

Retaining snap rings are compressed and placed in grooves in the pin bosses in the piston, Fig. 12-34. Some piston pins are a tight, press fit in the connecting rod, Fig. 12-35. The pin may turn in the piston bosses, bushings or needle bearings, depending on the type of construction used.

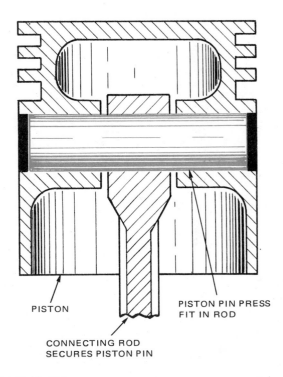

Fig. 12-33. A full floating piston pin used in a steel connecting rod requires use of a bushing.

Fig. 12-35. This piston pin is pressed into connecting rod, but is allowed to turn in pin bosses of piston.

REMOVING PISTON PINS

If snap rings are used, they must be removed first. Some are formed spring steel wire, others are stamped flat spring steel. Fig. 12-36 shows one method of removing the wire type snap rings with a screwdriver. Needle nose pliers can also be used. Removal of stamped snap rings usually requires a specially designed snap ring pliers. *Be careful when removing or replacing snap rings.* Should they slip out of the boss or out of

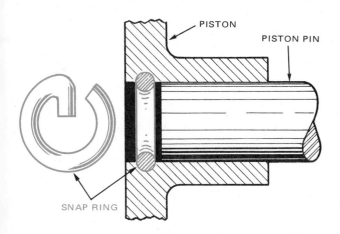

Fig. 12-34. Snap rings keep full floating pin in place in piston.

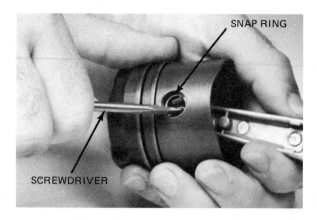

Fig. 12-36. Some rings being removed from their grooves as shown. Always wear safety glasses for this operation. (Jacobsen Mfg. Co.)

the jaws of the pliers, they can cause severe eye injuries. Wear safety glasses.

The piston pin may slide out easily, if it is full floating and somewhat worn. On the other hand, it may be quite snug. A soft-faced mallet and a dowel rod can be used to tap out the pin, Fig. 12-37. Be careful not to hit the piston. Let the pin fall into a soft cloth.

Press fitted pins must be removed with a mechanical or hydraulic press. Refer to the engine manual for the proper support of the piston. If needle bearings are used, be careful not to lose them or misplace them.

MEASURING PISTON PINS AND BOSSES

When the piston pin has been removed, measure its outside diameter with a micrometer. Make a note of the measurement. Next, measure the inside diameter of the bosses, using a small hole gage. Expand the gage in the boss until it gently contacts the inner surfaces. With-draw the gage and measure it with the micrometer. Then, subtract the pin diameter from the boss diameter. The difference must be within the limits specified for the engine being serviced.

If the pin-to-boss clearance is excessive, a new pin and bushing or an oversize pin and reaming of the bosses are needed. Most piston pins are a snug fit in the piston. If the pin turns by hand, inspect it for wear.

REVIEW QUESTIONS – CHAPTER 12

1. Explain why the piston head diameter may be less than the skirt diameter.
2. If a piston has too much clearance in the cylinder, it

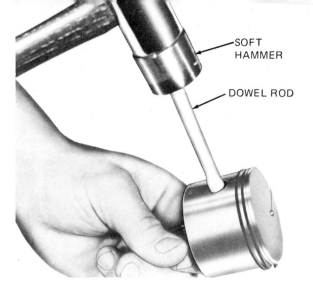

Fig. 12-37. Tapping out piston pin requires the use of a soft hammer and dowel rod to avoid damaging piston or pin boss. (Lawn-Boy Power Equipment, Gale Products)

will produce a motion and sound called:
   a. Knocking.   c. Pinging.
   b. Slapping.   d. Hammering.
3. Piston rings are made of:
   a. Cast iron.   d. Cast iron, aluminum
   b. Steel.   alloy.
   c. Aluminum alloy. e. Cast iron, steel.
4. Name the tool used to squeeze the piston rings together so they can be installed in the cylinder.
5. Briefly name the results of the following conditions:
   a. Excessive ring end gap.
   b. Lack of ring end gap.
6. Pinned piston rings would most likely be found in:
   a. Four cycle engines.
   b. Two cycle engines.
7. Piston rings can be one of two basic types. Can you name them?
8. Piston pins are prevented from contacting the cylinder wall by use of:
   a. Cotter pins.   c. Tapered pins.
   b. Setscrews.   d. Snap rings.

SUGGESTED ACTIVITIES

1. Remove the rings from a piston, clean the piston and measure it.
2. Measure ring end gap and side clearance.
3. Using an old ring, demonstrate method of dressing ring ends with a file to increase end gap.
4. Recondition piston bosses by reaming, and replace the worn piston pin with an oversize pin.
5. Replace piston rings with a ring expander.
6. Using ring compressor, replace reconditioned piston assembly in the cylinder.

Chapter 13

RODS, BEARINGS AND VALVES

Like pistons and piston rings, connecting rods, bearings and valves are used in areas of the engine which demand close "fits." Yet some extra clearance must be allowed for expansion of parts due to high temperatures. While there are differences between makes of engines, maintenance is much the same for all.

Special attention must be given to the four cycle engines because there are more parts involved that require service. Rod and bearing service is the same for both two cycle and four cycle types. The valve system of two cycle engines (major area of difference) will be treated alone near the end of the chapter.

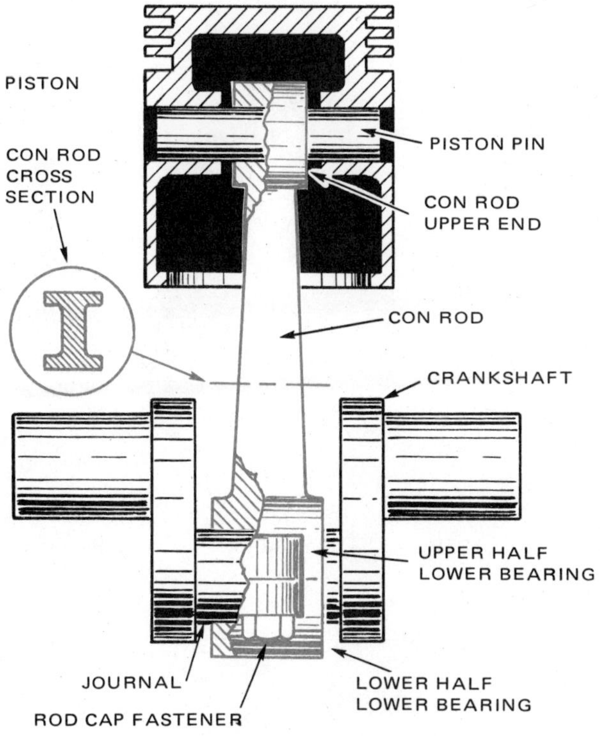

CONNECTING RODS AND BEARINGS

The connecting rod attaches the piston to the crankshaft, Fig. 13-1. The upper end has a hole through which the piston pin is passed. The lower end has a large bearing which fits around the crankshaft journal, Fig. 13-1.

The lower end of the connecting rod usually is split when friction type bearings are used. The place at which the halves separate is called the parting line. When needle or roller bearings are used, the rod end can be split or solid. See Fig. 13-2.

Fig. 13-1. Connecting rod attaches piston to crankshaft. Bearings are used at both ends of rod to reduce friction.

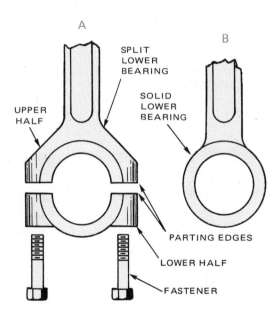

Fig. 13-2. Two types of connecting rod design, crankshaft end. A—Split construction. B—Solid construction.

A variety of connecting rods, both split and solid, are pictured in Fig. 13-3. Roller bearings, needle bearings and precision inserts are also shown.

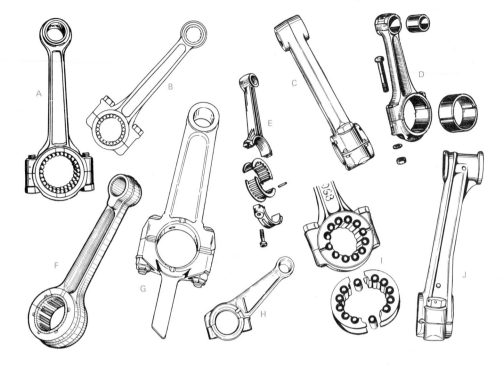

Fig. 13-3. Small engine rods and bearings. A and B—Split rod, roller bearings. C—Split rod, cast-in bearing. D—Split rod, precision inserts. E—Split rod, needle bearings. F—Solid rod, paired roller bearings. G—Split rod with dipper, precision inserts. H—Rod split at an angle, precision inserts. I—Split rod, roller bearings. J—Split rod with offset cap, cast-in bearing.

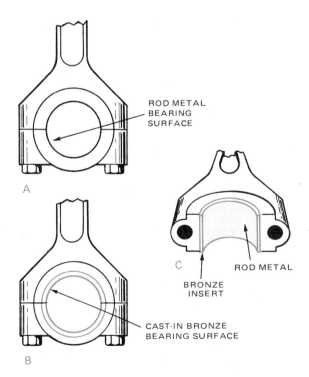

Fig. 13-4. Friction type connecting rod bearings. A—Rod metal forms bearing surface. B—Bronze bearing is cast into rod metal. C—Replaceable precision insert bearing.

FRICTION TYPE ROD BEARINGS

There are three types of friction bearings in common use in the big end of connecting rods. See Fig. 13-4.
A. Rod metal, when made of aluminum alloy.
B. Bearing bronze cast into rod end, bored and finished.
C. Removable precision insert bearing halves, using steel shells lined with various materials.

The thin lining material on inserts could be lead-tin babbitt, aluminum or copper-lead-tin. Fig. 13-5 shows an insert having a steel back (1) coated with cast babbitt (2). They are called "precision" inserts because they are made to an exact size for proper fit.

The insert is kept from turning in the rod end by a locating tab on the parting line edge of each insert. The tab fits into a slot in the rod itself. Fig. 13-6 illustrates this tab and slot arrangement.

BEARING SPREAD

The insert bearing halves are made with more "spread" than the curve machined into the rod and rod cap. To seat the insert, the ends must be forced down and snapped into place. *Never press down in the center*

Fig. 13-5. Construction of a typical precision insert bearing.
1—Steel back. 2—Cast babbitt about .004 in. thick.
(Clevite Corp.)

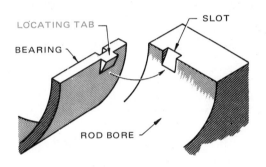

Fig. 13-6. Locating tabs prevent precision inserts from turning.

of the insert to seat it in the rod bore. The correct amount of bearing spread gives tight insert-to-bore contact around the entire bearing. It provides support, alignment and helps to carry heat away, through the rod and bearing cap. It also holds the bearing in place during assembly.

BEARING CRUSH

When precision inserts are snapped into the rod bore, the ends will protrude slightly above the parting surface. See A in Fig. 13-7. This built-in design feature is called bearing crush. The actual distance varies from a fraction of .001 in. to .002 in.

When the rod cap is installed and drawn in place, the insert ends meet first and force the insert halves tightly against the rod bore. This provides firm support for the insert. The forced fit makes it round and, through close metal-to-metal contact, allows heat to be carried away through the rod. View B in Fig. 13-7 shows how radial pressure is exerted against the rod bore.

INSERTS ARE MATCHED

Precision inserts must be kept in matched pairs. *Never mismatch bearing inserts, and use the exact size needed.* They cannot be made larger or smaller in the shop. Standard and various undersizes are available.

ANTIFRICTION BEARINGS

Many small gasoline engines use an antifriction bearing in the big end of the connecting rod. Often,

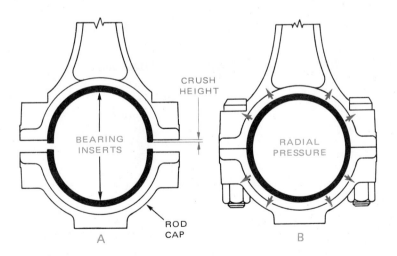

Fig. 13-7. Effect of bearing "crush." A—Rod and cap separated. B—Rod and cap drawn together, creating radial pressure on inserts. (Sunnen Products Co.)

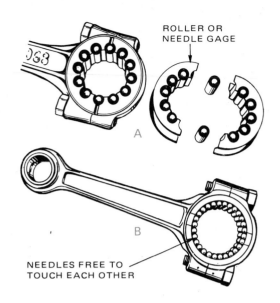

Fig. 13-8. Two types of connecting rod needle bearings.
A—Caged needles. B—Free needles. (Evinrude Motors)

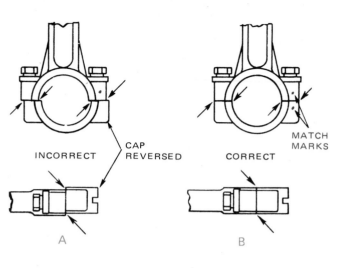

Fig. 13-9. Connecting rod cap installation. A—If cap is turned
180 deg., rod bore will be offset. B—Match marks on rod and
cap signal correct assembly.

needle rollers are used. These roller elements can be held together by a roller cage or separator. See A in Fig. 13-8. The rollers also can be left free as in view B. Antifriction bearing assemblies are hardened and ground to exact size. They must fit accurately, but still have some clearance for expansion.

During manufacture, the rod cap is bolted into position on the rod. Then the assembly is bored to exact size. *It is important, therefore, that the rod cap is always put back the same way it was removed.* If the cap is turned 180 deg., the upper and lower halves can be offset. This error in assembly will cause bearing and shaft failure. See A in Fig. 13-9.

Connecting rods are usually marked with either a line, punch mark, number, special boss or chamfered edge to show correct alignment during assembly. If a "mark" is not apparent, punch mark the rod end and cap as at B in Fig. 13-9 for later reference. NOTE: Caps must never be switched from one rod to another.

ROD BOLT LOCKING DEVICES

To stop connecting rod bolts or cap screws from loosening in service, locking devices are used. One common device is a thin sheet metal strip with locking tabs, Fig. 13-10. The cap screw is inserted through holes in the locking strip, holding it in place against the rod. After the cap screw is turned tight, the metal tabs are bent up against the flat sides of the screw head, Fig. 13-11.

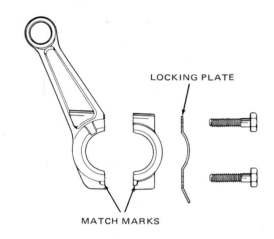

Fig. 13-10. Often a locking plate is used between connecting
rod cap and cap screws. (Tecumseh Products Co.)

Self-locking nuts, lock washers and special-shape cap screws are also used to prevent loosening. The final tightening of the cap screws is especially important. *Always tighten rod fasteners with a torque wrench, no more and no less than the amount of torque tightness specified.*

CRANKSHAFT

The crankshaft converts the reciprocating (back and forth) motion of the piston into rotary (circular) motion. It transmits engine torque to a pulley or gear so that some object may be driven by the engine. The crankshaft also drives the camshaft (on four cycle

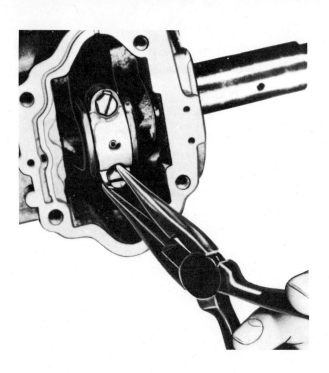

Fig. 13-11. After cap screws have been torque tightened, tab on locking plate is bent up against head of cap screw to lock it in place. (Lawn-Boy Power Equipment, Gale Products)

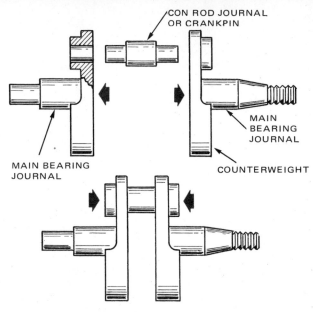

Fig. 13-13. Multi-piece crankshafts have various parts pressed together under heavy pressure.

engines), supports the flywheel and, in many engines, operates the ignition system.

Crankshafts can be made of cast or drop-forged steel. One and multi-piece crankshafts are used. Fig. 13-12 shows a typical one-piece small engine crankshaft. A multi-piece crankshaft is shown in Fig. 13-13.

The crankshaft "throw" is the offset portion of the shaft measured from the center line of the main bearing bore to the center line of the connecting rod journal. The connecting rod journal is commonly referred to as the "crank throw" or "crankpin."

CRANKSHAFT BALANCE

To help offset the unbalance created by the force of the reciprocating mass (connecting rod, piston and throw), counterweights are added to the crankshaft. By placing these weights opposite the crankpin, engine

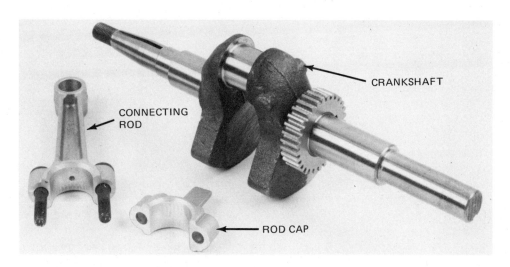

Fig. 13-12. Single-piece crankshafts are most popular in small gasoline engine applications.

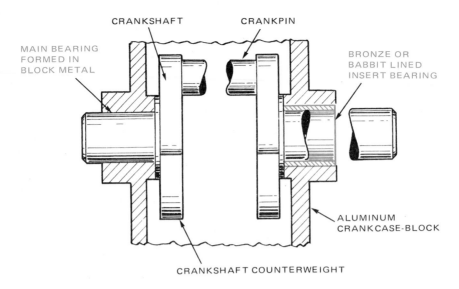

Fig. 13-14. Friction type crankshaft main bearings. Shaft at left uses bore in aluminum crankcase as bearing surface. Shaft at right uses a precision insert bearing.

vibration is greatly reduced. The counterweights usually are forged as an integral part of the crankshaft, Figs. 13-12 and 13-13.

CRANKSHAFT MAIN BEARINGS

The crankshaft is supported by one or more main bearings. Often, the main bearing journal surfaces are hardened by an induction hardening process to provide long service life. Three types of main bearings are used: the sleeve or bushing, Fig. 13-14; the roller bearing, Fig. 13-15; and the ball bearing, Fig. 13-16.

CRANKSHAFT CLEARANCES

To allow space for lubricant between the moving parts, as well as to provide room for expansion when heated, crankshaft bearings must have a slight end clearance, Fig. 13-17. Shaft movement from end to end is controlled by the bearing adjustment when tapered roller or ball bearings are used.

With friction bearings, a thrust surface on the shaft rubs against a similar surface on the crankcase. A precision insert main bearing may have a thrust flange for the crank to rub against. In some applications, a bronze thrust washer is used.

Clearances will vary with engine type, design and use. Bearing and thrust surface clearances are critical. They must be held to exact tolerances as recommended by the manufacturer.

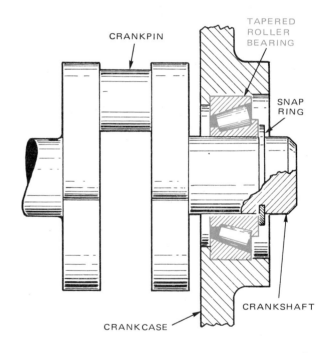

Fig. 13-15. Typical use of tapered roller bearing as crankshaft main bearing.

Fig. 13-18 pictures the method of measuring the bearing surfaces on a crankshaft with a micrometer. Measurement must be taken in at least two positions 90 deg. to each other. *If any of the dimensions are less than those specified, or if there are any scoremarks, the bearing surface should be reground.* Basically, wear and taper should not exceed .001 in.

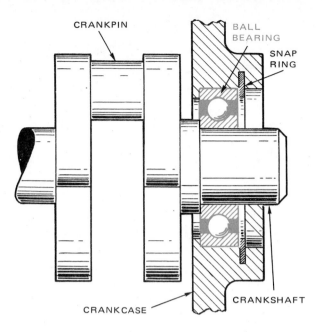

Fig. 13-16. A ball bearing also can be used as crankshaft main bearing.

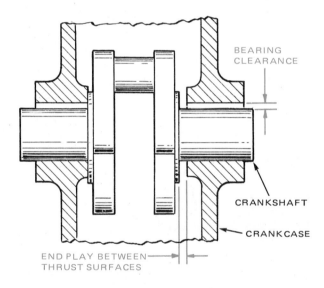

Fig. 13-17. Crankshaft bearings and thrust surfaces must have some clearance (end play) to provide space for lubricant and for heat expansion.

MEASURING BEARING CLEARANCE

Bearing clearance is the space left between the inner bearing surface and the crankshaft main or rod journal. When checking bearing clearance, use a special compressible plastic material called "Plastigage." The plastic material is color coded and selected according to the

Fig. 13-18. A micrometer is required to accurately measure bearing surface diameter on a crankshaft.

recommended clearance range. It comes in a thin, round strand stored in a paper package.

To use "Plastigage," select the correct color for the specified clearance. Cut a piece of plastic equal to the width of the bearing and lay it across the bearing surface. Torque tighten the bearing cap in place. Then remove the cap and compare the compressed width of the plastic with the comparison chart printed on the package. Clearance is given alongside the matching marks on the chart. In effect, the wider the plastic, the less clearance there is.

If bearing clearance is too great, undersize inserts will have to be used. If the crank journal is worn, it will require grinding to clean it up to a given undersize. Recheck the clearance with Plastigage.

On many small engines, the main bearings are simply machined bores in the crankcase halves or pressed inserts. Plastigage will not "measure" wear in these bearings. To check clearance, first measure the crankshaft diameter with a micrometer. Next, measure the inside diameter of the main bearing with a telescoping gage, Fig. 13-19. Lock the gage, remove it from the bearing and measure the setting with a micrometer. The difference between the shaft reading and the bearing reading is bearing clearance.

CRANKCASE SEALS

Crankcase seals prevent leakage of oil from the areas where the crankshaft and crankcase come together. The

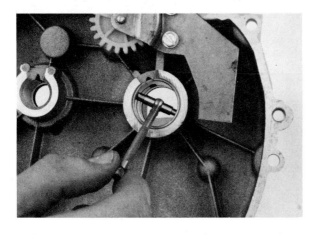

Fig. 13-19. Measuring diameter of a pressed main bearing with a telescoping gage. (Deere & Co.)

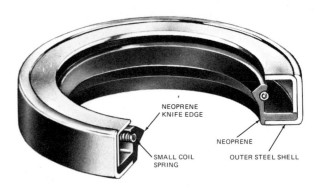

Fig. 13-21. This crankshaft oil seal has an outer steel shell with neoprene center. Small coil spring produces contact pressure. (Chicago Rawhide Mfg. Co.)

typical crankcase seal has a steel outer shell with a neoprene center. A small coil spring keeps the sealing lip in constant contact with the shaft it seals, Fig. 13-21.

Note in Fig. 13-20 that the sealing lip must face the fluid being sealed in. In this application, it faces the crankcase. In this way, the pressure of the oil will tend to force the lip against the shaft. *If the seal is installed backwards, oil pressure will force the sealing lip away from the shaft and oil leakage will occur.*

When removing the crankcase cover from the crankcase and crankshaft, as in Fig. 13-22, place tape over the keyway. This will keep the sharp keyway edges

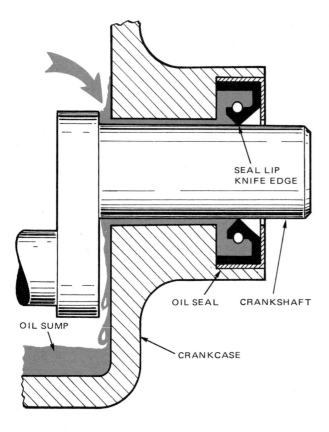

Fig. 13-20. Typical neoprene oil seal has sealing lip with sharp edge, providing increased pressure and reduced friction.

Fig. 13-22. Taped keyway edges will protect oil seal when cover is removed. (Deere & Co.)

shell of the seal makes fixed contact with the crankcase, while the knife edge of the sealing lip lightly rubs against the crankshaft. See Fig. 13-20.

Seals are made of neoprene, leather, graphite or other materials, depending on how the seals are used. A

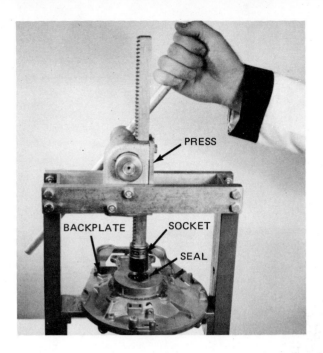

Fig. 13-23. On some engines, an arbor press can be used to push out old oil seals. Press with care to avoid damaging housing. (Jacobsen Mfg. Co.)

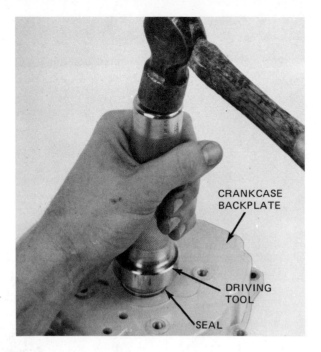

Fig. 13-25. Use a special driving tool to tap seal in place in housing. (Deere & Co.)

from cutting the neoprene oil seal. Fig. 13-23 shows how a press is used to push the old seal out of the backplate. In Fig. 13-24, the seal is being readied for installation. A liquid sealant is applied to the outside of the shell of the seal before pressing it in place in the backplate. Often, however, seals can be replaced by tapping them into the bore of the backplate with a special driving tool, Fig. 13-25.

VALVE SERVICE

Four cycle engine poppet valves are subjected to tremendous heat. Normal operating temperature of the exhaust valve exceeds 1000 deg. F. To withstand this, high quality, heat resistant steel must be used and correct operating clearances must be maintained.

REMOVING VALVE ASSEMBLY

The engine valve assembly includes the valve, valve spring and one or more retainers, Fig. 13-26. The

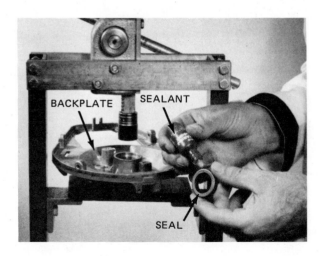

Fig. 13-24. Apply sealing compound to shell of seal before pressing seal into bore of housing.

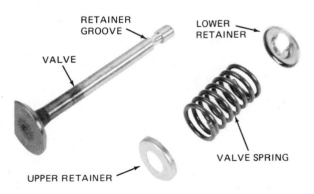

Fig. 13-26. Poppet valve assembly is used in four cycle engines.

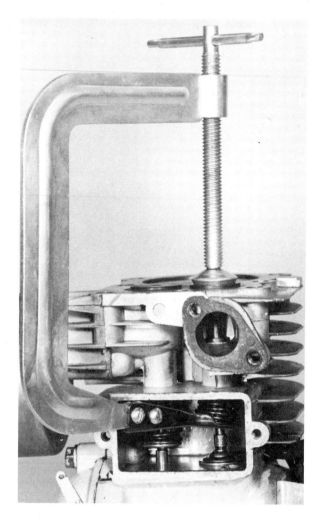

Fig. 13-27. A valve spring compressor "squeezes" spring to uncover keepers in valve stem.

Fig. 13-28. When spring is fully compressed, use a pliers to remove valve keepers. Then, release valve lifter and remove valve, spring and retainers.

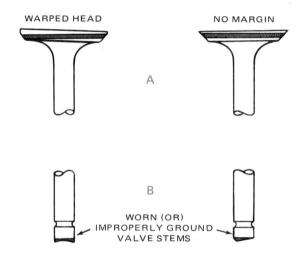

Fig. 13-29. Warped valve head, thin margin or worn stem may mean facing or replacement is needed.

locking types are called "keepers." To remove the valve, once the cylinder head has been removed, compress the valve spring with a compressor, Fig. 13-27. Remove the valve spring retainer from groove in the valve stem, using a pair of pliers as shown in Fig. 13-28. Then, slowly release pressure on the spring and remove the compressor. The valve can then be pulled out the top and the spring taken from the side.

INSPECTING VALVES AND SEATS

When the valves have been removed, clean them with a power-operated wire brush. Then inspect them for the following defects:

1. Eroded, cracked or pitted valve faces, heads or stems.
2. Warped head, as at A in Fig. 13-29.
3. Worn or improperly ground valve stems, as at B in Fig. 13-29.
4. Bent valve stems.
5. No margin, or less than 1/64 in. margin.
6. Partial seating.

Heavy carbon deposits on intake valves sometimes cause faulty valve operation by restricting the flow of fuel into the cylinder. If any serious defects are observed, the valve should be replaced. In any case, valve faces should be machined to a smooth, true finish.

INSPECTING VALVE SPRINGS

Valve springs, through overheating and long use, can lose their elasticity and become distorted (warped or bent). First, check each spring for squareness and length with a square and surface plate, Fig. 13-30. Replace all

Fig. 13-30. Use a square and a surface plate to check valve spring for proper length and squareness. (Deere & Co.)

Fig. 13-31. Special tester is available to test a valve spring for adequate tension.

springs that are badly distorted or reduced in length. Then, test spring tension, Fig. 13-31, and compare the reading with the specification in the engine manual. Lack of spring tension can cause valve "flutter" or incomplete closing and sealing of the valve.

VALVE GUIDES

Valve guides align and "steer" the valve so that it can open fully and close completely. Guide-to-valve stem clearance must not exceed tolerances, since this would allow the valve to tip. Tipping permits the valve face to strike the seat at an angle, permitting hot combustion gases to escape. Some clearance is required, however, to allow for heat expansion and lubrication. Guide-to-valve stem clearance should run about .002 to .003 in.

Valve guides can be a replaceable insert or integral with (part of) the block. See Fig. 13-32. Replaceable

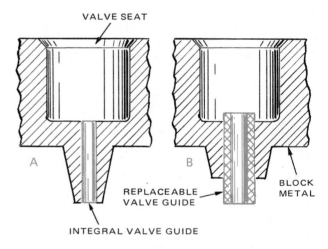

Fig. 13-32. Valve guides are one of two types. A—Bored in block. B—Pressed in block and can be replaced.

guides are cast iron. The old guide must be driven out and the new guide pressed in place. Integral guides can be reamed to fit a valve with an oversize stem.

INSPECTING VALVE GUIDES

Valve guides must be clean before inspection. A special cylindrical wire brush driven by a power drill is made for this job. After cleaning the guide, measure the bore with a small hole gage, Fig. 13-33. Expand the gage until it lightly touches the sides of the bore. Remove the gage and measure with a micrometer.

Fig. 13-33. Valve guide diameter can be measured with a small hole gage. (Deere & Co.)

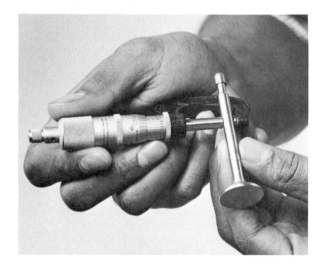

Fig. 13-34. Measure valve stem diameter to determine valve stem-to-guide clearance.

Next, measure the valve stem diameter with a micrometer, Fig. 13-34. Then, subtract the stem diameter from the guide diameter to find the precise amount of clearance. Compare this with clearance specified.

VALVE GUIDE REAMING

If the clearance between the stem and the guide exceeds the allowable limit, enlarge the guide with an adjustable reamer to the next oversize dimension. An adjustable reamer, Fig. 13-35, should be used. Select and install a valve with the correct oversize stem. *Do not enlarge the tappet guides because oversize tappet stems are seldom available.*

Fig. 13-35. Use an adjustable reamer to recondition and resize an integral valve guide. Replacement valve must have an oversize stem. (Deere & Co.)

VALVE SEAT ANGLE AND WIDTH

Valve seats are generally cut to 45 deg., although 30 deg. seats are used in a few engines. The width of the seat is important too. It must be wide enough to prevent cutting into the valve face. It also must provide enough contact area to provide for adequate heat dissipation.

Likewise, the seat must not be too wide. If it is, carbon will pack between the seat and the valve face, holding the valve off the seat. A valve that fails to seat produces a rough-running engine and will soon warp and burn. Specified seat widths, Fig. 13-36, range from .030 to .060 in. (1/32 to 1/16 in.).

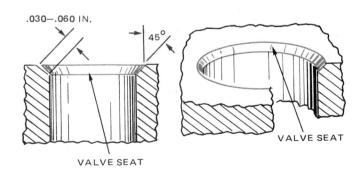

Fig. 13-36. A typical integral valve seat, in which hole is bored in block metal.

Some valve seats are ground to an angle of 44 deg. and the valve face is ground to 45 deg., or vice versa. The 1 deg. variation produces a hairline contact that gives fast initial seating. Some manufacturers believe that, upon heating, the valve will form a perfect seal. The difference in angle between valve face and valve seat is called an "interference fit," Fig. 13-37.

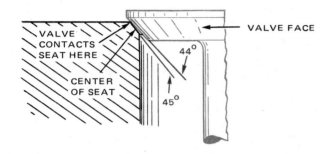

Fig. 13-37. A 1 deg. difference between valve face and valve seat provides better seating.

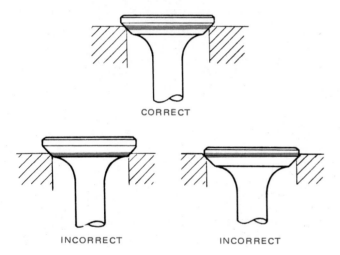

Fig. 13-38. Comparison of correct and incorrect location of seating area on valve face. (Deere & Co.)

Valve seat contact must be near the center of the valve face, Fig. 13-38. A valve seat cutter, Fig. 13-39, can be used to recondition the seat. After cutting the seat, the cutter can be turned over to produce the valve narrowing angle as in E, in Fig. 13-40.

LAPPING VALVES

Good used valves may be reseated by a hand-lapping process, if desired. Some manufacturers do not recom-

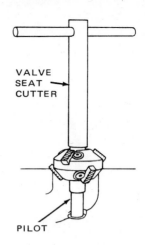

Fig. 13-39. Use a valve seat cutting tool to recondition seats. Turn cutter over to produce a seat narrowing angle.

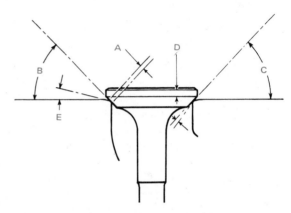

A—VALVE SEATING SURFACE (3/64 IN.)
B—VALVE SEAT ANGLE (46 DEG.)
C—VALVE FACE ANGLE (45 DEG.)
D—VALVE MARGIN (1/16 IN.)
E—VALVE NARROWING ANGLE (30 DEG.)

Fig. 13-40. Valve seat angles and dimensions are given for one specific engine. Angles shown are typical, but amounts may vary from one engine make to another.

mend lapping. When the valve expands, it does not seat in the same place as it does when it is cold. Thereby, any benefits from the lapping process would cancel out. Nevertheless, the procedure can be followed if you wish to seat valves this way.

Some engine manufacturers recommend hand lapping of the valve seats. Lapping compound is available from engine parts distributors. Some suppliers package it in a two-compartment canister. One contains a coarse silicon carbide abrasive combined with a special grease. The second contains a finer compound. Condition of valve will dictate which grade to use.

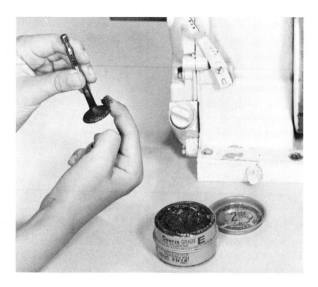

Fig. 13-41. Apply lapping compound to valve face before lapping face to valve seat.

If the coarse lapping compound is used, follow up with the finer compound. Fig. 13-41 shows how to apply the compound to the valve face only. *The compound should not be allowed to contact the valve stem or guide.* Next, a lapping tool is attached to the valve head by means of a suction cup, Fig. 13-42. The

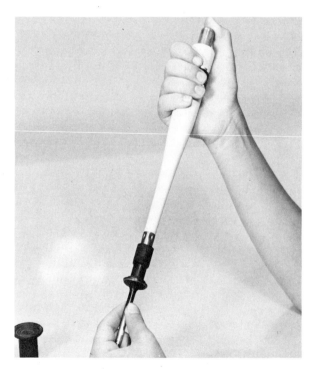

Fig. 13-42. To use a lapping stick, attach stick to valve head by suction. (Powr-Grip Co.)

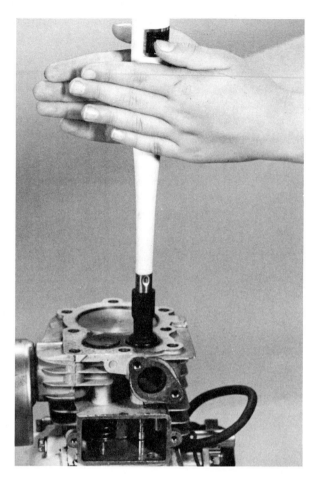

Fig. 13-43. Lap valve to seat by twirling lapping stick between palms of hands. Lift lapping stick and valve occasionally to increase cutting action of compound.

tool shown has a spring-loaded piston in the handle to create suction. With tool attached, valve is placed in guide and twirled back and forth, Fig. 13-43.

The lapping is completed when a dark gray, narrow band equal to the seat width can be seen all the way around the valve face. Do not lap more than is necessary to show a complete seat.

After lapping, thoroughly clean the valve and valve seat chamber so that none of the abrasive finds its way into the engine. The best way to get the cleaning job done is to turn the engine upside down and wash the chamber with solvent from the bottom.

VALVE LIFTER-TO-STEM CLEARANCE

Valve clearance refers to the space between the end of the valve stem and the top of the valve lifter when the valve is closed. The amount of clearance needed depends upon engine design and use. The exhaust valve, due to hotter operation, often requires more clearance than the intake valve. Clearances of around .008 in. for the intake valve and .012 in. for the exhaust valve are fairly common. *Follow manufacturer's specifications.*

When there is too little valve clearance, the valve may be held open when the valve stem heats up and lengthens (expands). As a result, engine performance is poor and both the valve face and valve seat will burn, Fig. 13-44. Insufficient clearance also alters valve timing, making it too far advanced.

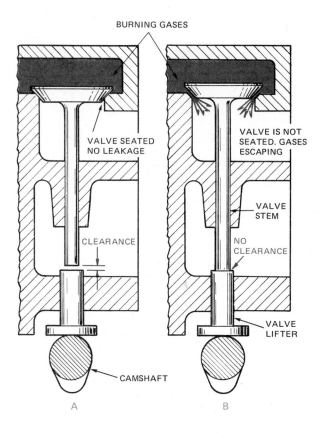

Fig. 13-44. Valve clearance setting is essential to good engine performance. A—Correct clearance permits valve to seat. B—Lack of clearance keeps valve open.

Too much valve clearance, on the other hand, will make valve timing late and reduce valve lift. This results in sluggish engine performance. It also can cause rapid lifter wear because of the pounding action involved. Under these conditions, the engine will be noisy and the valve could break. Fig. 13-45 shows the complete small engine valve train.

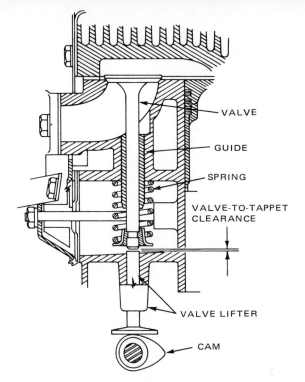

Fig. 13-45. Components of a complete valve train. (Kohler Co.)

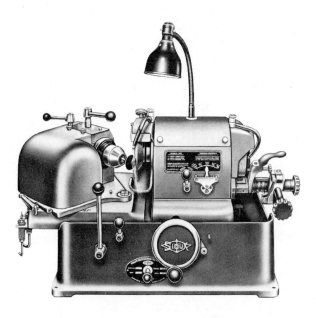

Fig. 13-46. During refacing operation, valve rotates slowly while being fed back and forth across abrasive wheel. (Sioux Tools, Inc.)

REFACING VALVES

Valve refacing is done on a specially designed grinder, Fig. 13-46. The valve is revolved while being fed over an abrasive wheel. The collet which holds the valve is adjusted to the face angle wanted. Coolant flows over the valve head during grinding to reduce heat and produce a good surface finish. The infeed wheel adjusts for the precise amount to be ground from the valve face.

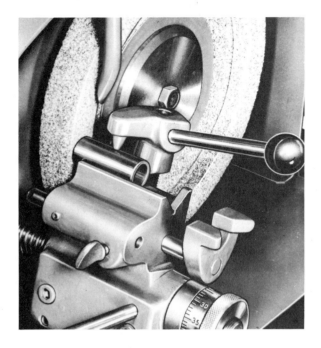

Fig. 13-47. Machine head of valve lifter as shown.

When valve lifters become concave or deformed, they can be reground flat as in Fig. 13-47. Likewise, the end of the valve stem can be dressed and shortened to produce correct valve clearance, Fig. 13-48.

Fig. 13-48. Grinding end of valve stem trues end surface and removes stock for clearance adjustment.

ADJUSTING VALVE CLEARANCE

When a valve has been refaced, it rides lower in the guide, and valve-to-tappet clearance is reduced. If the engine does not have adjustable tappets, the end of the valve stem must be ground to obtain correct clearance. To check clearance, turn the camshaft until the lobe is away from the tappet. Hold the valve against its seat while testing clearance with a thickness gage, Fig. 13-49. If there is too little clearance, remove the valve

Fig. 13-49. Valve stem-to-lifter clearance is checked with a thickness gage.   (Deere & Co.)

and grind .001 or .002 in. off the stem. Repeat the clearance check and grinding operation until clearance is correct.

After the valves and seats have been properly reconditioned, place each valve in its respective guide. Use a valve compressor to compress the spring, then install the keepers. Make sure that the oil drainback holes are open, Fig. 13-50, then reinstall the valve breather cover.

PORTS, REEDS AND ROTARY VALVES

Two cycle engines generally use "porting" of the cylinder wall (instead of poppet valves) in the fuel feed

Fig. 13-50. Make sure that all drainback holes are open before installing breather cover.

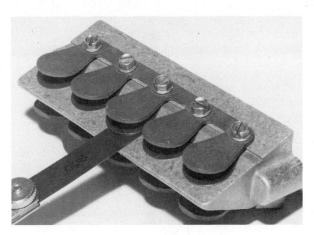

Fig. 13-51. Reed valves not apt to stick shut if they are adjusted to stay open from .005 to .010 in. when at rest.

and exhaust systems. Porting, basically, is a hole in the cylinder wall that admits the air-fuel mixture, and another hole that allows exhaust gases to escape.

Ports have no service requirements other than keeping them free of carbon. Unlike poppet valves, they cannot be replaced since they are part of the cylinder. Reed valves are used in two cycle engines to control fuel flow from carburetor to crankcase, which serves as a second combustion chamber. A reed valve, operating on vacuum, opens during the intake stroke of the piston and closes before the start of the power stroke.

A reed valve usually is mounted on a plate to separate it from the carburetor. Fig. 13-51 shows a reed valve assembly used in an outboard engine.

The openings through the reed plate and the reeds must be kept clean. Surfaces of the plate must be smooth so there is a good seal when the reed closes against it. If the valve has lost its "springiness" or tension, install a new one. High-speed two cycle engines use reed stops to prevent distortion and damage to the reeds, Fig. 13-52. If they are bent, they must be replaced.

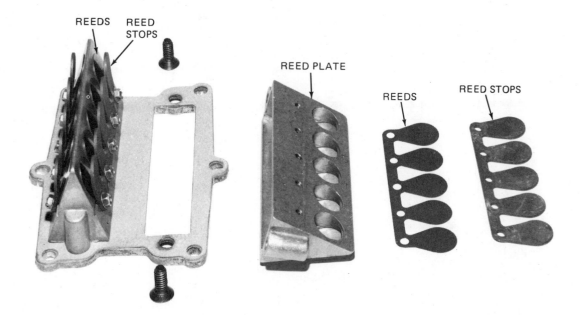

Fig. 13-52. A partially disassembled reed valve assembly shows ease with which parts may be replaced.

Some two cycle engines use rotary valves (instead of reeds) to control air-fuel intake. These generally are attached to the end of the crankshaft, although some run on a separate shaft geared off the crankshaft. In operation, a valve rotates against a wear plate. Both the valve and plate have holes in them. When the holes align, an air-fuel charge enters the crankcase.

Inspect rotary valve ports for wear or damage. Replace defective parts. Keep openings clean, and see that the surface of the wearplate is smooth and flat. If the wearplate or the mating surface of the valve is pitted or bent, replace the assembly. If a spring holds the parts in contact, check it against specifications for proper length and tension.

CAMSHAFTS AND GEARS

The camshaft of an engine is designed to operate the valves. A single camshaft is used in most small engines, with a cam (lobe) for each valve. When the camshaft rotates, the lobe of the cam lifts the valve from its seat.

Camshafts are made of steel or cast iron. The surface of the shaft is hardened to improve wearability. The ends of the camshaft may turn in bearings or in the block metal. See A in Fig. 13-53. Some small engine

camshafts are hollow with another shaft installed within it, as in view B. With this setup, the inner shaft is fixed and the hollow camshaft revolves on it.

Most small gasoline engines use gears to turn the camshaft. A gear on the crankshaft meshes with and drives a gear on the camshaft. Since the camshaft gear is exactly twice the size of the crankshaft gear, it runs at half crankshaft speed, Fig. 13-54. Also note that the

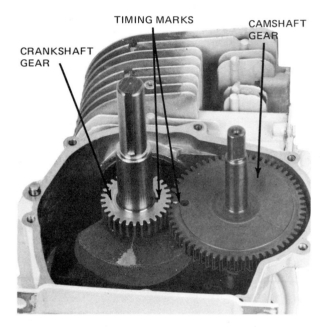

Fig. 13-54. Camshaft gear is meshed with crankshaft gear so that timing marks are aligned. Camshaft turns at half crankshaft speed. (Deere & Co.)

timing mark on the cam gear is aligned with the keyway on the crankshaft. This is correct procedure for timing valve operation to the crankshaft on this engine.

AUTOMATIC COMPRESSION RELEASE

To make hand cranking less of an effort, some small engines have an automatic compression release mechanism on the camshaft. This device lifts the exhaust valve a small amount during cranking and releases part of the compression pressure.

One manufacturer's release mechanism is pictured in Fig. 13-55. In view A, the camshaft is at rest and springs are holding the flyweights in. In this position, the tab on the larger flyweight sticks up above the base circle of the exhaust cam, holding the exhaust valve partially

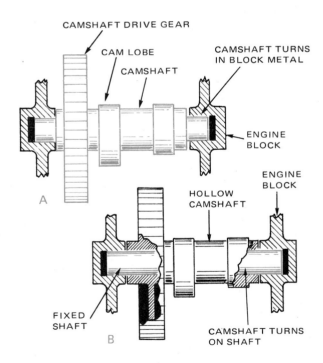

Fig. 13-53. Typical small engine camshafts. A—Solid camshaft. B—Hollow camshaft turning on a fixed shaft.

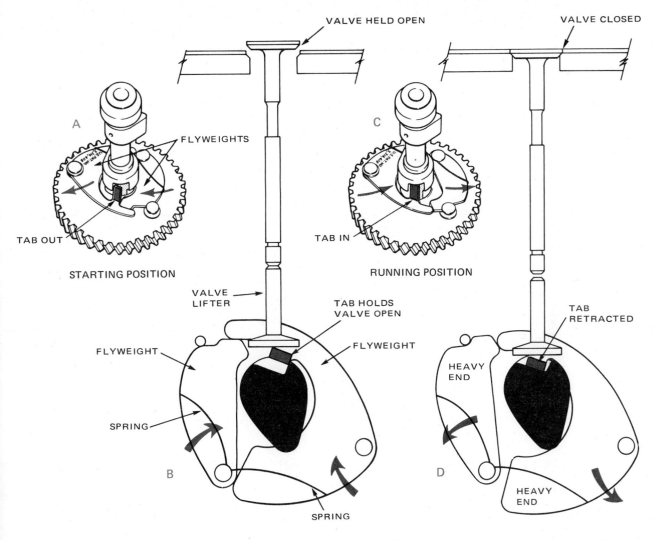

Fig. 13-55. Automatic compression release makes cranking easier. A—Tab is out, preventing valve from closing completely. B—When engine starts and reaches 600 rpm, flyweights move out, tab retracts and valve functions normally. (Kohler Co.)

open. In view B, the tab prevents the exhaust lifter from resting on the cam.

After the engine starts and its speed reaches about 600 rpm, centrifugal force overcomes spring pressure and the flyweights move outward. Movement of the flyweights causes the tab to be retracted, and the exhaust valve fully seats. See views C and D in Fig. 13-55. The flyweights remain in this position until the engine is stopped.

Fig. 13-56 shows the compression release mechanism in starting position (A) and running position (B). Compression release is one of many advances in small engines that eases the chore of engine start-up.

Fig. 13-56. Automatic compression release. A—Starting position. B—Running position.

REVIEW QUESTIONS – CHAPTER 13

1. Properly fitted friction bearing ends protrude slightly above the parting surface of the connecting rod cap. This characteristic produces what is commonly called:
   a. Bearing crush.
   b. Bearing spread.
   c. Bearing seat.
   d. Bearing swell.
2. Bearing caps must never be _____ _____ when being replaced on the rods.
3. What tool must always be used to tighten rod caps?
4. Name three types of crankshaft main bearings.
5. What do you call the special plastic substance used to measure bearing clearance?
6. When removing the crankcase cover from the engine:
   a. Always pry it loose with screwdrivers.
   b. Pry out the oil seal first.
   c. Place tape over the keyway to protect the oil seal.
   d. Hold the cover while hammering on the crankshaft end.
7. When replacing oil seals, the knife edge of the seal lip should face:
   a. Fluid being sealed.
   b. Away from fluid being sealed.
8. Valve margins should not be allowed to be less than:
   a. 1/64 in.
   b. 1/32 in.
   c. 1/16 in.
   d. 3/32 in.
9. Valve seats that are too wide will: (Select correct answers.)
   a. Cause valves to stick closed.
   b. Cause valves to stick open.
   c. Transfer too much heat to the block.
   d. Warp and burn.
10. What is the 1 deg. difference between the valve face and valve seat angles called?
11. Name the process of placing abrasive compound on the valve face and twirling it back and forth in the valve seat.
12. Too little valve clearance will cause the valve to:
    a. Break.
    b. Be noisy.
    c. Burn.
    d. Open late.
13. The camshaft revolves at:
    a. Twice crankshaft speed in all engines.
    b. Four times crankshaft speed in four cycle engines.
    c. One-half crankshaft speed in four cycle engines.
    d. One-half crankshaft speed in two cycle engines.
14. An automatic compression release is used on some engines to:
    a. Make cranking easier.
    b. Prevent knocking due to too high compression.
    c. Control speed.
    d. Prevent overheating.

SUGGESTED ACTIVITIES

1. Measure crankshaft bearing clearances with Plastigage and telescoping gages.
2. Install new main and rod bearing inserts. Observe rules of cleanliness and torque tighten rod bolts to specified value.
3. Replace oil seals in the crankcase.
4. Cut new valve seats and reface valves. Lap valves.
5. Test valve springs for length, straightness and tension.
6. Ream valve guides to fit an oversize valve stem with proper clearance.
7. Adjust valve lifter-to-valve clearance by grinding valve stems or adjusting tappets.
8. Time the camshaft to the crankshaft.

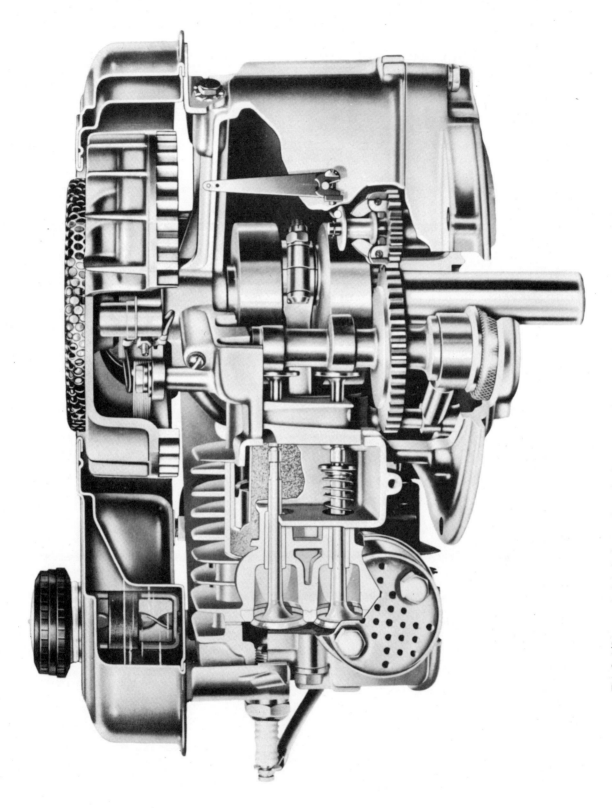

This 3 1/2 hp, vertical shaft, four cycle engine has poppet valves, pressed-in valve guides and mushroom-type nonadjustable lifters. Clearance of .010 in. must be obtained by grinding ends of valve stems. (Tecumseh Products Co.)

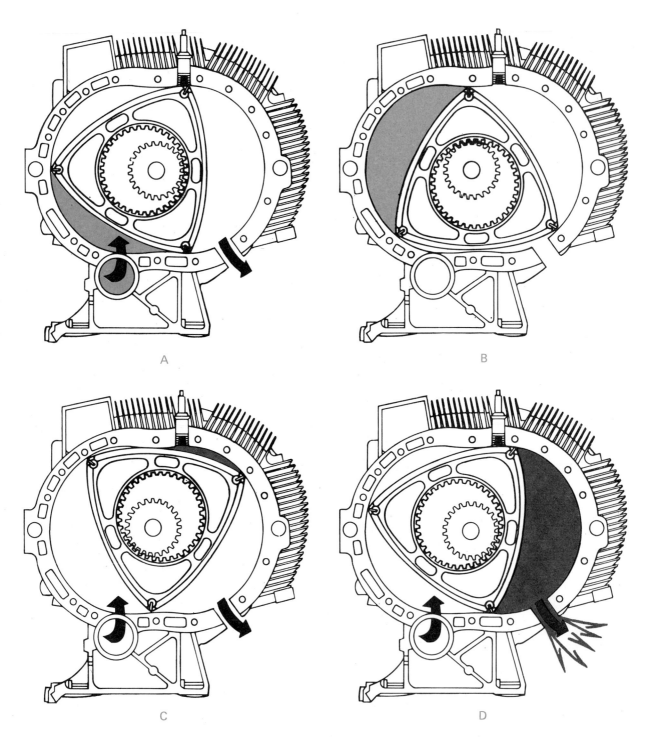

Fig. 14-1. Four cycle sequence of rotary engine. A—Intake stroke begins when rotor tip uncovers intake port. It continues until trailing rotor tip closes intake port, ending one intake cycle and beginning another. B—Compression starts as intake port is closed and reaches highest point in front of spark plug. C—Combustion takes place when charge is most compressed. Ignited air-fuel mixture expands and pushes against turning rotor to keep it rotating. D—Exhaust begins as rotor tip passes exhaust port. Rotor motion provides complete scavenging, pushing spent gases out port before trailing rotor tip closes it. (Outboard Marine Corp.)

Chapter 14

ROTARY ENGINES

Most small gasoline engines in use today are of the reciprocating piston type. It is unchallenged as a power source wherever light air-cooled engines are needed.

Now a challenge comes from the Wankel rotary engine. As a design, the Wankel has been around for over 30 years. Named for its inventor, Dr. Felix Heinrich Wankel, it was developed out of efforts to create a rotary type valve for a motorcycle engine.

Wankel did succeed in designing a rotary valve which was used on both German Messerschmidt fighters and Junker bombers during World War II. After the war, he resumed research on the rotary engine and, in 1951, opened his own laboratory with NSU, a small motorcycle manufacturer.

While Wankel's idea was to be used as a supercharger on the motorcycle, that soon changed. By 1954, NSU and Wankel decided it was possible to build a four-stroke cycle rotary automotive engine. By 1957, the first Wankel engine was being tested. It was small and had a number of imperfections.

One of the main problems was in the seals used on the rotors. Often, just a few minutes of running would wear them out. But these problems were solved and rotary engines are available today in automobiles, and various small engine applications.

Curtiss-Wright, one of the oldest aerospace companies in America, purchased the first rights to manufacture the Wankel in the United States. At this time, there are a dozen or more licenses to manufacture them here. Americans use them in powering snowmobiles, outboard and marine engines, motorcycles and small electric generators.

Wankel engines are known by other names: rotary engines, rotating combustion engines or, simply, the R.C. engine. There have been other rotary designs but, in our discussions, we will be dealing only with the Wankel type engine.

FEWER PARTS

Beyond the rotary principle, there are great differences between the rotary and the reciprocating engine. The rotary is much lighter and has only a very few moving parts. A six cylinder automotive piston engine has over 230 basic parts, 166 of them moving. A two-rotor Wankel has about 70; only three move.

This should mean that less engine fuel is being used to overcome friction and drive the engine parts. Too, there are fewer parts to manufacture and replace.

HOW THE ROTARY OPERATES

You are familiar with the four-stroke Otto cycle of the reciprocating piston engine – intake, compression, power, exhaust. The rotary engine has the same familiar four strokes. Only, it completes four strokes three times for each revolution of the rotor. Also, the shaft that is driven by the rotor revolves three times for each revolution of the rotor.

Fig. 14-1 illustrates the cycle of events as they take place through one revolution of the rotor. Don't forget that each side of the rotor is a separate and independent chamber. Therefore, one rotor does the same work as a three-cylinder reciprocating engine. Two rotors would increase the engine potential to at least six cylinders, three rotors to nine, and so on.

Beginning with A in Fig. 14-1, the rotor is turning clockwise. The chamber is exposed to the intake port at position A. The increasing volume draws in the air-fuel mixture until, at position B, the trailing apex seal moves past the intake port and compression starts. When the

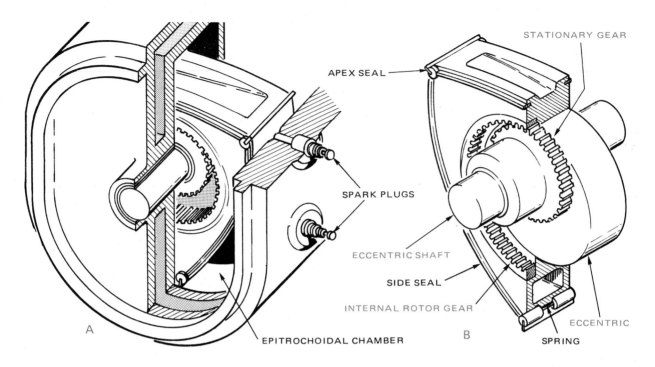

APEX SEAL

SPARK PLUGS

ECCENTRIC SHAFT

SIDE SEAL

INTERNAL ROTOR GEAR

ECCENTRIC

SPRING

A

EPITROCHOIDAL CHAMBER

B

Fig. 14-2. As rotor turns, eccentric is driven by rotor. Stationary gear keeps rotor in position and does not transmit torque.

rotor reaches full compression as in C, the intake port is fully closed to that chamber and the volume of the combustion chamber is at its smallest. Then the spark plug ignites the charge to supply force to continue the rotation. When the rotor reaches the point shown in D, the spent gases are pushed out the exhaust port.

Two spark plugs are used in some rotaries to improve combustion efficiency. The plugs fire one after the other, about 10 deg. apart, igniting the fuel mixture. Remember that while tracing the action of one chamber through one revolution, the other two chambers are following the same sequence 120 deg. and 240 deg. behind. Actually, three intake, three compression, three power and three exhaust actions take place during the one revolution of the rotor.

Notice, though, that the eccentric portion of the shaft rotates the shaft three times. This means there are three power pulses per rotor revolution. If there were two rotors overlapping, there would be six power pulses per shaft rotation and, greater power produced.

THE DRIVING MECHANISM

Fig. 14-2 shows in greater detail, the way in which the eccentric shaft is driven by the rotor. First, the stationary gear and internal rotor gear do not transmit torque. The purpose of the gears is to control the motion and position of the rotor while maintaining proper ignition timing since the ignition distributors are driven from the eccentric shaft. Torque is transmitted from the rotor by the eccentric as shown in B. The eccentric shaft replaces the familiar crankshaft used in the reciprocating engine.

SEALS

Instead of piston rings, the rotor requires side seals and apex seals to prevent leakage of compression and combustion gases, Fig. 14-2. One of the major problems encountered with early rotary engines was extreme wear of the apex seals which slide over the inner surface of the epitrochoidal chamber.

Although various materials are being tried, current rotary engine seals are not as durable as conventional piston rings. Hard carbon and aluminum impregnated carbon have provided the best results so far. New sealing materials are now developed which can provide greater than 100,000 miles of service in automobiles.

Side seals B, Fig. 14-2, are just as important as apex seals but do not provide any particular technical difficulty. Fig. 14-3 shows spring loading and relative position of the apex and side seals.

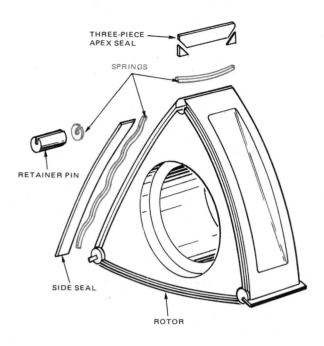

Fig. 14-3. Seals are an important part of rotor. Springs are used to keep seals in sliding contact with chamber walls.

WEARABILITY

Due to the sliding friction created by the seals, it is necessary to use some material which will resist wear to both seals and the inner walls they slide on. The inner walls of the epitrochoid chamber are faced with tungsten carbide, hard chrome or nickel chromium carbide.

COOLING

The rotary can be either air-cooled or water-cooled. In applications requiring small rotaries, air cooling is used. Heat is concentrated in the combustion area of the chamber around the spark plug and in the rotor.

Cooling of the rotor is difficult. It is completely enclosed in the chamber and does not get any cooling from the crankcase as in the reciprocating engine. Heat is controlled by circulating cool oil from the sump.

LUBRICATION

Lubricating oils in rotaries are not contaminated by blow-by. Thus, periodic oil changes are not needed. Oil should be added as needed to replace that which is lost to lubricate the rotor seals and housing.

Bearings which surround the output shaft are lubricated by oil from the sump under pressure from a gear type oil pump. In some cases, oil has been mixed with the fuel to provide lubrication of seals, bearings and chambers. More recent designs, however, have an independent feed system which introduces measured amounts of oil at the intake ports.

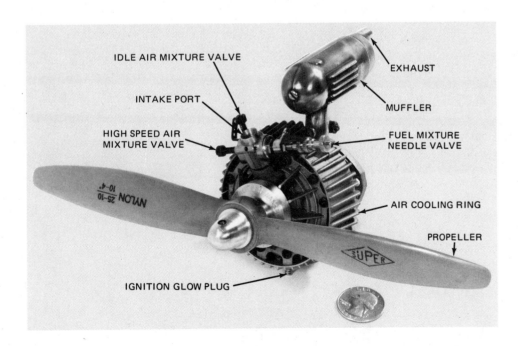

Fig. 14-4. A small Wankel model airplane engine manufactured by NSU operates with a glow plug.

SERVICE

Rotary engines have no valves to burn or stick. There are no rings, but there are seals that need replacing occasionally. These should be just as durable as rings. Carburetors and ignition systems are very accessible and should be easy to service. Ignition units and carburetors are of conventional design.

IMPLICATIONS FOR SMALL ENGINES

The rotary engine can be produced in a wide range of sizes. One small enough to be held in the palm of the hand is available for model airplane use, Fig. 14-4. Currently, rotary engines are available in some snowmobiles. Outboard engines and motorcycles are on the market and new ones are being developed.

ROTARY ENGINES

ADVANTAGES
1. Though following the four-stroke cycle, there is a power pulse for each rotation of the crankshaft.
2. Far fewer moving parts.
3. Less power lost to friction.
4. About half the size and weight of an equal horsepower piston engine.
5. Comparatively simple.
6. Almost vibrationless.
7. Easier to muffle than two cycle engines.
8. Operates on low octane and unleaded fuels.
9. Quieter running.
10. Greater acceleration and top speed.
11. Permits effective use of anti-pollution devices.
12. Efficient lubrication without oil contamination.

DISADVANTAGES
1. Newness of the engine.
2. Requires major expenditures for production.
3. Rapid wear of apex seals (currently being solved).
4. Poor fuel economy in early Wankel engine automobiles was a common complaint. Dual spark plugs and improved seals have reduced problem.
5. Early engines released more unburned hydrocarbons. It has been found that it is easier to control emissions from the R.C., particularly oxide of nitrogen which has plagued the piston engine.

REVIEW QUESTIONS — CHAPTER 14

1. What company was first licensed in the United States to produce Wankel engines?
2. A two-rotor engine would produce power pulses equivalent to a:
   a. Single cylinder reciprocating engine.
   b. Two cylinder reciprocating engine.
   c. Four cylinder reciprocating engine.
   d. Six cylinder reciprocating engine.
3. When the rotor revolves once, the eccentric shaft revolves:
   a. Once.   c. 120 deg.
   b. Three times.   d. Two times.
4. Why do some rotary engines use two spark plugs?
5. The main purpose of the stationary and internal rotor gears is to:
   a. Drive the shaft.   c. Control rotor position.
   b. Operate the valves.   d. Transmit torque.
6. Apex seals are made of hard carbon and carbon impregnated with what other material?

SUGGESTED ACTIVITIES

1. Assemble a plastic model of a Wankel engine.
2. Design an epitrochoid and rotor for a rotary engine.
3. After designing the epitrochoid and rotor, transfer this to acrylic sheets and make a working demonstration model.
4. Visit a local dealer of rotary engines, such as an outboard engine, snowmobile or automobile dealer. Ask to look over the engine and have a demonstration. Write a report on your observations.
5. Test a small rotary engine on a classroon dynamometer. Chart the results.

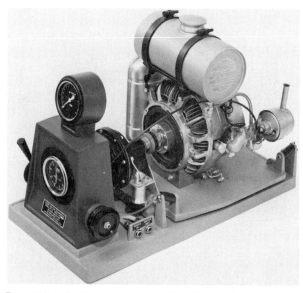

Eccentric shaft of rotary engine positioned for testing on a water brake dynamometer.

Chapter 15

SMALL GAS ENGINE APPLICATIONS

This chapter will examine a few of the more common small gas engine powered implements. It will discuss purchasing considerations, safety features, service warnings, and maintenance methods. Since there are hundreds of specialized powered implements, you should always study the owner's manual before operating and servicing these devices.

LAWNMOWER SAFETY

Lawnmowers have caused a great number of injuries. The more severe injuries were lacerations to hands and feet, and injuries from objects thrown from under the mower housing or out the discharge opening.

Recently, laws have been enacted forcing manufacturers to provide certain safety devices on lawnmowers to help prevent accidents.

Like many other implements, there is not an age, skill, nor intelligence requirement for using a lawnmower. However, young children should never be allowed to operate a power mower. The human body is no match for a sharp steel blade tip rotating at hundreds of miles per hour. Objects have been reported to have been thrown a distance of one quarter of a mile by a power mower.

A grass discharge chute guard is shown in Fig. 15-1. It is an important safety device that can prevent lawnmower injuries.

PURCHASING A LAWNMOWER

Because of the large variety available, certain considerations should be made when purchasing a power mower. The operator's strength may determine whether the mower should be the push-type or self-propelled type. Rotary mowers (those having horizontally rotating blades) for large areas have larger diameter blades with larger, heavy engines and housings. Also, safety and ac-

cessory items may add to mower weight. Remember, a mower may roll easily on a smooth floor. However, it may be difficult to push on a deep, rough lawn with the grass catcher loaded with grass.

If you have a super-quality, "showpiece" lawn that is very level, you may want to consider a reel-type mower (blades rotate down vertically). This is the type used on golf course greens. They produce a high quality job but are more difficult and expensive to maintain. Special equipment is needed to sharpen the reel. Adjustments of the reel blades to the cutter bar are critical and need occasional readjustments as the blades wear. Cutter height is controlled by raising or lowering rollers.

For the average size yard, a rotary-type mower with blade diameter of 22 inches is usually satisfactory. For small yards, a 20 inch mower is more maneuverable and takes less storage space. The length the grass is to be cut

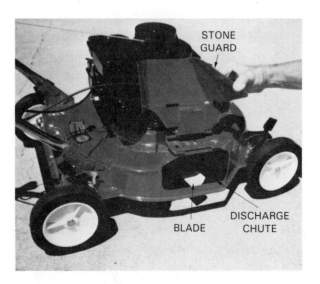

Fig. 15-1. Grass discharge chute guards, also called stone guards, are designed to deflect objects toward ground. Guards must always be kept in place and in good working order.

Fig. 15-2. Each wheel on this mower has an adjuster to raise or lower housing and blade for desired length of grass. Always stop engine before making these adjustments.

is controlled by raising or lowering the entire mower through adjustments on the wheels. See Fig. 15-2.

PUSH-TYPE OR SELF-PROPELLED

Selecting a push-type or self-propelled mower is a matter of personal preference. Engines for self-propelled mowers have to rotate the blade and drive mechanism to the wheels, Fig. 15-3. This requires additional engine horsepower and a drive mechanism to the wheels. In some mowers, transmission to the wheels is through shafts and gears; others use a belt and pulley system, Fig. 15-4. Greasing bearings and gears and replacing belts are additional maintenance requirements for a self-propelled mower.

Fig. 15-3. Self-propelled mowers are commonly driven by friction rollers that engage lawnmower wheels.

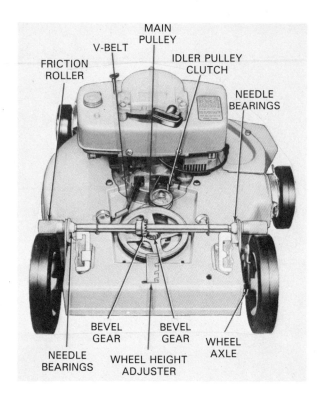

Fig. 15-4. This self-propelled mower uses a V-belt pulley system with an idler pulley clutch to tighten belt. Main pulley drives bevel gears to turn drive shaft. Friction rollers turn mower wheels.

TWO-CYCLE OR FOUR-CYCLE ENGINE

As discussed earlier in this text, two-cycle engines have fewer moving parts than four-cycle engines and require a proper mixture of oil with the gasoline. See Chapter 9, "Mixing Oil and Fuel." Four-cycle engines retain the oil in the crankcase and only gasoline is placed in the fuel tank. Crankcase oil must be drained and replaced at recommended intervals.

Both types of engines are quite reliable but be aware of the type you are purchasing. If it is a two-cycle engine, purchase some two-cycle oil, Fig. 15-5A. You should also have an approved type gasoline can labeled so you do not confuse it with another can of plain gasoline. Without the oil in the fuel, the two-cycle engine will be ruined in only a few moments of running.

OPTIONAL FEATURES

One optional feature for mowers is side-bagging, rear-bagging, or no bagging at all. Grass bags may be fabric or molded plastic containers. Side-bagging mowers tend to extend the width of the mower but may be a little easier to remove and install. If mowing in nar-

Fig. 15-5. A—Two-cycle engine oil. B—Fuel stabilizer for extended storage. Both can be obained from local implement dealers.

row quarters, a rear bag may be slightly more maneuverable, particularly if the mower must be backed out of the space.

Using either a side or rear bag reduces thatch (compacted dead grass cuttings) in the lawn. Thatch, if allowed to accumulate, tends to smother the lawn and create growth problems.

WARNING! To prevent serious injury, always stop engine or blade when emptying grass catcher. Never start engine or blade without grass catcher in place, Fig. 15-6.

Grass cuttings are excellent additions for the compost pile or garden. Some prefer to use a bagless mower

and rake the grass cuttings after mowing. Again, the cuttings must be removed. Some mowers have blades and housings designed to cut and mulch (cut into fine particles) grass and leaves for composting.

Another optional accessory is a de-thatcher blade, Fig. 15-7. This can be purchased to fit any mower and is installed in place of the cutting blade. There are spring-like fingers attached to the blade that reach down into the lawn. As the blade turns, the fingers rake the thatch up to the surface. The thatch is then collected in the grass bag, or hand raked later.

CONVENIENCES

Most manufacturers try to design bag removal and replacement as easy as possible. The bag usually must be removed a number of times before mowing is completed. This means stopping the mower, removing the bag and grass, emptying the bag, replacing the bag, restarting the mower. This is enough work without having to "wrestle" with a clumsy bag connection. The bag should also be durable and not wear on chaffing points.

Ease of cleaning the mower, particularly underneath, is another convenience. It is good practice to let the mower cool after mowing; then tip it on its side and wash out the residue of grass with a garden hose. See Fig. 15-8. If grass is allowed to build up on the blade and in the housing, it will dry and become very difficult to remove. Also, cutting efficiency will be lessened. The upper parts of the mower and engine should also be kept clean to maintain engine cooling and proper functioning of the carburetor and governor parts.

Fig. 15-6. Never have engine or blade running when grass catcher is removed. On this mower, blade brake and clutch stops blade while allowing engine to continue running.

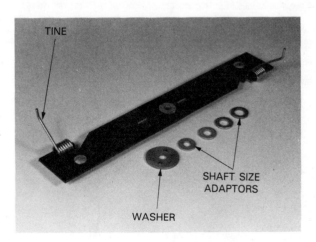

Fig. 15-7. A de-thatcher blade uses spring tines (pointed prongs) to rake out dead grass. Adaptors with various hole sizes permit bar to fit on any mower shaft diameter.

Fig. 15-8. Bottom of mower should be kept clean for efficient cutting. Steel housings should be cleaned and painted occasionally to prevent rust.

Air filter maintenance (cleaning and oiling) should be easily done. See Figs. 15-9 and 15-10. Also, the spark plug should be easily reached for cleaning or changing as needed, Fig. 15-11. Mufflers, like those on any internal combustion engine, eventually deteriorate to the point where they must be replaced. Ease and cost of replacement may vary, Fig. 15-12.

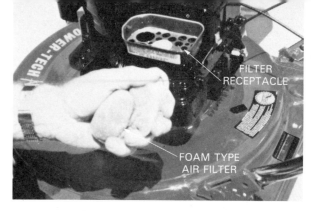

Fig. 15-10. Foam type air filter can be easily washed in detergent and water and squeezed dry. Saturate in oil and squeeze out excess before replacing.

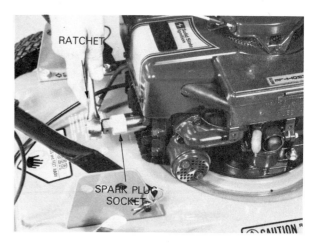

Fig. 15-11. Spark plug should be serviced or replaced at least once each season. A spark plug socket should be used to prevent damage to porcelain insulator.

Fig. 15-9. Air filter must allow large amounts of air to enter engine easily, while trapping dirt and debris. It should be easily removed for service.

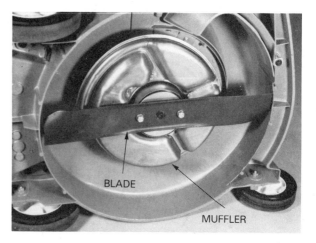

Fig. 15-12. Muffler on this mower requires removal of blade for replacement. It is tuned to engine for quietness and efficiency.

The "business end" of the mower is the cutting blade. After prolonged use, it will become nicked and dull. There are many styles of blades. To remove the blade for sharpening, the mower is tipped on its side and several bolts are removed, Fig. 15-13. The mower should be set on its side so that it cannot fall upside down.

WARNING! Always remove the spark plug wire and tie it back before tipping the mower. Wear gloves or wrap a cloth around the blade to protect hands from sharp edges.

Storage and portability may be important. Today, many mowers are designed with handles that fold down so that little space is needed for storage, Fig. 15-14. If transporting the mower in a car, this may also be convenient so you can shut and latch the trunk lid.

ENGINE STARTING

Various mechanical means have been devised for starting small engines. The simplest method has been the independent rope start. A rope with a knot in one end is wrapped around the flywheel pulley and given a quick pull. This design is no longer in use.

The recoil starter is common today. It utilizes a rope, a ratchet mechanism, and a rewind spring. When the rope is pulled, the ratchet engages the flywheel and rotates the crankshaft. When the engine starts, the

Fig. 15-14. If storage space is limited, mower handle should fold down or be removed easily.

ratchet disengages from the flywheel. The rewind spring retracts and recoils the rope for the next starting. The proper technique for starting is shown in Fig. 15-15.

Inertia starters are also available. The inertia starter utilizes a coil spring attached to a ratchet mechanism and a crank or lever. The coil spring is wound tightly with the crank and held with a locking pin. When the control knob or lever is released, the coil spring engages the ratchet with the flywheel to start engine, Fig. 15-16.

Fig. 15-13. Before removing a blade, ALWAYS remove spark plug wire and tie it away from plug. Hands and knuckles should be protected with gloves while loosening or tightening bolts.

Fig. 15-15. When hand starting vertical pull engine, place one foot on deck and other foot away from mower. Pull rope briskly.

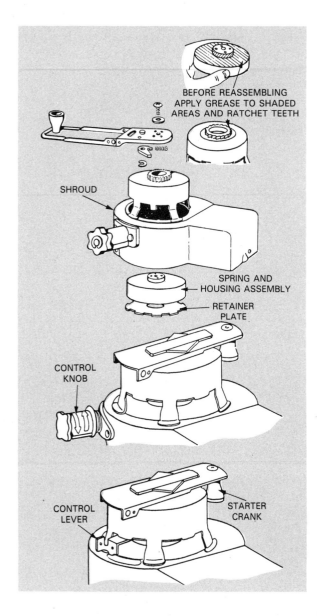

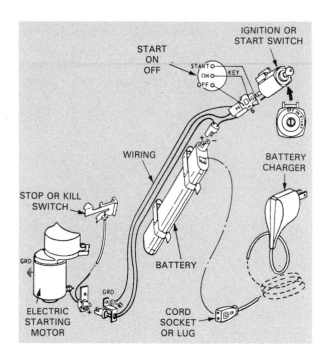

Fig. 15-17. An electric start system has an electric motor, battery, and switch like an automobile. Some mowers have a hand start backup system in case battery is discharged.

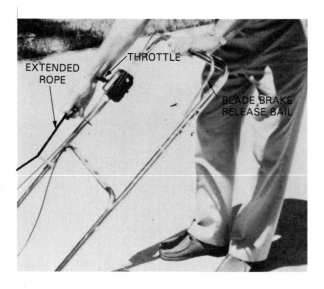

Fig. 15-18. Extended rope starting system helps keep hands and feet away from mower while starting.

Fig. 15-16. Inertia starter is used by winding a coil spring with starter crank. When ready to start engine, turn control knob or move control lever. Spring will then rotate engine flywheel for starting.

Electric start mowers are also available. The electric start mower has the added weight of a starter motor, switch, battery, and wiring, Fig. 15-17. A key is required to turn on the switch. A small battery charger is also needed to energize the battery after motor operation.

A recent innovation for convenience and safety is the "extended rope" start. The advantage to this system is the operator does not have to bend over as far and his or her feet are clear of the blade housing. See Fig. 15-18. Also, the blade brake must be released with one hand while the opposite hand pulls the rope.

BLADE BRAKES

One of the major safety features of every mower manufactured today is that the blade must automatically stop within three seconds after the operator's hands leave the handle. This requires a highly efficient braking system. There are basically two kinds.

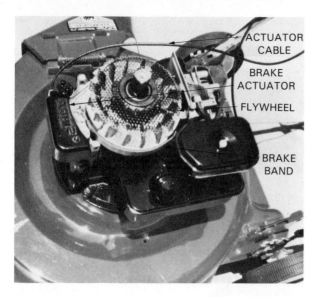

Fig. 15-19. To stop engine and blade within three seconds of release of handle, this engine has a flywheel brake. Cover has been removed for clarity to expose brake band around flywheel.

In one system, a brake band wraps around and grips the flywheel to stop the engine and blade, Fig. 15-19. The second system, as shown in Fig. 15-20, has a clutch release that allows the engine to remain running while a brake stops the blade. Both systems utilize a "bail" (hand lever) hinged to the handle. In the position shown in Fig. 15-21, the brake is engaged and the blade will not turn. When the bail is pulled into the handle, the blade begins to spin.

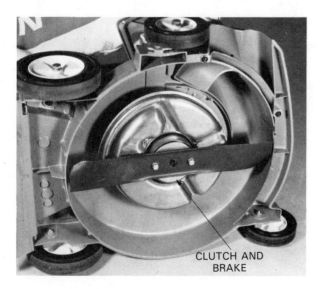

Fig. 15-20. This mower has a clutch that disengages engine from blade while brake stops blade. Engine can safely remain running when operator releases handle and bail.

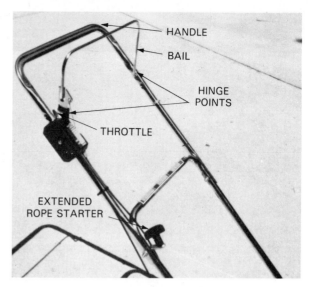

Fig. 15-21. Bail must be held against handle to release engine or blade brade.

WARNING! In all cases, operators should thoroughly study the owner's manual and be familiar with all safety precautions and operating procedures before using a power mower.

PROCEDURE FOR STARTING ENGINE

The following is a basic procedure for starting a cold engine:
1. Fill fuel tank with proper fuel for engine type. Refer to Fig. 15-22.
2. Check the oil level and condition of oil. Add or change if necessary, Fig. 15-23.

Fig. 15-22. Check fuel level in tank before starting.

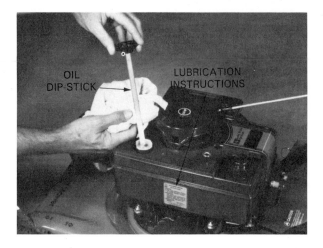

Fig. 15-23. This engine has an automotive type dip-stick for checking oil level. Oil is added through dip-stick opening.

3. Prime engine with fuel, Fig. 15-24, or close choke.
4. Turn key on or advance throttle to START position, Fig. 15-25.
5. a. Vertical or Horizontal Rope Start: Place foot on mower deck and pull rope, Fig. 15-15.
   b. Extended Rope Start: Hold bail to handle and pull rope. Refer to Fig. 15-21.
   c. Electric Start: Turn key in switch past RUN to START. Release key to RUN position as soon as engine starts, Fig. 15-26.
6. When engine starts, open choke and adjust throttle to operating speed (about 1/2 to 2/3 maximum speed). Refer to Fig. 15-25.

NOTE! For re-starting a warm engine, priming or choking should not be necessary. A cold two-cycle engine may need several prime pump strokes after the engine starts to keep it running.

Fig. 15-24. Depressing primer forces fuel into cylinder for quick starting a cold engine. Engines without a primer pump utilize a choke.

Fig. 15-25. Throttles have stop, start, and fast positions. Throttle lever is being moved to start position.

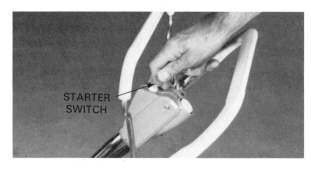

Fig. 15-26. Electric start mowers have a key type switch. To start, insert key and turn to start position until engine begins to run. Release key immediately upon starting.

MINOR CHECKS

If engine idles too fast or slow, adjust idle speed screw on carburetor with a screwdriver. The engine should idle slow but smooth. Refer to Fig. 15-27.

If the engine will not start, remove the spark plug as shown in Fig. 15-28. Test for spark as described in Chapter 10, "Ignition Testing Procedure."

Fig. 15-27. An idle speed adjustment screw located on carburetor can be turned with a screwdriver to increase or decrease rpm.

Fig. 15-28. If engine will not start, remove spark plug. With spark plug wire about 3/16 inch from engine or mower, a blue spark should jump gap when engine is cranked. No spark indicates ignition system problem.

If spark jumps the gap (3/16 inch) from spark plug wire to engine when the engine is cranked, replace the wire on the spark plug and ground the base of the plug against the engine (not close to gasoline tank), Fig. 15-29. Have someone hold the bail and crank the engine. Watch for spark between the electrodes. If there is no spark, service or replace the spark plug.

If the engine still does not start, refer to the troubleshooting chart near the end of this chapter.

GENERAL MAINTENANCE

As mentioned earlier, cleaning the mower after each use is an important maintenance procedure. Additional maintenance procedures will be described here.

Fig. 15-29. If a spark jumps gap described in Fig. 16-28, connect wire to spark plug and ground plug to engine. A bright blue spark should jump electrode gap. If no spark is present, replace spark plug.

BLADE SHARPENING

When mower blades become dull and nicked, it makes the engine work harder and the lawn is cut poorly. Blades can be sharpened by clamping them to a table or in a vise and filing. Always retain the same edge angle. Blades can also be sharpened on a grinding wheel, as in Fig. 15-30.

WARNING! When grinding a blade, always wear safety goggles or glasses with side shields.

When sharpening the blade, balance it at the same time. Grind or file a small amount of the heavy end until the blade balances horizontally. Test the blade by using a blade balancer, Fig. 15-31, or balance on a sharp edge, as in Fig. 15-32. An unbalanced blade will create damaging engine vibration.

Hitting solid objects with the blade, such as large rocks, cement edges, pipes, etc., can be very damaging

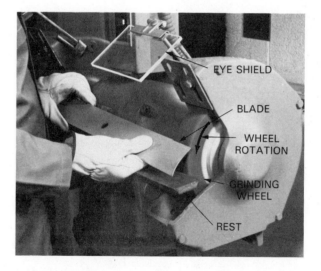

Fig. 15-30. Proper procedure for grinding cutting edges of blade. Note wheel rotation direction and position of blade. Always wear safety glasses with side shields.

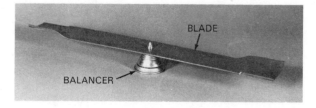

Fig. 15-31. This blade balancer is being used to test blade balance. An unbalanced blade can cause vibration and shorten engine life.

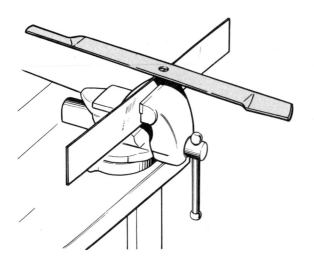

Fig. 15-32. A sharp edged object can also be used for blade balancing. Center blade hole over sharp edge.

to a mower, Fig. 15-33. Besides damage to blade, engine crankshaft can be bent, crankcase housings distorted, connecting rods bent, and flywheel keys sheared.

SPARK PLUG

A spark plug is normally easy to change. The spark plug should be carefully removed with a spark plug socket and ratchet wrench so the porcelain insulator is not damaged, Fig. 15-11.

Spark plugs can be cleaned and gapped as described in Chapter 10. When replacing the spark plug, clean the seat and do not overtighten the plug. Spark plugs should be serviced about every one-hundred hours of operation.

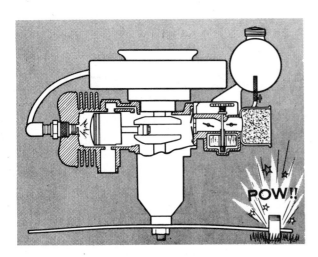

Fig. 15-33. Hitting solid objects with the blade can severely damage an engine.

When replacing the spark plug, use only the kind specified for the engine with the proper reach (thread length) and heat range (insulator tip length).

AIR CLEANERS

Engines consume tremendous amounts of air which passes through the air cleaner. When mowing, dust and dry grass materials are trapped by the filter. When too much debris collects in the air filter, it can restrict air flow. Therefore, the air cleaner element should be cleaned frequently, about every twenty-five hours of operation. Lawnmowers generally use the oil-foam type element which is a spongy plastic saturated with oil. Refer to Figs. 15-10 and 15-11.

To clean the oil-foam filter:
1. Remove filter cover.
2. Remove foam element from base.
3. Wash element in kerosene or liquid detergent and water. Follow the steps shown in Fig. 15-34.
4. Wrap foam in cloth and squeeze dry.
5. Saturate foam with engine oil.
6. Squeeze out excess oil.
7. Replace foam and filter cover.

For other type filters, see Chapter 4, "Air Cleaners, Air Filters."

CHANGING OIL

Four-cycle engines should have the oil changed about every twenty-five hours of operation. To drain oil from

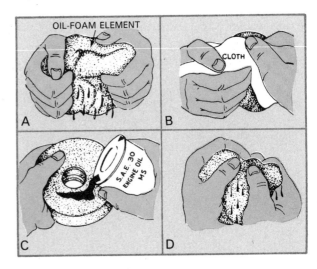

Fig. 15-34. Cleaning the oil-foam air filter: A—Wash foam element in kerosene or liquid detergent and water. B—Wrap foam in cloth and squeeze dry. C—Saturate foam with clean oil. D—Squeeze out excess oil.

lawnmower engines:
1. Place a pan under the mower housing.
2. Locate and remove oil drain plug.
3. Drain all oil and replace the drain plug.

To replace oil:
1. Obtain the kind of oil recommended by the manufacturer of the engine.
2. Pour oil into oil-fill neck that is provided and labeled. Fill to level indicated by the engine manufacturer. Some engines have dip-sticks. Fill to "Full" line, as shown in Fig. 15-23.
3. Replace filler cap.

MUFFLER REPLACEMENT

Engine mufflers get extremely hot and are subjected to the acidic products of combustion. They eventually need replacement, Fig. 15-35. There are many styles and shapes of mufflers.

Fig. 15-12 shows a rather large, extensive muffler that requires removal of the blade before its removal.

The mufflers in Fig. 15-35 simply require unscrewing of the tapered pipe threads. Coat threads with anti-seize compound prior to replacement. Other mufflers are removed by loosening two bolts with a wrench.

V-BELTS

V-belts are a widely used means of power transmission, particularly on riding type mowers. See Fig. 15-36. There may be four or five separate belts on a small tractor mower. V-belts may look very much alike but there are many unseen differences which may affect their

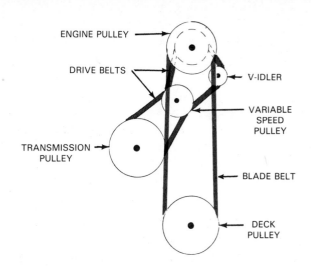

Fig. 15-36. This V-belt system for a riding mower illustrates multiple belting and pulleys.

quality and life span.

Some belts have a fabric cover; others to do. The cords imbedded inside the belt may be rayon, polyester, or Kevlar. Rayon will not last as long as the tremendously strong Kevlar. Polyester tends to shrink as it gets hot. Polyester could present a danger on a clutch, causing unintentional engagement due to shrinkage.

WARNING! Use only the type and size belt specified by the manufacturer. V-belt failures are usually a result of some other failure, such as those shown in Fig. 15-37.

CAUSE OF FAILURE	CORRECTION
1. Normal wear.	Replace belt.
2. Poor operating habits.	Do not engage and disengage clutch excessively.
3. Damaged or worn idler pulleys.	Replace idler, frozen bearings and belt.
4. Incorrectly positioned belt guards.	Realign guards. Replace damaged guards.
5. Misaligned pulleys.	Replace pulleys.
6. Damaged or worn pulleys.	Align pulleys (except where an offset system is used with special pulleys).
7. Incorrect tensions	Replace belt. Check idler spring tension. Lubricate idler brackets.
8. Oil and grease damage.	Replace belts. Eliminate oil leakage. Use oil resistant belts if possible.
9. Heat damage (140° or higher).	Use heat resistant belt. Avoid polyester belts. Shield belts from heat source.
10. Incorrect installation.	Install with care. Never use force. Recheck belts after 48 hours of use.

Fig. 15-37. Study typical causes and corrections for V-belt failure.

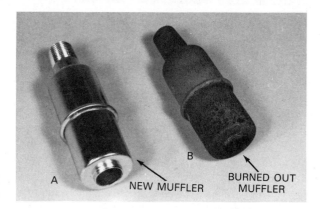

Fig. 15-35. Most mufflers are inexpensive and easy to replace. Combustion products and heat cause mufflers to corrode.

WARNING! Always keep hands and objects away from any exposed belts when engine is running. Remove spark plug wire when servicing belts. Always keep belt guards in space when operating implements.

STORING POWER LAWNMOWER

Several things can be done when preparing to store a lawnmower for extended periods. Proper preparation will help ensure long mower and engine life. It will also ensure easy starting the following season.

1. Clean the grass bag and hang it in a dry location.
2. Clean the mower of grass cuttings, mud, etc.
3. Avoid storing gasoline for long periods of time. Store only in approved "safe" containers. Never store fuel or mower in an enclosure where there is an open flame. If fuel must be stored, add a stabilizer to it. A fuel stabilizer is available from implement dealers.
4. Try to plan ahead and run the engine dry of fuel at the last use.
5. Drain the oil from the crankcase. Do not refill it now. See Chapter 16 "Changing Oil." Place a tag on the engine that says NO OIL.
6. Rotate the engine so the piston is at bottom of cylinder. Remove the spark plug. Spray storage oil through the spark plug hole, or squirt about one tablespoon of clean motor oil through the spark plug hole with an oil can. Rotate the engine slowly several times to distribute the oil on the cylinder walls. Replace the spark plug.
7. Leave the spark plug lead disconnected. Using the pull rope, rotate the engine slowly until compression resistance is felt. Then rotate the engine an additional one-quarter turn to close off its ports. This seals the cylinder and prevents moisture entry.
8. Leave the throttle in the off position (closed) and close the choke.
9. Lubricate the mower, drive system, etc., as described by the manufacturer.
10. Coat the cutting blade with chassis grease to prevent rusting.
11. Store the mower in a dry, clean area.

REMOVING MOWER FROM STORAGE

1. Replace the grass bag.
2. Fill crankcase with new oil, or make new mixture for two-cycle engine. See Chapter 9, "Mixing Oil and Fuel."
3. Remove the spark plug. Using the pull rope or starter, spin the engine rapidly to remove excess oil from cylinder. Clean or replace spark plug.
4. Clean and oil the air filter if necessary.
5. Fill fuel tank.

6. Start engine and idle until engine is warm. Adjust idle speed if necessary.
7. Increase engine speed in normal manner.
8. Make a brief test run while listening to engine and watching condition of all parts.
9. If engine does not start, review the troubleshooting chart in Fig. 15-38 and other chapters of this text.

CHAIN SAWS

Small gasoline engine powered chain saws, Fig. 15-39, have become very popular for cutting firewood and trimming trees. Like any cutting tool, they work best when properly maintained. To avoid hazards, all safety devices should be in place and safe operating procedures must be carefully followed.

PURCHASING CONSIDERATIONS

When investing in a chain saw, there are a number of things to consider so that you will be satisfied with its performance and operation. Chain saws are manufactured in a number of sizes from about 10 inches to over 40 inches of blade length. The type of work to be done is the best indicator for size selection. A smaller saw works well for cutting branches, small trees, and fireplace logs. A large professional model would be for big trees and continuous rugged work.

The lighter a chain saw is, the easier it is to handle, control, and carry. Chain saws range in weight from less than 10 pounds to over 25 pounds. Weight is a consideration that should be planned according to the type of work to be done. When trimming tree branches from a ladder, a lighter saw will be less tiring.

Chain saws also come with different size engines. With more rugged work and for long continuous cutting, a more powerful engine should be selected.

Some states have laws that require certain safety devices to be used on every chain saw. Check which devices are required and be sure the saw is properly equipped.

SAFETY FEATURES

Safe operaton of a chain saw comes from a good knowledge of correct operating procedure. However, there are a few safety features on many chain saws that are very important.

When the unshielded nose of a chain saw hits a solid surface, the spinning chain may cause the saw to fly back

TROUBLESHOOTING CHART

PROBLEM	CAUSE	REMEDY
1. Engine fails to start.	A. Blade control handle disengaged.	A. Engage blade control handle.
	B. Check fuel tank for gas.	B. Fill tank if empty.
	C. Spark plug lead wire disconnected.	C. Connect lead wire.
	D. Throttle control lever not in the starting position.	D. Move throttle lever to start position.
	E. Faulty spark plug.	E. Spark should jump gap between center electrode and side electrode. If spark does not jump, replace the spark plug.
	F. Carburetor improperly adjusted, engine flooded.	F. Remove spark plug. Dry the plug. Crank engine with plug removed, and throttle in off position. Replace spark plug and lead wire and resume starting procedures.
	G. Old, stale gasoline.	G. Drain and refill with fresh gasoline.
	H. Engine brake engaged.	H. Follow starting procedure.
2. Hard starting or loss of power.	A. Spark plug wire loose.	A. Connect and tighten spark plug wire.
	B. Carburetor improperly adjusted.	B. Adjust carburetor. See separate engine manual.
	C. Dirty air cleaner.	C. Clean air cleaner as described in separate engine manual.
3. Operation erratic.	A. Dirt in gas tank.	A. Remove the dirt and fill tank with fresh gas.
	B. Dirty air cleaner.	B. Clean air cleaner as described in separate engine manual.
	C. Water in fuel supply.	C. Drain contaminated fuel and fill tank with fresh gas.
	D. Vent in gas cap plugged.	D. Clear vent or replace gas cap.
	E. Carburetor improperly adjusted.	E. Adjust carburetor. See separate engine manual.
4. Occasional skip (hesitates) at high speed.	A. Carburetor idle speed too slow.	A. Adjust carburetor. See separate engine manual.
	B. Spark plug gap too close.	B. Adjust to .030''.
	C. Carburetor idle mixture adjustment improperly set.	C. Adjust carburetor. See separate engine manual.
5. Idles poorly.	A. Spark plug fouled, faulty, or gap too wide.	A. Reset gap to .030'' or replace spark plug.
	B. Carburetor improperly adjusted.	B. Adjust carburetor. See separate engine manual.
	C. Dirty air cleaner.	C. Clean air cleaner as described in separate engine manual.
6. Engine overheats.	A. Carburetor not adjusted properly.	A. Adjust carburetor. See separate engine manual.
	B. Air flow restricted.	B. Remove blower housing and clean as described in separate engine manual.
	C. Engine oil level low.	C. Fill crankcase with the proper oil.
7. Excesssive vibration.	A. Cutting blade loose or unbalanced.	A. Tighten blade and adapter. Balance blade.
	B. Bent cutting blade.	B. Replace blade.

Fig. 15-38. Study troubleshooting chart for engine problems.

toward the operator. This is known as KICKBACK and may be very dangerous. Some chain saws have a tip guard device that is attached to the end of the blade that helps prevent kickback. It can be installed or removed very quickly.

Another safety feature on many chain saws is the quick-stop device. Illustrated in Fig. 15-40, this device stops the chain to reduce the possibility of injury. With a sudden kickback, the operator's left hand moves forward to make contact with the front hand guard and the

chain is stopped. The front hand guard is the quick-stop activating lever.

When carrying the chain saw by hand, the engine must be stopped and the saw must be in the proper position. Grip the front handle and place the muffler at the side away from the body. The chain guide bar should be behind you.

Some chain saws come with a case, Fig. 15-41, that protects the saw during transportation and storage.

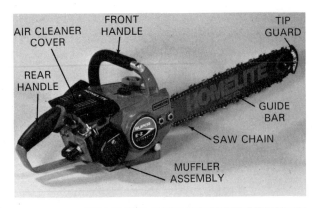

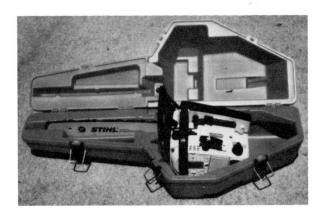

Fig. 15-41. This chain saw is well protected in a tough plastic carrying case.

Fig. 15-39. Study parts of gasoline engine powered chain saw.

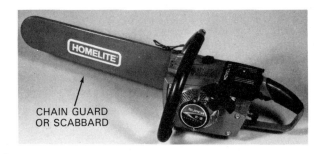

Fig. 15-42. A scabbard (chain guard) protects saw blade during transportation and storage.

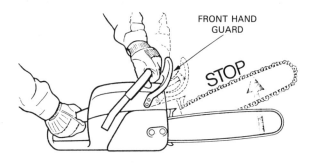

Fig. 15-40. When operator's left hand makes contact with front hand guard, it will activate quick-stop to stop chain and reduce risk of injury.

Fig. 15-43. Correct position of hands when operating chain saw. Note use of heavy gloves for protection.

Another safety feature is the chain guard (scabbard) shown in Fig. 15-42. The chain guard protects the operator from the sharp blade when not in use and protects the blade from moisture and rusting.

Hold the chain saw with both hands when cutting. See Fig. 15-43. You should also wear heavy gloves to protect your hands and dampen vibration.

Another safety feature required by some states is the spark arrestor. The spark arrestor is built into the ex-

haust system to prevent sparks from the exhaust system from catching dry grass or wood chips on fire. These devices are sometimes known as fire arrestor screens.

RULES FOR SAFE OPERATION

The following rules for the operation of a chain saw should be followed carefully.

1. Never operate a chain saw when you are tired.
2. Use safety footwear, snug-fitting clothing, and eye, hearing, and head protection devices.
3. Always use caution when handling fuel. Move the chain saw at least 10 feet (3 m) from the fueling point before starting the engine.
4. Do not allow other persons to be near the chain saw when starting or cutting. Keep bystanders and animals out of the work area.
5. Never start cutting until you have a clear work area, secure footing, and a planned retreat path from the falling tree.
6. Always hold the chain saw firmly with both hands when the engine is running. Use a firm grip with thumb and fingers encircling the chain saw handles.
7. Keep all parts of your body away from the saw chain when the engine is running.
8. Before you start the engine, make sure the saw chain is not contacting anything.
9. Always carry the chain saw with the engine stopped, with the guide bar and saw chain to the rear, and the muffler away from your body.
10. Never operate a chain saw that is damaged, improperly adjusted, or is not securely assembled. Be sure that the saw chain stops moving when the throttle control trigger is released.
11. Always shut off the engine before setting the chain saw down.
12. Use extreme caution when cutting small size brush and saplings because slender material may catch the saw chain. This could fling the saw toward you or pull you off balance.
13. When cutting a limb that is under tension, be alert for spring back so that you will not be struck when the tension in the wood fibers is released.
14. Keep the handles dry, clean, and free of oil or fuel mixture.
15. Do not operate the chain saw with a deteriorated or removed muffler system. Fire preventing mufflers (fire arrester screen types) should be used in dry areas.
16. Operate the chain saw only in well ventilated areas.
17. Do not operate a chain saw in a tree unless specially trained to do so.
18. Guard against kickback. Kickback can lead to severe injuries.

TO AVOID KICKBACK:

1. Hold the chain saw firmly with both hands. Do not reach too far.
2. Do not let the nose of the guide bar contact a log, branch, the ground, or any other obstruction.

3. Cut at high engine speeds.
4. Do not cut above shoulder height.
5. Follow manufacturer's sharpening and maintenance instructions for the saw chain.

CHAIN SAW MAINTENANCE

Careful maintenance of the chain saw engine and chain provide for long lasting service life and safe use. Never operate a chain saw that is damaged, improperly adjusted, or not completely assembled.

WARNING! Always stop the engine and be sure that the chain is stopped before doing any maintenance or repair work on the saw.

FUEL AND CARBURETOR

Always use the correct gasoline and oil mixture for a two-stroke engine as recommended by the manufacturer. Before refueling, carefully clean the filler cap and the area around it to ensure that no dirt falls into the tank. See Fig. 15-44.

Do not adjust the carburetor unless it is necessary. The high speed and low speed carburetor adjustments on a chain saw engine are very critical. Incorrect settings of these speeds can cause serious damage to the engine. If adjustments become necessary, follow the manufacturer's recommendations.

If the engine stops while idling or if the exhaust smokes or engine does not run smooth, try adjusting the carburetor, Fig. 15-45. Normally you must turn the adjusting screw clockwise when the idle setting is too lean and counterclockwise when the setting is too rich.

Fig. 15-44. Clean area around filler cap before refueling engine.

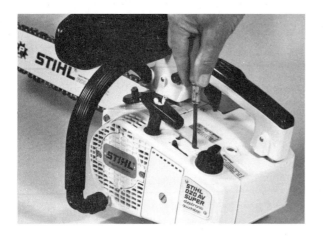

Fig. 15-45. Turn idle speed adjusting screw with a screwdriver to obtain correct idle.

Apart from minor adjustments, carburetor repairs should be made by a trained technician who has all the necessary service tools and equipment.

CYLINDER FINS

Check the cylinder fins periodically since clogged fins result in poor engine cooling. Remove the dirt and dust from between the fins to allow cooling air to pass freely. This can be done with a small stick, brush, or compressed air, Fig. 15-46.

AIR CLEANER (FILTER)

The function of the air filter is to catch dust and dirt in the inlet air and reduce wear on engine components. Clogged air cleaners cut down on engine power, increase fuel consumption, and make starting more difficult. The air cleaner should be cleaned every day in very dusty operating conditions.

Fig. 15-47. Move choke into closed position so dirt cannot enter carburetor.

Before removing the air cleaner, close the choke valve so that no dirt can get into the carburetor, Fig. 15-47. Unscrew the air cleaner cover and remove the element, Fig. 15-48. Lightly brush off dust or wash cleaner element in cleaning solvent if extremely dirty. It should be dried completely before replacing it in the engine.

FUEL FILTER

Check the fuel filter periodically. A clogged fuel filter will cause engine trouble, such as hard starting. Remove the fuel filler cap and fish for the flexible pick-up tube with a hook. Pull the fuel pick-up out, Fig. 15-49. Remove the old filter sleeve and slide a new sleeve in place. Return the flexible pick-up tube to the gas tank.

LUBRICATION

The saw chain and guide bar must be continuously lubricated during operation to protect them from excessive wear. This is provided for by the automatic chain oiling system. Clean the lubricating oil supply hole dai-

Fig. 15-46. Cylinder fins can be cleaned with a small wood stick.

Fig. 15-48. This air filter cover is removed and air filter is ready for cleaning.

Fig. 15-49. Flexible fuel pick-up tube is removed through fuel filler hole.

CYLINDER EXHAUST PORT

Fig. 15-51. Clean cylinder exhaust port with a wood stick or dowel.

ly. It is located at the base of the blade. Always fill the oil tank with chain oil each time the engine is refueled, Fig. 15-50. NEVER use waste oil for this purpose.

MUFFLER

The muffler should be kept clean and open. Do not run a chain saw without the muffler. If regulations require the use of a spark arrestor, check its condition periodically. Carbon deposits in the muffler and cylinder exhaust port will cause lower engine power output and sparking from the muffler. If needed, remove the muffler and clean the cylinder exhaust port, Fig. 15-51. Use a wood stick or dowel to protect the metal surfaces.

SPARK PLUG

If the engine does not start, it may be due to a wet, fouled, or faulty spark plug. Check the spark plug periodically and clean or install a new one as necessary. Adjust the spark gap if wider or narrower than the standard gap. Be sure the stop-start switch is in the "off" position when checking the spark plug, Fig. 15-52.

GUIDE BAR

Clean the guide bar daily or before each use of the chain saw. Remove any burrs that may be found along

Fig. 15-52. Set engine stop-start switch in "off" position when servicing.

Fig. 15-53. Lubricate roller nose bearing with a grease pump as shown.

Fig. 15-50. Removing oil tank cap to fill tank with oil.

the bar rails. On most chain saws, the roller nose bearing must be lubricated. Place the chain saw on its side so that the bar nose is firmly supported. Clean the grease hole and pump in grease as shown in Fig. 15-53.

STORAGE AFTER USE

Inspect and make adjustments of every part of the chain saw for future use before storage. Clean all parts and apply a thin coating of oil to all metal surfaces to prevent rust. Remove chain and guide bar. Apply a sufficient oil coating and wrap them up in a plastic bag. Drain the fuel tank and pull the starter a few times to drain the carburetor.

Pour a small amount of oil in the spark plug hole and replace the spark plug. Slowly pull the starter to crank the engine a couple of revolutions. Place the saw in its case and store in a dry, dust-free area.

EDGERS AND TRIMMERS

The combination edgers and trimmers powered by small gasoline engines are versatile units for lawn care maintenance. Since the edger has an exposed, fast spinning blade, it is very important that safety devices designed for the unit are always in place. Use proper operating procedures to assure safe, dependable service.

PURCHASING CONSIDERATIONS

When purchasing of a combination edger/trimmer, Fig. 15-55, a number of factors should be considered. Edgers come in a variety of sizes with respect to the construction and power of the unit. Small engines in the range of two to five horsepower are generally used to drive a belt to the blade.

Commercial edgers for heavy-duty work are built more rugged. They incorporate a larger, more powerful engine, and are considerably more expensive. The most popular edgers are smaller units designed for home lawn care and weed cutting.

Some edgers are single units designed for edging along sidewalks and driveways. Others are designed so the blade can be tilted from vertical to horizontal for trimming long grass or weeds under fences and close to buildings. Consider the type of work the edger will be expected to perform when planning a purchase.

SAFETY FEATURES AND ADJUSTMENTS

The blade of a gasoline engine driven edger is usually belt driven from the engine. The blade clutch should always be disengaged, Fig. 15-55, when starting the engine or when doing maintenance work on the unit. This loosens the belt drive to the blade, Fig. 15-56, and allows the engine shaft to turn freely.

DANGER! Never attempt to make any adjustments on an edger/trimmer while the engine is running. Serious injury could result.

The guide wheel can normally be adjusted horizontally by loosening the lever on the front of the frame and moving the wheel to either side of the frame as need-

Fig. 15-55. Disengage clutch lever when you are not trimming or edging.

Fig. 15-54. This is a combinaton edger/trimmer set for edging operation.

Fig. 15-56. This blade drive belt is in loosened position.

Fig. 15-57. Locking front guide wheel in a newly adjusted position.

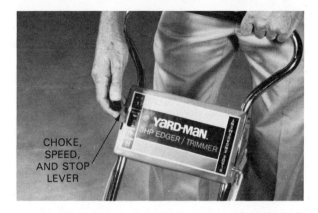

Fig. 15-59. Adjusting choke, speed, and stop lever.

ed. See Fig. 15-57. The lever should then be tightened to hold the wheel firmly in place.

With some designs, by raising and lowering the guide wheel, the depth of the edger blade can be controlled. The height of the wheel is adjusted by placing the lever on the right of the frame in the desired notch, Fig. 15-58.

One lever on the handle of the edger usually operates the choke, regulates engine speed from slow to fast, and stops the engine, Fig. 15-59. When starting the engine, place this lever in the choke position and pull the starter handle rapidly, Fig. 15-60. Grip the handle firmly and place your foot behind the rear wheel so the unit will not move during starting.

The blade guard should always be in place during use. The cam lever releases the blade guard, Fig. 15-61, so the guard can be rotated to cover the blade in any edging or trimming position. Tighten the cam lever securely before operating the edger.

Fig. 15-60. Pulling starter handle to start engine. Note belt is in loose position.

Fig. 15-58. Raising front guide wheel by placing adjusting lever in correct notch.

Fig. 15-61. Releasing cam lever so blade guard can be set at a new position.

For edging along a sidewalk or driveway, the blade is in the vertical position. When the unit is to be used for trimming, the blade is set horizontally, as shown in Fig. 15-62. It can also be set at an angle for special edging jobs, Fig. 15-63. This adjustment is made by loosening the belt with the blade clutch lever and releasing the lever on the pivot bracket and rotating the spindle housing. See Fig. 15-64. The notches in the bracket will hold the blade firmly in any position.

RULES FOR SAFE OPERATION

1. Thoroughly inspect the area where the equipment is to be used and remove all stones, sticks, wire, bones, and other foreign objects.
2. Do not operate equipment when barefoot or when wearing open sandals. Always wear substantial footwear.
3. Check the fuel before starting the engine. Do not fill the gasoline tank indoors, when the engine is running, or while the engine is still hot. Wipe off any spilled gasoline before starting the engine.

Fig. 15-62. Blade is set horizontally for trimming.

Fig. 15-63. Blade is set at an angle for special trimming jobs.

Fig. 15-64. Lever on pivot bracket is placed in correct notch for trimming.

4. Disengage the blade clutch before starting the engine.
5. Never attempt to make a wheel adjustment while the engine is running.
6. Never operate the equipment in wet grass. Always be sure of your footing; keep a firm hold on the handle.
7. Do not change the engine governor settings or overspeed the engine.
8. Do not put hands or feet near or under rotating parts. Keep clear of the discharge opening at all times.
9. Stop the blade when crossing a gravel drive, walks, or roads.
10. After striking a foreign object, stop the engine. Remove the wire from the spark plug. Thoroughly inspect the edger for any damage, and repair the damage before restarting and operating the edger.
11. If the equipment should start to vibrate abnormally, stop the engine and check for the cause. Vibration is generally a warning of trouble.
12. Stop the engine whenever you leave the equipment, before cleaning the guard assembly, and when making any repairs or inspections.
13. When cleaning, repairing, or inspecting, make certain the blade and all moving parts have stopped. Disconnect the spark plug wire, and keep the wire away from the plug to prevent accidental starting.
14. Do not run the engine indoors.
15. Shut the engine off and wait until the blade comes to a complete stop before unclogging guard assembly.
16. Safety glasses or other eye protection should always be used when operating an edger.

EDGER AND TRIMMER MAINTENANCE

The standard edger blade shown in Fig. 15-65 is 10 inches long and notched on the ends. Since the blade scrapes the edges of driveways or sidewalks during operation, it wears down quickly and should be replaced as needed.

To change the blade, raise the front wheel and loosen the nut on the drive shaft, Fig. 15-66. Remove the old blade, and replace it with a new one. Be sure the blade nut is tightened properly. Always wear a glove to hold the blade to prevent injury to your hand.

Edger blades are never sharpened. They are just replaced when they become too short to make good contact with a surface being edged.

To replace a worn belt, remove the belt guard on the engine pulley and the belt guard on the spindle housing. Remove the belt. Replace with a proper size V-belt and secure the belt guards.

Check the engine oil level before starting the engine and after every six hours of use. Add oil as necessary

Fig. 15-65. This is a standard edger blade with lock washer, plain washer, and nut.

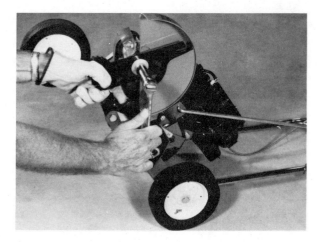

Fig. 15-66. Loosening blade nut on drive shaft. Hold blade with a heavy glove to protect hand.

to keep level on full. Before removing filler plug, clean the area around the plug to prevent dirt from entering the oil fill opening.

Change the oil after 30 hours of operation by draining oil through the lower oil drain plug. Refill with the correct amount and weight of fresh oil.

Lubricate all moving parts of the edger with engine oil periodically. Check and clean or replace spark plug each operating season. Remove and clean the air filter at recommended intervals.

EDGER AND TRIMMER STORAGE

The following steps should be taken to prepare edgers and trimmers for storage.
1. Clean and lubricate the unit thoroughly.
2. Loosen the belt so it will not stretch during storage.
3. Coat the cutting blade with oil to prevent rusting.
4. Remove the spark plug and pour a tablespoon of clean engine oil into the spark plug hole. Rotate the crankshaft a few times and replace the spark plug.
5. Check the blade and engine mounting bolts for proper tightness.
6. Never store the edger with gasoline in the tank when inside of a building where fumes may reach a spark or open flame.
7. Store the edger in a dry, clean area.

REVIEW QUESTIONS—Chapter 15

1. Name five safety features used on power lawnmowers.
2. List eight considerations for purchasing a power lawnmower.
3. List five safe operating practices when using a power lawnmower.
4. Name seven lawnmower conveniences.
5. List the procedures for preparing an engine before starting.
6. List the cold engine starting procedure for an extended rope type lawnmower with blade brake.
7. If the engine idles too fast, you can adjust the idle speed _____ on the _____.
8. If the engine will not start, though proper procedures have been used, the first trouble check should be the _____ _____.
9. List five maintenance steps that will keep the power lawnmower in good working condition.
10. When the unshielded nose of the chain saw hits a solid surface, it may jump toward the operator. This is called _____.

11. The device often built into the muffler of a chain saw to prevent sparks from causing a fire is called a _____ _____.
12. The _____ _____ should always be in place when operating an edger.
13. A standard edger blade is _____ long.
14. A small chain saw works well for cutting _____, small _____, and _____ _____.

SUGGESTED ACTIVITIES

1. Change the oil in a power lawnmower.
2. Service an air filter.
3. Clean and gap a used spark plug.
4. Sharpen and balance a lawnmower blade.
5. Demonstrate proper engine starting procedures.
6. Adjust engine idle setting.
7. Replace a burned out muffler.
8. Go to a local implement store and compare lawnmowers as though you were planning to purchase one.
9. From literature obtained from dealers, make a list of chain saws produced by different manufacturers. Describe the advantages and disadvantages of each in terms of operator safety.
10. Replace a worn out blade on an edger.
11. Make a list of maintenance features to be done before placing a chain saw in storage.
12. Check and change, if necessary, the gas line filter in a chain saw fuel tank.
13. Check the muffler of a chain saw to see if it has a spark arrestor device.

Chapter 16

CAREER OPPORTUNITIES

The small gas engine field offers career opportunities for men and women in three different areas: manufacturing, sales and service. Through training, study and work experience, you can become an engine mechanic, service manager, sales manager or owner/manager of a small engine service center. Or, you can be a manufacturer's technician, service representative or engineer.

ENGINE MECHANIC

Implement sales facilities and equipment rental centers need engine mechanics to do tune-ups, service equipment and make repairs. Often, the quality of

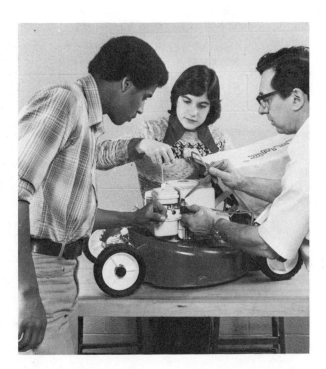

Fig. 16-1. A small gas engine mechanic must know engine construction and principles of operation. Mechanics must be proficient in troubleshooting, maintenance, service and repair.

workmanship and reliability of the mechanic who services the customers' equipment directly affects the reputation and sales volume of the business.

Small gas engine mechanics must be able to diagnose engine trouble and make appropriate repairs and/or parts replacement, Fig. 16-1. They should be able to analyze the mechanical condition and performance of any particular engine and make proper recommendations to the owner. They should be competent in the use of test equipment and thoroughly familiar with manufacturer's manuals.

Good mechanics keep their tools and equipment in first class condition and organized for convenient use. They must have specialty tools (pullers, drivers, etc.) that are available from the engine manufacturers or tool manufacturers. Small gas engine mechanics must know how to use and read micrometers and dial indicators. Also, some basic machining and welding experience is desirable. Engine mechanics generally receive their knowledge and skill at technical career centers or vocational schools. Some engine manufacturers have their own training programs for service personnel.

SERVICE MANAGER

Men and women are needed as service managers in small gas engine shops with more than one mechanic. They are responsible for quality workmanship and satisfactory shop operation. They must plan and supervise the activities of all service department employees.

Service managers discuss service problems with customers, make recommendations, write job tickets and assign work to the mechanics. They handle customer complaints, are responsible for the training of apprentices and generally inspect all finished repair work. They report directly to the general manager or owner of the service facility.

SALES MANAGER

Sales managers are needed to sell or rent implements and vehicles that utilize small gasoline engines. Some sales managers handle a wide variety of products such as yard and farm equipment, marine and sports vehicles, construction and emergency repair or rescue equipment. Others specialize in one field or sell a few closely related products such as motorcycles, all terrain vehicles (ATV) and snowmobiles.

The geographical location of a sales or rental business often determines the type of vehicle or kind of equipment that is most in demand by the consumer. Gasoline engine-driven applications are numerous, and more new uses are being developed each year.

OWNER/MANAGER

Owning or managing a successful small engine sales and service business requires experience and education. Keeping accurate sales and service records, budgeting, promoting sales, meeting payrolls, providing benefits for employees, maintaining adequate tools, parts, supplies and accessories are management responsibilities.

The owner/manager must have a sincere, personal commitment to provide fair, quality service to customers. Often, a mechanic or sales person works up to this position through years of work experience and post high school courses in business and management.

TECHNICIAN

Small gas engine manufacturers need men and women to develop prototype engines or engine parts and test new design theories. Technicians need to be skilled in the use of tools, materials and machine processes in order to produce a special part or engine unit. They usually are required to run exhaustive tests using dynamometers and other specialized testing equipment, observing and recording test results.

Technicians generally are involved with experiments, tests and analysis of various engine systems and designs in the plant and in actual field use. When testing is completed, they present the test results and recommend changes. Therefore, they must be able to communicate in clear technical language to engineers and others involved with the project. Their observations of experiments may be presented orally or in writing to

engineers and others involved in the project. This requires adequate communicative skills combined with technical talents.

Many colleges, technical institutes and universities offer programs for technicians. Engine mechanics may become technicians if they have the desire to further their education through evening courses and service training programs.

SERVICE REPRESENTATIVE

Small gas engine manufacturers may train certain employees with broad service experience to become service representatives for the company. These representatives are required to work in close cooperation with service managers and mechanics in the field to catch and correct chronic service problems. In some cases, service representatives write and distribute service bulletins concerning these problems. They also meet with and report findings to company engineers involved in engine design.

ENGINEER

Manufacturers need engineers to design engines that will perform satisfactorily under certain environmental conditions. For example, a small gas engine designed for use in a garden tractor is quite different from one intended for use in a chain saw. Engineers must use their knowledge of scientific principles to design and create engines that will meet all specified operating requirements.

Engineers usually have college engineering degrees. Sometimes, however, an engineer's license can be obtained by passing special examinations. Engineers must have a strong background in science, mathematics and many specialized technical subjects. They must be analytical and creative with a practical knowledge of manufacturing processes and materials.

EXECUTIVE

Any of the careers outlined in this chapter could serve as stepping stones to high level management positions in the small gas engine field. Most successful executives in this field began their careers in engine production, design, sales or service. Almost invariably, the key to their success is learning to do the job at hand to the best of their ability.

USEFUL INFORMATION

GENERAL SMALL ENGINE TIPS

FUEL RECOMMENDATION

Premium high octane or high test gasolines are of no particular advantage in engines other than those with extremely high compression ratios. Premium gasolines have more tetraethyl lead and can cause depost buildup in engines with low compression ratios. Always be sure to follow the vehicle and/or engine manufacturer's recommendation for type and grade of fuel to be used.

COMPLAINT, CAUSE AND CORRECTIVE ACTION

TOO MUCH FUEL

Open choke and throttle completely. Crank engine over a couple of times. If it does not start, remove spark plug. With plug out, crank engine several times to clean out excess fuel. Reinstall cleaned plug and crank again.

NO FUEL

Check fuel lines for icing or obstruction. Clean fuel filter, carburetor filter and fuel lines.

NO IGNITION SPARK

Remove spark plug, replace boot and contact plug body with engine housing. If no spark at electrode while turning engine over, ignition device is defective. Check ignition system.

SPARK PLUG FOULING

Replace with new spark plug or clean plug carefully and reinstall. To prevent fouling, avoid extended periods of engine idling.
1. White or light gray color on insulator nose, badly eroded electrodes: Spark plug is too hot. Check heat range chart and try one heat range colder. If same results, repeat procedure. White color can also be caused by a lean carburetor setting or over-advanced ignition timing.
2. Black, damp, oily carbon film over electrode and insulator nose: Spark plug is too cold. Check heat range chart and try one heat range hotter. If same results, repeat procedure. (Other possible causes could be improper fuel-to-oil ratio or an overly rich carburetor setting.)
3. Brown or light tan color on insulator, electrodes not burned: You are using correct plug. Change plugs at recommended intervals, using same heat range.

DETONATION (SPONTANEOUS IGNITION OF UNBURNED PORTIONS OF AIR/FUEL CHARGE AFTER SPARK PLUG HAS FIRED)

Detonation shock and pressure waves can cause severe stress on engine parts and could cause damage to piston rings and ring lands, overheating of piston and scoring of cylinder walls. Causes are overadvanced timing, low grade of fuel, too lean an air-fuel mixture, and increased compression ratio due to a buildup of deposits in combustion chamber.
1. Deposits on combustion chamber surface and piston head should be cleaned away.

2. Carburetor set too lean (adjust). Start with rich mixture and lean out carburetor 1/16 turn at a time.
3. Using too cold a spark plug. Try one heat range hotter.
4. Use grade of fuel and octane rating recommended by the engine manufacturer and mix with proper amount, type and grade of lubricant.
5. Over-advanced ignition timing is the main cause of detonation. Have qualified person periodically check or adjust the timing approximately once a year.

EXCESSIVE ENGINE VIBRATION
1. See if engine mounts are secure.
2. Check drive clutch assembly for faulty clutch halves. pressure levers or centrifugal governor, and replace faulty parts.
3. Check magneto generator assembly.
4. Check crankshaft for trueness.
   Remember, all small engines will vibrate somewhat at low rpm, since most are balanced at high rpm.

CYLINDER HEAD NUTS/BOLTS

Always check whether cylinder head nuts are firmly tightened after operating engine for first time or after operating at full load.

IGNITION (MACHINE DOES NOT START OR RUNS UNEVENLY)
1. Check ignition to be sure it is in "on or run" position. Check wire leading to switch.
2. Inspect spark plug wire for worn insulation or ·breaks. Replace if needed.
3. Engine has been too long at idle speed (plug probably fouled). Replace with new plug or clean plug carefully and reinstall.
4. Carburetor could also be set too rich (adjust).

GAP SETTING

Check gap on new plugs to make sure they match specifications for vehicle. (Check manual.) For easier cold weather starting, maintain electrode gap to minimum recommended spacing. If engine misses at high speeds, as a result of lack of high voltage, problem may be relieved by setting plug gap slightly closer.

ENGINE LUBRICATION

In four cycle engines, crankcase acts as a storage tank for oil needed to lubricate internal engine moving parts. (Check manual for type and grade.) Crankcase of a two cycle engine is part of induction system, so oils specially prepared for two cycle, air-cooled engines are highly recommended. (They reduce exhaust port deposits, piston and cylinder wall scuffing, and spark plug fouling.) A standard grade non-detergent SAE 30 automotive oil may be substituted if two cycle, air-cooled engine oils are not available. Oils for outboards are not recommended unless specifically formulated for both water-cooled and air-cooled, two cycle engines. Low-priced SA or SB (light duty) oils or multiple viscosity 10W30 should be avoided.

(A. C. Spark Plug)

CHECK SHEET FOR DISASSEMBLY
4 CYCLE ENGINE RECONDITIONING

ENGINE: MAKE _____   MODEL NO. _____

HP _____   SERIAL NO. _____

_____ 1. Remove all gasoline from engine.

_____ 2. Inspect engine for broken or missing parts.

_____ 3. Record all important data on an information sheet.

_____ 4. Remove spark plug. Check gap and condition of electrodes.

_____ 5. Take compression reading. Record on information sheet.

_____ 6. Check ignition output with spark tester.

_____ 7. Drain oil from crankcase.

_____ 8. Disconnect all linkage from remote throttle assembly to engine.

_____ 9. Remove engine from its mountings.

_____ 10. Clean engine housing or mounting area.

_____ 11. Remove blower housing from engine.

_____ 12. Remove carburetor and carburetor linkage.

_____ 13. Remove governor air vane and governor linkage.

_____ 14. Remove muffler.

_____ 15. Remove valve chamber cover.

_____ 16. Remove cylinder head and head gasket.

_____ 17. Measure bore and stroke. Record measurements.

_____ 18. Check and record valve clearance.

_____ 19. Remove air deflector shields.

_____ 20. Check and record armature air gap.

_____ 21. Remove starter clutch and flywheel nut.

_____ 22. Remove flywheel.

_____ 23. Remove ignition breaker point cover.

_____ 24. Check and record ignition point gap setting.

_____ 25. Remove ignition points, condenser and ignition cam.

_____ 26. Remove magneto assembly.

_____ 27. Remove all rust and burrs from end of crankshaft.

_____ 28. Remove mounting flange, if any.

_____ 29. Check timing marks.

_____ 30. Remove camshaft and oil pump.

_____ 31. Remove valve tappets.

_____ 32. Remove piston and rod assembly. Mark rod first.

_____ 33. Remove piston rings from piston.

_____ 34. Check and record ring end gap.

_____ 35. Clean ring grooves in piston.

_____ 36. Check and record piston ring-to-land clearance.

_____ 37. Remove crankshaft and inspect it.

_____ 38. Remove intake and exhaust valves.

_____ 39. Wash and clean all parts that will not be damaged by solvent.

_____ 40. Inspect engine block for scores or imperfections.

_____ 41. Check all bearings and oil seals for possible damage.

_____ 42. Recondition, then reassemble engine.

CHECK SHEET FOR REASSEMBLY
4 CYCLE ENGINE RECONDITIONING

ENGINE: MAKE _____   MODEL NO. _____

HP_____   SERIAL NO. _____

_____ 1. Clean valves and valve seats with wire wheel and brush.
_____ 2. Have instructor check valve parts after they are cleaned.
_____ 3. Lap valves against valve seats, using lapping compound.
_____ 4. Install valve assemblies.
_____ 5. Install crankshaft.
_____ 6. Fit rings on piston in proper order.
_____ 7. Oil cylinder wall. Install piston and rod assembly in proper direction.
_____ 8. Torque rod bolts to specifications. Bend up retainer clips.
_____ 9. Install tappets, camshaft and oil pump.
_____ 10. Align timing marks on camshaft and crankshaft.
_____ 11. Bolt mounting flange (crankcase cover) onto engine. Check for proper fit.
_____ 12. Check and record valve clearance measurements.
_____ 13. Assemble valve cover and breather and bolt to engine block.
_____ 14. Install cylinder head and head gasket. Torque to specifications.
_____ 15. Reassemble and install ignition system. Set point gap to specifications.
_____ 16. Install ignition point cover.
_____ 17. Install flywheel key and flywheel. Torque flywheel nut.
_____ 18. Set armature air gap to specifications.
_____ 19. Fasten governor air vane to engine block.
_____ 20. Mount carburetor on engine block.
_____ 21. Install muffler.
_____ 22. Install blower housing.
_____ 23. Connect fuel lines and valve breather tube.
_____ 24. Mount engine in implement or equipment.
_____ 25. Connect engine to drive train.
_____ 26. Connect all linkage between remote throttle and engine.
_____ 27. Tighten oil plug and fill engine with proper oil to oil level.
_____ 28. Put oil in air cleaner to proper level, if oil bath type.
_____ 29. Check compression of engine.
_____ 30. Clean and set gap of spark plug electrodes. Install spark plug.
_____ 31. Check to be sure all components are tight and properly adjusted.
_____ 32. Fill fuel tank with clean regular gasoline.
_____ 33. Turn on exhaust fan (if indoors) and wear goggles.
_____ 34. Engage carburetor choke and start engine.
_____ 35. Adjust carburetor.

Keep a record on file of all reconditioning repairs, part replacements and engine identification information.

GENERAL TORQUE SPECIFICATIONS IN CONSIDERATION OF FASTENER QUALITY

THE FOLLOWING RULES APPLY TO THE CHART:

1. CONSULT MANUFACTURERS SPECIFIC RECOMMENDATIONS WHEN AVAILABLE.
2. THE CHART MAY BE USED DIRECTLY WHEN ANY OF THE FOLLOWING LUBRICANTS ARE USED:
   a. NEVER-SEEZ COMPOUND, MOLYKOTE, FEL-PRO C-5, GRAPHITE AND OIL OR SIMILAR MIXTURES.
3. INCREASE THE TORQUE BY 20% WHEN USING ENGINE OIL OR CHASSIS GREASE AS A LUBRICANT.
   (THESE LUBRICANTS ARE NOT GENERALLY RECOMMENDED FOR FASTENERS)
4. REDUCE TORQUE BY 20% WHEN PLATED BOLTS ARE USED.
5. INCREASE TORQUE BY 20% WHEN MULTIPLE TAPERED TOOTH LOCKWASHERS ARE USED.

	MOST USED	MUCH USED	USED AT TIMES	USED AT TIMES	RECOMMENDED FOR COMPETITION AND CRITICAL USE
CURRENT AUTOMOTIVE USAGE					
MIN. TENSILE STRENGTH	64,000 P.S.I.	105,000 P.S.I.	133,000 P.S.I.	150,000 P.S.I.	160,000 P.S.I.
MATERIAL	LOW CARBON STEEL	MEDIUM CARBON STEEL TEMPERED	MEDIUM CARBON STEEL QUENCHED/TEMP.	MED. CARBON ALLOY STEEL QUENCHED/TEMP.	SPECIAL ALLOY STEEL QUENCHED/TEMP.
DEFINITION	INDETERMINATE QUALITY	MINIMUM COMMERCIAL QUALITY	MEDIUM COMMERCIAL QUALITY	BEST COMMERCIAL QUALITY	BEST QUALITY SUPERTANIUM
GRADE MARKINGS	(hexagon)	(cube)	(hexagon +)	(hexagon *)	(hexagon)
BOLT SIZE	S.A.E. GRADE 1 OR 2	S.A.E. GRADE 5	S.A.E. GRADE 6	S.A.E. GRADE 8	EXCEEDS ALL S.A.E. GRADES
1/4	5	7	10	10.5	11
5/16	9	14	19	22	24
3/8	15	25	34	37	40
7/16	24	40	55	60	65
1/2	37	60	85	92	97
9/16	53	88	120	132	141
5/8	74	120	167	180	192
3/4	120	200	280	296	316
7/8	190	302	440	473	503
1	282	466	660	714	771

NOTE: THE TORQUE SPECIFICATIONS ARE GIVEN IN FOOT-POUNDS. INCH-POUND EQUIVALENT MAY BE OBTAINED BY MULTIPLYING BY 12.

NOTE: TO CONVERT FROM FOOT-POUNDS TO NEWTON-METRES, MULTIPLY FOOT-POUNDS (ACTUALLY POUND FEET) BY .7376.

(P. A. Sturtevant)

TORQUE WRENCH SELECTOR CHART

TORQUE APPLICATIONS ARE MUCH EASIER ACCOMPLISHED WHEN USING THE CORRECT TORQUE WRENCH FOR THE JOB.

THE CHART SHOWN BELOW INDICATES AT A GLANCE WHICH TORQUE WRENCH TO USE WHEN CONSIDERING BOLT DIAMETER AND TIGHTENING SPECIFICATIONS.

FOOT POUND SCALE	INCH POUND SCALE	TYPICAL BOLT SIZE	RECOMMENDED TORQUE WRENCH
5	60		
10	120	1/4"	
15	180	5/16"	
20	240		
25	300	3/8"	
30	350		
35	420		
40	480	7/16"	
45	540		
50	600		
55		1/2"	
60			
70			
80			
90		9/16"	
100			
110			
120		5/8"	
130			
140			
150			
160			
180			
200		3/4"	
220			
240			
260			
280			
300		7/8"	

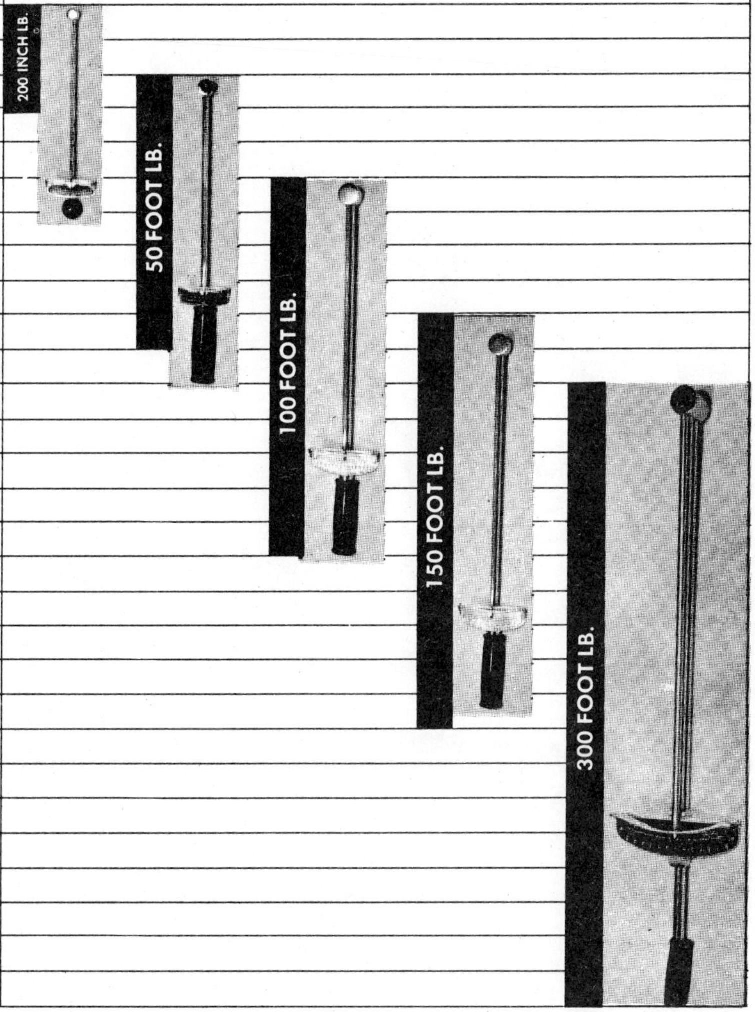

(Wrench scale labels shown in the chart: 200 INCH LB., 50 FOOT LB., 100 FOOT LB., 150 FOOT LB., 300 FOOT LB.)

BOLT SIZES SHOWN HEREIN ARE OF AVERAGE QUALITY AND CONFORM TO. THE THREE LINE TYPE SPECIFICALLY.

(P. A. Sturtevant)

The Metric System of Measurement

Measures of Length

1 Millimetre (mm) = ...0.03937079 inch, or about 1/25 inch
10 Millimetres = 1 Centimetre (cm) = ...0.3937079 inch
10 Centimetres = 1 Decimetre (dm) = ...3.3937079 inch
10 Decimetres = 1 Metre (m) = ...39.37079 inches, 3.2808992 feet, or 1.09361 yards
10 Metres = 1 Dekametre (dam) = ...32.808992 feet
10 Dekametres = 1 Hectometre (hm) = ...19.927817 rods
10 Hectometres = 1 Kilometre (km) = ...1093.61 yards, or 0.6213824 mile
10 Kilometres = 1 Myriametre (mym) = ...6.213824 miles

1 inch = 2.54 cm, 1 foot = 0.3048 m, 1 yard = 0.9144 m, 1 rod = 0.5029 m,
1 mile = 1.6093 km

Measures of Weight

1 Gram (g) = 15.4324874 gr. Troy, or 0.03215 oz. Troy, or .. .03527398 oz. avoirdupois
10 Grams = 1 Dekagram (dag) = ...0.3527398 oz. avoirdupois
10 Dekagrams = 1 Hectogram (hg) = ...3.527398 oz. avoirdupois
10 Hectograms = 1 Kilogram (kg) = ...2.20462125 lbs.
1000 Kilograms = 1 Ton (t) = 2204.62125 lbs., or 1.1023 tons of 2000 lbs., or 0.9842 ton
of 2240 lbs., or 19.68 cwts.

1 grain = 0.0648 g, 1 oz. avoirdupois = 28.35 g, 1 lb. = 0.4536 kg,
1 ton (2000 lbs.) = 0.9072 t, 1 ton (2240 lbs.) = 1.016 t, or 1016 kg

Measures of Capacity

1 Litre (L) = 1 cubic decimetre = 61.0270515 cubic in., or 0.03531 cu. ft., or 1.0567 liquid
qts., or 0.908 dry qt., or 0.26417 Amer. gal.
10 Litres = 1 Decalitre (dal) = 2.6417 gal., or 1.135 pk.
10 Decalitres = 1 Hectolitre (hl) = 2.8375 bu.
10 Hectolitres = 1 Kilolitre (kl) = 61027.0515 cu. in., or 28.375 bu.
1 cu. foot = 28.317 Litres, 1 gallon (American) = 3.785 Litres,
1 gallon (British) = 4.543 Litres

(L.S. Starrett)

Metric Conversion Table

Millimetres	×	.03937	=	Inches
Millimetres	=	25.400	×	Inches
Metres	×	3.2809	=	Feet
Metres	=	.3048	×	Feet
Kilometres	×	.621377	=	Miles
Kilometres	=	1.6093	×	Miles
Square centimetres	×	.15500	=	Square inches
Square centimetres	×	6.4515	=	Square inches
Square metres	×	10.76410	=	Square feet
Square metres	=	.09290	×	Square feet
Square kilometres	×	247.1098	=	Acres
Square kilometres	×	.00405	=	Acres
Hectares	×	2.471	=	Acres
Hectares	=	.4047	×	Acres
Cubic centimetres	=	.061025	×	Cubic inches
Cubic centimetres	=	16.3866	×	Cubic inches
Cubic metres	=	35.3156	×	Cubic feet
Cubic metres	=	.02832	×	Cubic feet
Cubic metres	×	1.308	=	Cubic yards
Cubic metres	×	.765	=	Cubic yards
Litres	×	61.023	=	Cubic inches
Litres	=	.01639	×	Cubic inches
Litres	×	.26418	=	U.S. gallons
Litres	=	3.7854	×	U.S. gallons
Grams	=	15.4324	×	Grains
Grams	=	.0648	×	Grains
Grams	×	.03527	=	Ounces, avoirdupois
Grams	×	28.3495	=	Ounces, avoirdupois
Kilograms	×	2.2046	=	Pounds
Kilograms	=	.4536	×	Pounds
Kilograms per square centimetre	×	14.2231	=	Pounds per square inch
Kilograms per square centimetre	=	.0703	×	Pounds per square inch
Newton-metres	×	1.3558	=	Pound feet
Newton-metres	×	.7376	=	Pound feet
Metric tons (1,000 kilograms)	×	1.1023	=	Tons (2,000 pounds)
Metric tons	×	.9072	=	Tons (2,000 pounds)
Kilowatts	×	1.3405	=	Horsepower
Kilowatts	×	.746	=	Horsepower
Calories	×	3.9683	=	B.T. units
Calories	=	.2520	×	B.T. units

Decimal Equivalents of Millimetres

mm	Inches	mm	Inches	mm	Inches	mm	Inches	mm	Inches
.01	.00039	.41	.01614	.81	.03189	21	.82677	61	2.40157
.02	.00079	.42	.01654	.82	.03228	22	.86614	62	2.44094
.03	.00118	.43	.01693	.83	.03268	23	.90551	63	2.48031
.04	.00157	.44	.01732	.84	.03307	24	.94488	64	2.51968
.05	.00197	.45	.01772	.85	.03346	25	.98425	65	2.55905
.06	.00236	.46	.01811	.86	.03386	26	1.02362	66	2.59842
.07	.00276	.47	.01850	.87	.03425	27	1.06299	67	2.63779
.08	.00315	.48	.01890	.88	.03465	28	1.10236	68	2.67716
.09	.00354	.49	.01929	.89	.03504	29	1.14173	69	2.71653
.10	.00394	.50	.01969	.90	.03543	30	1.18110	70	2.75590
.11	.00433	.51	.02008	.91	.03583	31	1.22047	71	2.79527
.12	.00472	.52	.02047	.92	.03622	32	1.25984	72	2.83464
.13	.00512	.53	.02087	.93	.03661	33	1.29921	73	2.87401
.14	.00551	.54	.02126	.94	.03701	34	1.33858	74	2.91338
.15	.00591	.55	.02165	.95	.03740	35	1.37795	75	2.95275
.16	.00630	.56	.02205	.96	.03780	36	1.41732	76	2.99212
.17	.00669	.57	.02244	.97	.03819	37	1.45669	77	3.03149
.18	.00709	.58	.02283	.98	.03858	38	1.49606	78	3.07086
.19	.00748	.59	.02323	.99	.03898	39	1.53543	79	3.11023
.20	.00787	.60	.02362	1.00	.03937	40	1.57480	80	3.14960
.21	.00827	.61	.02402	1	.03937	41	1.61417	81	3.18897
.22	.00866	.62	.02441	2	.07874	42	1.65354	82	3.22834
.23	.00906	.63	.02480	3	.11811	43	1.69291	83	3.26771
.24	.00945	.64	.02520	4	.15748	44	1.73228	84	3.30708
.25	.00984	.65	.02559	5	.19685	45	1.77165	85	3.34645
.26	.01024	.66	.02598	6	.23622	46	1.81102	86	3.38582
.27	.01063	.67	.02638	7	.27559	47	1.85039	87	3.42519
.28	.01102	.68	.02677	8	.31496	48	1.88976	88	3.46456
.29	.01142	.69	.02717	9	.35433	49	1.92913	89	3.50393
.30	.01181	.70	.02756	10	.39370	50	1.96850	90	3.54330
.31	.01220	.71	.02795	11	.43307	51	2.00787	91	3.58267
.32	.01260	.72	.02835	12	.47244	52	2.04724	92	3.62204
.33	.01299	.73	.02874	13	.51181	53	2.08661	93	3.66141
.34	.01339	.74	.02913	14	.55118	54	2.12598	94	3.70078
.35	.01378	.75	.02953	15	.59055	55	2.16535	95	3.74015
.36	.01417	.76	.02992	16	.62992	56	2.20472	96	3.77952
.37	.01457	.77	.03032	17	.66929	57	2.24409	97	3.81889
.38	.01496	.78	.03071	18	.70866	58	2.28346	98	3.85826
.39	.01535	.79	.03110	19	.74803	59	2.32283	99	3.89763
.40	.01575	.80	.03150	20	.78740	60	2.36220	100	3.93700

Decimal Equivalents of 8ths, 16ths, 32nds, 64ths

8ths

$\frac{1}{8}$	= .125
$\frac{1}{4}$	= .250
$\frac{3}{8}$	= .375
$\frac{1}{2}$	= .500
$\frac{5}{8}$	= .625
$\frac{3}{4}$	= .750
$\frac{7}{8}$	= .875

16ths

$\frac{1}{16}$	= .0625
$\frac{3}{16}$	= .1875
$\frac{5}{16}$	= .3125
$\frac{7}{16}$	= .4375
$\frac{9}{16}$	= .5625
$\frac{11}{16}$	= .6875
$\frac{13}{16}$	= .8125
$\frac{15}{16}$	= .9375

32nds

$\frac{1}{32}$	= .03125
$\frac{3}{32}$	= .09375
$\frac{5}{32}$	= .15625
$\frac{7}{32}$	= .21875
$\frac{9}{32}$	= .28125
$\frac{11}{32}$	= .34375
$\frac{13}{32}$	= .40625
$\frac{15}{32}$	= .46875
$\frac{17}{32}$	= .53125
$\frac{19}{32}$	= .59375
$\frac{21}{32}$	= .65625
$\frac{23}{32}$	= .71875
$\frac{25}{32}$	= .78125
$\frac{27}{32}$	= .84375
$\frac{29}{32}$	= .90625
$\frac{31}{32}$	= .96875

64ths

$\frac{1}{64}$	= .015625	$\frac{33}{64}$	= .515625
$\frac{3}{64}$	= .046875	$\frac{35}{64}$	= .546875
$\frac{5}{64}$	= .078125	$\frac{37}{64}$	= .578125
$\frac{7}{64}$	= .109375	$\frac{39}{64}$	= .609375
$\frac{9}{64}$	= .140625	$\frac{41}{64}$	= .640625
$\frac{11}{64}$	= .171875	$\frac{43}{64}$	= .671875
$\frac{13}{64}$	= .203125	$\frac{45}{64}$	= .703125
$\frac{15}{64}$	= .234375	$\frac{47}{64}$	= .734375
$\frac{17}{64}$	= .265625	$\frac{49}{64}$	= .765625
$\frac{19}{64}$	= .296875	$\frac{51}{64}$	= .796875
$\frac{21}{64}$	= .328125	$\frac{53}{64}$	= .828125
$\frac{23}{64}$	= .359375	$\frac{55}{64}$	= .859375
$\frac{25}{64}$	= .390625	$\frac{57}{64}$	= .890625
$\frac{27}{64}$	= .421875	$\frac{59}{64}$	= .921875
$\frac{29}{64}$	= .453125	$\frac{61}{64}$	= .953125
$\frac{31}{64}$	= .484375	$\frac{63}{64}$	= .984375

Rules Relative to the Circle

To Find Circumference—
Multiply diameter by 3.1416 . . . . Or divide diameter by 0.3183

To Find Diameter—
Multiply circumference by 0.3183 . . . . Or divide circumference by 3.1416

To Find Radius—
Multiply circumference by 0.15915 . . . . Or divide circumference by 6.28318

To Find Side of an Inscribed Square—
Multiply diameter by 0.7071
Or multiply circumference by 0.2251 . . . . Or divide circumference by 4.4428

To Find Side of an Equal Square—
Multiply diameter by 0.8862 . . . . Or divide diameter by 1.1284
Or multiply circumference by 0.2821 . . . . Or divide circumference by 3.545

Square—
A side multiplied by 1.4142 equals diameter of its circumscribing circle.
A side multiplied by 4.443 equals circumference of its circumscribing circle.
A side multiplied by 1.128 equals diameter of an equal circle.
A side multiplied by 3.547 equals circumference of an equal circle.
Square inches multiplied by 1.273 equals circle inches of an equal circle.

To Find the Area of a Circle—
Multiply circumference by one-quarter of the diameter.
Or multiply the square of diameter by 0.7854
Or multiply the square of circumference by .07958
Or multiply the square of ½ diameter by 3.1416

(L. S. Starrett)

Tap Drill Sizes

For Machine Screw Threads

75% Depth of Thread

A bolt inserted in an ordinary nut, which has only one-half of a full depth of thread, will break before stripping the thread. Also a full depth of thread, while very difficult to obtain, is only about 5% stronger than a 75% depth.

These tables give the exact size of the hole, expressed in decimals, that will produce a 75% depth of thread, and also the nearest regular stock drill to this size. Holes produced by these drills are considered close enough for any commercial tapping.

$$\text{Diameter of Tap, Minus } \frac{.974}{\text{No. threads per Inch}} = \text{Diameter of Hole}$$

Tap Size	Threads per Inch	Diameter Hole	Drill
0	80	.048	3/64
1	72	.060	53
1	64	.058	53
2	64	.071	50
2	56	.069	50
3	56	.082	45
3	48	.079	47
4	48	.092	42
4	40	.088	43
4	36	.085	44
5	44	.103	37
5	40	.101	38
5	36	.098	40
6	40	.114	33
6	36	.111	34
6	32	.108	36
7	36	.124	1/8
7	32	.121	31
7	30	.119	31
8	36	.137	29
8	32	.134	29
8	30	.132	30
9	32	.147	26
9	30	.145	27
9	24	.136	29
10	32	.160	21
10	30	.158	22
10	28	.155	23
10	24	.149	25
12	28	.181	14
12	24	.175	16
14	24	.201	7
14	20	.193	10
16	22	.224	2
16	20	.219	7/32
16	18	.214	3
18	20	.245	D
18	18	.240	B
20	20	.271	I
20	18	.266	17/64
22	18	.292	L
22	16	.285	9/32
24	18	.318	O
24	16	.311	5/16
26	16	.337	R
26	14	.328	21/64
28	16	.363	23/64
28	14	.354	T
30	16	.389	25/64
30	14	.380	V

Tap Drill Sizes *For Fractional Size Threads*

75% Depth of Thread

AMERICAN NATIONAL THREAD FORM

Tap Size	Threads per Inch	Diam. Hole	Drill
1/16	72	.049	3/64
1/16	64	.047	3/64
1/16	60	.046	56
5/64	72	.065	52
5/64	64	.063	1/16
5/64	60	.062	1/16
5/64	56	.061	53
3/32	60	.077	5/64
3/32	56	.076	48
3/32	50	.074	49
3/32	48	.073	49
7/64	56	.092	42
7/64	50	.090	43
7/64	48	.089	43
1/8	48	.105	36
1/8	40	.101	38
1/8	36	.098	40
1/8	32	.095	3/32
9/64	40	.116	32
9/64	36	.114	33
9/64	32	.110	35
5/32	40	.132	30
5/32	36	.129	30
5/32	32	.126	1/8
11/64	36	.145	27
11/64	32	.141	9/64
3/16	36	.161	20
3/16	32	.157	22
3/16	30	.155	23
3/16	24	.147	26
13/64	32	.173	17
13/64	30	.171	11/64
13/64	24	.163	20
7/32	32	.188	12
7/32	28	.184	13
7/32	24	.178	16
15/64	32	.204	6
15/64	28	.200	8
15/64	24	.194	10
1/4	32	.220	7/32
1/4	28	.215	3
1/4	27	.214	3
1/4	24	.209	4
1/4	20	.201	7
5/16	32	.282	9/32
5/16	27	.276	J
5/16	24	.272	I
5/16	20	.264	17/64
5/16	18	.258	F
3/8	27	.339	R
3/8	24	.334	Q
3/8	20	.326	21/64
3/8	16	.314	5/16
7/16	27	.401	Y
7/16	24	.397	X
7/16	20	.389	25/64
7/16	14	.368	U
1/2	27	.464	15/32
1/2	24	.460	29/64
1/2	20	.451	29/64
1/2	13	.425	27/64
1/2	12	.419	27/64
9/16	27	.526	17/32
9/16	18	.508	33/64
9/16	12	.481	31/64
5/8	27	.589	19/32
5/8	18	.571	37/64
5/8	12	.544	35/64
5/8	11	.536	17/32
11/16	16	.627	5/8
11/16	11	.599	19/32
3/4	16	.689	11/16
3/4	12	.669	43/64
3/4	10	.653	21/32
13/16	12	.731	47/64
13/16	10	.715	23/32
7/8	27	.839	27/32
7/8	18	.821	53/64
7/8	14	.805	13/16
7/8	12	.794	51/64
7/8	9	.767	49/64
15/16	12	.856	55/64
15/16	9	.829	53/64
1	14	.930	15/16
1	27	.964	31/32
1	12	.919	59/64
1	8	.878	7/8
1 1/16	8	.941	15/16
1 1/8	12	1.044	1 3/64
1 1/8	7	.986	63/64
1 3/16	7	1.048	1 3/64
1 1/4	12	1.169	1 11/64
1 1/4	7	1.111	1 7/64
1 5/16	7	1.173	1 11/64
1 3/8	12	1.294	1 19/64
1 3/8	6	1.213	1 7/32
1 1/2	12	1.419	1 27/64
1 1/2	6	1.338	1 11/32
1 5/8	5 1/2	1.448	1 29/64
1 3/4	5	1.555	1 9/16
1 7/8	5	1.680	1 11/16
2	4 1/2	1.783	1 25/32
2 1/8	4 1/2	1.909	1 29/32
2 1/4	4 1/2	2.034	2 1/32
2 3/8	4	2.131	2 1/8
2 1/2	4	2.256	2 1/4
2 5/8	4	2.381	2 3/8
2 3/4	4	2.506	2 1/2
2 7/8	3 1/2	2.597	2 19/32
3	3 1/2	2.722	2 23/32
3 1/8	3 1/2	2.847	2 27/32
3 1/4	3 1/2	2.972	2 31/32
3 3/8	3 1/4	3.075	3 1/16
3 1/2	3 1/4	3.200	3 3/16
3 5/8	3 1/4	3.325	3 5/16
3 3/4	3	3.425	3 7/16
4	3	3.675	3 11/16

(L. S. Starrett)

No.	Size of Drill in Inches	No.	Size of Drill in Inches	No.	Size of Drill in Inches	No.	Size of Drill in Inches
1	.2280	21	.1590	41	.0960	61	.0390
2	.2210	22	.1570	42	.0935	62	.0380
3	.2130	23	.1540	43	.0890	63	.0370
4	.2090	24	.1520	44	.0860	64	.0360
5	.2055	25	.1495	45	.0820	65	.0350
6	.2040	26	.1470	46	.0810	66	.0330
7	.2010	27	.1440	47	.0785	67	.0320
8	.1990	28	.1405	48	.0760	68	.0310
9	.1960	29	.1360	49	.0730	69	.0292
10	.1935	30	.1285	50	.0700	70	.0280
11	.1910	31	.1200	51	.0670	71	.0260
12	.1890	32	.1160	52	.0635	72	.0250
13	.1850	33	.1130	53	.0595	73	.0240
14	.1820	34	.1110	54	.0550	74	.0225
15	.1800	35	.1100	55	.0520	75	.0210
16	.1770	36	.1065	56	.0465	76	.0200
17	.1730	37	.1040	57	.0430	77	.0180
18	.1695	38	.1015	58	.0420	78	.0160
19	.1660	39	.0995	59	.0410	79	.0145
20	.1610	40	.0980	60	.0400	80	.0135

Decimal Equivalents of Letter Size Drills

Letter	Size of Drill in Inches	Letter	Size of Drill in Inches
A	.234	N	.302
B	.238	O	.316
C	.242	P	.323
D	.246	Q	.332
E	.250	R	.339
F	.257	S	.348
G	.261	T	.358
H	.266	U	.368
I	.272	V	.377
J	.277	W	.386
K	.281	X	.397
L	.290	Y	.404
M	.295	Z	.413

Suggested Cutting Speeds for Drills

A feed per revolution of .004 to .007 for drills ¼ inch and smaller, and from .007 to .015 for larger drills is about all that should be required.

This feed is based on a peripheral speed of a drill equal to:
30 feet per minute for steel; 35 feet per minute for iron; 60 feet per minute for brass.

It may also be found advisable to vary the speed somewhat as the material to be drilled is more or less refractory.

We believe that these speeds should not be exceeded under ordinary circumstances.

Drill Diam., Inches	FEET PER MINUTE										
	15	20	25	30	35	40	45	50	60	70	80
	REVOLUTIONS PER MINUTE										
1/16	917.	1223.	1528.	1834.	2140.	2445.	2751.	3057.	3668.	4280.	4891.
1/8	459.	611.	764.	917.	1070.	1222.	1375.	1528.	1834.	2139.	2445.
3/16	306.	408.	509.	611.	713.	815.	917.	1019.	1222.	1426.	1630.
1/4	229.	306.	382.	458.	535.	611.	688.	764.	917.	1070.	1222.
5/16	183.	245.	306.	367.	428.	489.	550.	611.	733.	856.	978.
3/8	153.	204.	255.	306.	357.	408.	458.	509.	611.	713.	815.
7/16	131.	175.	218.	262.	306.	349.	393.	437.	524.	611.	699.
1/2	115.	153	191.	229.	268.	306.	344.	382.	459.	535.	611.
5/8	91.8	123.	153.	184.	214.	245.	276.	306.	367.	428.	489.
3/4	76.3	102.	127.	153.	178.	203.	229.	254.	306.	357.	408.
7/8	65.5	87.3	109.	131.	153.	175.	196.	219.	262.	306.	349.
1	57.3	76.4	95.5	115.	134.	153.	172.	191.	229.	267.	306.
1 1/8	51.0	68.0	85.0	102.	119.	136.	153.	170.	204.	238.	272.
1 1/4	45.8	61.2	76.3	91.8	107.	123.	137.	153.	183.	214.	245.
1 3/8	41.7	55.6	69.5	83.3	97.2	111.	125.	139.	167.	195.	222.
1 1/2	38.2	50.8	63.7	76.3	89.2	102.	115.	127.	153.	178.	204.
1 5/8	35.0	47.0	58.8	70.5	82.2	93.9	106.	117.	141.	165.	188.
1 3/4	32.7	43.6	54.5	65.5	76.4	87.3	98.2	109.	131.	153.	175.
1 7/8	30.6	40.7	50.9	61.1	71.3	81.5	91.9	102.	122.	143.	163.
2	28.7	38.2	47.8	57.3	66.9	76.4	86.0	95.5	115.	134.	153.
2 1/4	25.4	34.0	42.4	51.0	59.4	68.0	76.2	85.0	102.	119.	136.
2 1/2	22.9	30.6	38.2	45.8	53.5	61.2	68.8	76.3	91.7	107.	122.
2 3/4	20.8	27.8	34.7	41.7	48.6	55.6	62.5	69.5	83.4	97.2	111.
3	19.1	25.5	31.8	38.2	44.6	51.0	57.3	63.7	76.4	89.1	102.

(L. S. Starrett)

SAFETY HINTS

SNOWMOBILES

1. Wear protective head gear and goggles, and the proper winter clothing.

2. Before starting engine check your spring loaded throttle to be sure it moves freely and returns to the correct position.

3. Check your lights before any night running.

4. Never add fuel while the machine is running.

5. Always use the "buddy system" and take a reserve supply of fuel, on cross country trips.

6. Never leave unattended an idling machine.

7. Young children should ride in front of the driver so they can be seen at all times.

8. Use only a "stiff hitch tow bar" if pulling a sled, skier or toboggan.

9. Stop, look and listen when crossing roadways and railroad tracks.

10. Always carry a tool kit and extra Spark Plugs.

MOTORCYCLES, MOTOR BIKES AND ALL-TERRAIN VEHICLES

A. Wear clothing of heavy material with long sleeves and pants, heavy boots, gloves, and protective head gear and goggles.

B. Check to make sure front and rear brakes are in good operating condition at all times.

C. Use rear brake when stopping on loose surfaces for better control.

D. Allow a greater distance in which to stop when carrying a passenger.

E. Do not drag feet on road surface when turning.

F. Shift to a lower gear when stopping.

G. When following a vehicle, ride in a position best to be seen by the other vehicle operator.

H. Do not use high speed highways unless you can keep up with the flow of traffic.

I. Always carry a tool kit.

(A. C. Spark Plug)

HELI-COIL THREAD REPAIR KITS

INSERT INSTALLATION DATA — TOOLING DATA

	THREAD SIZE	KIT PART NO.	INSERT QTY. EA. STYLE & LENGTH	INSERT SIZE DESIG-NATION	INSERT LENGTH "L" DIAMETERS 1½	2	DRILLING AND TAPPING* DRILL SIZE	HOLE LIMITS MIN.	HOLE LIMITS MAX.	DRILL DEPTH ALLOWANCE** PLUG TAP	BOTT. TAP	TAP PART NO.***	INSERT TOOL PART NO.	TANG BREAK-OFF TOOL PART NO.	EXTRACTING TOOL PART NO.
UNIFIED COARSE STANDARD (FREE-RUNNING) AND SCREW-LOCK (SPECIFY 4131 FOR UNIFIED COARSE KITS; SPECIFY 1185 FOR STANDARD (FREE-RUNNING) AND 3585 FOR SCREW-LOCK)	2 (.086)-56	4131-02-1	15	-02CN	.129	.172	3/32	.0899	.0961	.150	.070	02CPA	551-02	4156-02	1227-02
	3 (.099)-48	-031-1	15	-031CN	.148	.198	#37	.1036	.1104	.175	.083	031CPA	551-031	4156-03	1227-06
	4 (.112)-40	-04-1	15	-04CN	.168	.224	#31	.1175	.1252	.206	.100	04CPB	7551-04	4156-04	1227-06
	5 (.125)-40	-05-1	15	-05CN	.188	.250	#29	.1305	.1373	.212	.100	05CPB	7551-05	4156-05	1227-06
	6 (.138)-32	-06-1	10	-06CN	.207	.276	#25	.1448	.1527	.256	.125	06CPB	7551-06	4156-06	1227-06
	8 (.164)-32	-2-1	10	-2CN	.246	.328	#17	.1708	.1781	.270	.125	2CPB	7551-2	4156-2	1227-06
	10 (.190)-24	-3-1	10	-3CN	.285	.380	13/64	.1990	.2080	.345	.167	3CPB	7551-3	4156-3	1227-6
	12 (.216)-24	-1-1	10	-1CN	.324	.432	#2	.2250	.2340	.358	.167	1CPB	7551-1	4156-1	1227-6
	1/4 (.2500)-20	-4-1	10	-4CN	.375	.500	17/64	.2608	.2704	.425	.200	4CPB	7551-4	4156-4	1227-6
	5/16 (.3125)-18	-5-1	8	-5CN	.469	.625	Q	.3245	.3342	.488	.222	5CPB	7551-5	4156-5	1227-6
	3/8 (.3750)-16	-6-1	8	-6CN	.562	.750	X	.3885	.3987	.375	.250	6CPB	7551-6	4156-6	1227-6
	7/16 (.4375)-14	-7-1	6	-7CN	.656	.875	29/64	.4530	.4639	.430	.286	7CPB	7551-7	4156-7	1227-16
	1/2 (.5000)-13	-8-1	6	-8CN	.750	1.000	33/64	.5166	.5273	.462	.308	8CPB	7551-8	4156-8	1227-16
	9/16 (.5625)-12	-9-1	5	-9CN	.844	1.125	37/64	.5806	.5918	.500	.333	187-9	3724-9		1227-16
	5/8 (.6250)-11	-10-1	5	-10CN	.938	1.250	21/32	.6447	.6564	.545	.363	8187-10	3724-10	FOR	1227-16
	3/4 (.7500)-10	-12-1	4	-12CN	1.125	1.500	25/32	.7716	.7838	.600	.400	8187-12	3724-12	SIZES	1227-16
	7/8 (.8750)-9	-14-1	4	-14CN	1.312	1.750	29/32	.8990	.9119	.667	.445	8187-14	3724-14	9/16 &	1227-16
	1 (1.0000)-8	-16-1	4	-16CN	1.500	2.000	1 1/32	1.0271	1.0421	.750	.500	8187-16	3724-16	OVER	1227-16
	1 1/8 (1.1250)-7	-18-1	3	-18CN	1.688	2.250	1 3/16	1.1559	1.1730	.857	.570	8187-18	3724-18	USE	1227-24
	1 1/4 (1.2500)-7	-20-1	3	-20CN	1.875	2.500	1 5/16	1.2809	1.2980	.857	.570	8187-20	3724-20	LONG	1227-24
	1 3/8 (1.3750)-6	-22-1	3	-22CN	2.062	2.750	1 7/16	1.4110	1.4310	1.000	.667	8187-22	3724-22	NOSE	1227-24
	1 1/2 (1.5000)-6	-24-1	3	-24CN	2.250	3.000	1 9/16	1.5360	1.5560	1.000	.667	8187-24	3724-24	PLIERS	1227-24
UNIFIED FINE STANDARD (FREE-RUNNING) AND SCREW-LOCK (SPECIFY 4132 FOR UNIFIED FINE KITS; SPECIFY 1191 FOR STANDARD (FREE RUNNING) AND 3591 FOR SCREW-LOCK)	4 (.112)-48	4132-041-1	15	-041CN	.168	.224	#31	.1166	.1229	.180	.083	041FPA	7552-041	4156-04	1227-06
	6 (.138)-40	-06-1	10	-06CN	.207	.276	#26	.1435	.1503	.219	.100	06FPB	7552-06	4156-06	1227-06
	8 (.164)-36	-2-1	10	-2CN	.246	.328	#17	.1701	.1771	.250	.111	2FPB	7552-2	4156-2	1227-06
	10 (.190)-32	-3-1	10	-3CN	.285	.380	#7	.1968	.2041	.282	.125	3FPB	7552-3	4156-3	1227-6
	1/4 (.2500)-28	-4-1	10	-4CN	.375	.500	G	.2577	.2646	.339	.143	4FPB	7552-4	4156-4	1227-6
	5/16 (.3125)-24	-5-1	8	-5CN	.469	.625	21/64	.3215	.3288	.406	.167	5FPB	7552-5	4156-5	1227-6
	3/8 (.3750)-24	-6-1	8	-6CN	.562	.750	25/64	.3840	.3910	.250	.167	6FPB	7552-6	4156-6	1227-6
	7/16 (.4375)-20	-7-1	6	-7CN	.656	.875	29/64	.4483	.4561	.300	.200	7FPB	7552-7	4156-7	1227-16
	1/2 (.5000)-20	-8-1	6	-8CN	.750	1.000	33/64	.5108	.5186	.300	.200	8FPB	7552-8	4156-8	1227-16
	9/16 (.5625)-18	-9-1	5	-9CN	.844	1.125	37/64	.5745	.5826	.333	.222	38193-9	535-9		1227-16
	5/8 (.6250)-18	-10-1	5	-10CN	.938	1.250	41/64	.6370	.6451	.333	.222	8193-10	535-10	FOR	1227-16
	3/4 (.7500)-16	-12-1	4	-12CN	1.125	1.500	49/64	.7635	.7720	.375	.250	8193-12	535-12	SIZES	1227-16
	7/8 (.8750)-14	-14-1	4	-14CN	1.312	1.750	57/64	.8905	.8994	.430	.286	8193-14	535-14	9/16 &	1227-16
	1 (1.0000)-14	-16-1	4	-16CN	1.500	2.000	1 1/32	1.0181	1.0281	.430	.286	8193-16	535-16	OVER	1227-16
	1 (1.0000)-12	-161-1	4	-161CN	1.500	2.000	1 1/32	1.0155	1.0243	.500	.333	8193-161	535-161	USE	1227-16
	1 1/8 (1.1250)-12	-18-1	3	-18CN	1.688	2.250	1 5/32	1.1431	1.1531	.500	.333	8193-18	535-18	LONG	1227-24
	1 1/4 (1.2500)-12	-20-1	3	-20CN	1.875	2.500	1 9/32	1.2681	1.2781	.500	.333	8193-20	535-20	NOSE	1227-24
	1 3/8 (1.3750)-12	-22-1	3	-22CN	2.062	2.750	1 13/32	1.3931	1.4031	.500	.333	8193-22	535-22	PLIERS	1227-24
	1 1/2 (1.5000)-12	-24-1	3	-24CN	2.250	3.000	1 17/32	1.5181	1.5281	.500	.337	8193-24	535-24		1227-24
COARSE STUD LOCK (SPECIFY 3758 FOR STUD LOCK INSERTS)	10 (.190)-24	4151-3-1	25	-3CN	.285	.380	13/64	.1990	.2080	.345	.167	3CBB	5551-3	4156-3	1227-6
	1/4 (.2500)-20	-4-1	25	-4CN	.375	.500	17/64	.2608	.2704	.425	.200	4CBB	5551-4	4156-4	1227-6
	5/16 (.3125)-18	-5-1	25	-5CN	.469	.625	Q	.3245	.3342	.488	.222	5CBB	5551-5	4156-5	1227-6
	3/8 (.3750)-16	-6-1	20	-6CN	.562	.750	X	.3885	.3987	.375	.250	6CBB	5551-6	4156-6	1227-6
	7/16 (.4375)-14	-7-1	20	-7CN	.656	.875	29/64	.4530	.4639	.430	.286	7CBB	5551-7	4156-7	1227-16
	1/2 (.5000)-13	-8-1	15	-8CN	.750	1.000	33/64	.5166	.5273	.462	.308	8CBB	5551-8	4156-8	1227-16

*Full tapped thread depth equals insert length "L"
**Drill depth min. equals insert length "L" plus drill depth allowance
***Kits 1/4" & smaller contain 2 taps

Complete insert part number — TYPE SIZE LENGTH
example - 10-24 x 1-1/2 dia. std. 1185 - 3CN x .285

1½ DIA. LENGTH STANDARD INSERTS ONLY
2 DIA. LENGTH NOT FURNISHED IN KITS

CONDITION	POSSIBLE CAUSE	CORRECTION
1. Carbon Fouling	Overrich carburetion. Clogged air cleaner can restrict air flow to carburetor causing rich mixture. Poor ignition output can reduce voltage and cause misfiring. Excessive idling, slow speeds under light load also can keep spark plug temperatures so low that normal combustion deposits are not burned off.	After cause has been eliminated, spark plugs may be cleaned, re-gapped, and reinstalled. If carbon deposits are not burned off, a hotter spark plug will better resist carbon deposits.
2. Oil Fouling.	Oil fouling may be caused by oil pumping past worn rings. "Break-in" of a new or recently overhauled engine before rings are fully seated may also result in this condition.	Usually, oil fouled spark plugs can be degreased, cleaned, and reinstalled. While hotter type spark plugs will reduce oil deposits, an engine overhaul may be necessary in severe causes to obtain satisfactory service.
3. Deposit Fouling "A"	Deposits which accumulate on insulator are by-products of combustion and come from fuel and lubricating oil, both of which today generally contain additives. Most powdery deposits have no adverse effect on spark plug operation; however, they may cause intermittent missing under severe operating conditions, especially at high speeds and heavy load.	If insulator is not too heavily coated, spark plugs may be cleaned, re-gapped, and reinstalled.
4. Deposit Fouling "B"	Deposits are similar to those identified as "Deposit Fouling-A." These deposits are by-products of combustion and come from fuel and lubricating oil. Excessive valve stem clearances will allow excessive lube oil to enter combustion chamber with fuel. Deposits will accumulate on spark plug and will be heaviest on side facing intake valve. Defective seals should be suspected when condition is found in only one or two cylinders on multi-cylinder engines.	After cause has been eliminated, spark plugs may be cleaned, re-gapped, and reinstalled.
5. Deposit Fouling "C"	Most powdery deposits, as in "A," have no adverse effect on operation of spark plug as long as they remain in powdery state. However, under certain conditions, these deposits melt and form a shiny glaze coating on insulator which, when hot, acts as a good electrical conductor. This allows current to follow deposits instead of jumping gap, thus shorting out spark plug.	Glazed deposits can be avoided by not applying sudden load, such as wide open throttle acceleration, after sustained periods of low speed and idle operation. Glazed spark plugs must be replaced.
6. Detonation	Overadvanced ignition timing, or use of low octane fuel will result in detonation, commonly referred to as engine knock or "ping." This causes severe shock inside combustion chamber, resulting in damage to adjacent parts which include spark plugs. A common result of prolonged detonation is to have sidewire of a spark plug torn off.	After cause has been eliminated, replace damaged plug with new spark plug.
7. Preignition	Preignition indicates excessive overheating. Cooling system stoppage, sticking valves, or overlean air-fuel mixtures are common causes of preignition. Spark plugs which are wrong (too hot) heat range, or not properly installed are also a possible cause. Sustained high speed, heavy load service can produce high temperatures which will cause preignition.	After cause has been eliminated, replace damaged plugs with new spark plugs. For sustained high speeds or heavy loads, install colder spark plugs.
8. Heat Shock Failure	Overadvanced ignition timing and low grade fuel are usually responsible for heat shock failures. Rapid increase in tip temperature under severe operating conditions causes heat shock and a fracture results. Another common cause of chipped or broken insulator tips is carelessness in regapping by either bending centerwire to adjust gap, or allowing gapping tool to exert pressure against tip of center electrode or insulator when bending side electrode to adjust gap.	After cause has been eliminated, replace damaged plugs with new spark plugs.

CONDITION	POSSIBLE CAUSE	CORRECTION
9. Insufficient Installation Torque	Poor contact between spark plug and entire seat. Lack of proper heat transfer causes excessive overheating of spark plug and, in many cases, severe damage. Dirty threads in an engine head can also result in plug seizing before it is seated.	Replace damaged spark plugs with new spark plugs. Observe torque specifications.
10. Proper Gasket Compression (For Applications Requiring a Gasket Seat Spark Plug)	This gasket has conformed to plug and engine seat to allow maximum heat transfer from plug to engine and to form a gas tight seal. (A properly compressed 14MM gasket will be .045'' to .050'' thick. A new gasket is .070'' to .080''.)	No correction required.
11. No Gasket Compression (For Applications Requiring a Gasket Seat Spark Plug)	Little or no force has been applied to gasket. In actual installation, it is necessary to almost completely compress gasket to allow proper heat transfer to engine block.	Observe proper torque specifications.
12. Thread Seizure	Scaled and deposit filled threads have not allowed proper gasket compression (or seating). CAUTION: Proper installation torque may be attained without compressing engine seat gasket for applications requiring a gasket seat spark plug. For taper seat applications, proper installation torque may be obtained without taper seat engagement in cylinder head. Operation of an engine with this type of installation can result in plug overheating and destruction of either spark plug or engine, or both.	Insure cylinder head and spark plug threads are free of deposits, burrs, and scales. Replace damaged plugs with new spark plugs.
13. Manganese Deposits or Rust Colored Tip Deposits	Manganese additive in some fuels. These deposits of manganese dioxide (MnO_2) do not adversely affect operation of spark plugs.	Spark plugs may be cleaned, re-gapped, and reinstalled.
14. Improper Cleaning	Overblasting and not rocking spark plugs during cleaning results in dirty, damaged insulator tips. Eroded tips change heat range and conductive paths mean fouled spark plugs, both resulting in poor engine operation.	Replace improperly cleaned spark plugs with new spark plugs.
15. Proper Cleaning		Follow recommended cleaning procedure of short abrasive blasts accompanied by rocking of spark plug in adapter, results in a properly cleaned spark plug ready to give many additional miles of service.
16. Normal Operation	Brown to grayish-tan deposits and slight electrode wear indicate correct spark plug heat range and mixed periods of high and low speed.	Spark plug may be cleaned, re-gapped, and reinstalled. When reinstalling spark plugs that have been cleaned and regapped, be sure to use a new engine seat gasket in each case where gasket is required.

DICTIONARY OF TERMS

ABRASION: Wearing or rubbing away.

ADDITIVE: As used with reference to automotive oils, a material added to the oil to give it certain properties. Example: a material added to engine oil to lessen its tendency to congeal or thicken at low temperature.

AEA: Automotive Electric Association.

AIR: A gas containing approximately 4/5 nitrogen, 1/5 oxygen and some carbonic gas.

AIR BLEED: A tube in a carburetor through which air can pass into fuel moving through a fuel passage.

AIR CLEANER: A device for filtering, cleaning, and removing dust from the air admitted to an engine.

AIR-COOLED ENGINE: An engine cooled by air.

AIR-FUEL RATIO: Ratio, by weight, of the fuel as compared to air in carburetor mixture.

AIR GAP: The space between the spark plug electrodes.

AIR HORN: Part of air passage in carburetor which is on the atmospheric side of the venturi. The choke valve is located in the air horn.

AIR LOCK: A bubble of air trapped in a fluid circuit which interferes with normal circulation of the fluid.

ALIGNMENT: An adjustment to bring related components into a line.

ALLEN WRENCH: A hexagonal (six-sided) wrench which fits into a recessed hexagonal hole. Used commonly in set screws.

ALLOY: A mixture of different metals such as solder, an alloy of lead and tin.

ALTERNATING CURRENT: An electrical current alternating flow back and forth in a circuit.

ALTERNATOR: A generator which produces alternating current.

ALUMINUM: A metal noted for its lightness often alloyed with small quantities of other metals.

AMMETER: An instrument for measuring the flow of electric current.

AMPERE: Unit of measurement for flow of electric current.

AMPERE-HOUR CAPACITY: A term to indicate the capacity of a storage battery. Example: the delivery of a certain number of amperes for a certain number of hours.

ANTIFREEZE: A material such as ethylene glycol added to water to lower its freezing point.

ANTIFRICTION BEARING: A bearing constructed with balls or rollers between journal and bearing surface to provide rolling instead of sliding friction.

APEX SEALS: The spring loaded seals located at the apexes (points or tips) of the triangular rotor in the Wankel engine.

API: American Petroleum Institute.

ARC: A discharge of electric current across a gap as between electrodes.

ARMATURE: Part of an electrical motor or generator which includes the main current-carrying winding.

ASME: American Society of Mechanical Engineers.

ATMOSPHERIC PRESSURE: The weight of the air at sea level (about 14.7 psi; less at higher altitudes).

BABBITT: An alloy of tin, copper and antimony having good antifriction properties. Used as a facing for bearings.

BACKFIRE: Ignition of the mixture in the intake manifold by flame from the cylinder such as might occur from a leaking or open intake valve.

BACKLASH: The clearance or "play" between two parts, such as meshed gears.

BACK PRESSURE: A resistance to free flow, such as a restriction in the exhaust line.

BAFFLE OR BAFFLE PLATE: An obstruction for checking or deflecting the flow of gases or sound.

BALL BEARING: An anti-friction bearing consisting of a hardened inner and outer race with hardened steel balls set between the two races.

BATTERY: Any number of complete electrical cells assembled in one housing or case.

BCI: Battery Council International.

BDC: Bottom dead center.

BEARING: A part in which a journal, pivot or the like turns or moves.

BEVEL: The angle that one surface makes with another when they are not at right angles.

BHP: Brake horsepower. A measurement of power developed by an engine in actual operation.

BLOW-BY: A leakage or loss of pressure, often used with reference to leakage of compression past piston ring between piston and cylinder.

BOILING POINT: The temperature at atmospheric pressure at which bubbles or vapors rise to the surface of a liquid and escape.

BORE: The diameter of a hole, such as a cylinder; also to enlarge a hole as distinguished from making a hole with a drill.

BOSS: An extension or strengthened section, such as projections in a piston which support piston pin or piston pin bushings.

BOUNCE: Applied to engine valves, a condition where the valve is not held tightly to its seat when cam is not lifting it. In ignition distributor, a condition where breaker points make and break contact when they should remain closed.

BREAKER ARM: Movable part of a pair of contact points in an ignition distributor or magneto.

BREAKER POINTS: Two separable points usually faced with silver, platinum or tungsten which interrupt primary circuit in distributor or magneto for purpose of inducing a high tension current in ignition systems.

BREAK-IN: The process of wearing into a desirable fit between surfaces of two new or reconditioned parts.

BRUSHES: The bars of carbon or other conducting material which contact the commutator of an electric motor or generator.

BTU (British Thermal Unit): A measurement of the amount of heat required to raise the temperature of 1 lb. of water, 1 deg. F.

BUSHING: A removable liner for a bearing.

BYPASS: An alternate path for a flowing substance.

CALIBRATE: To determine or adjust the graduation or scale of any instrument giving quantitative measurements.

CAM OR BREAKER CAM: The lobed cam rotating in the ignition system which interrupts the primary circuit to induce a high tension spark for ignition.

CAM ANGLE: Applied to an ignition distributor or magneto; number of degrees of rotation of distributor shaft during which contact points are closed.

CAM GROUND PISTON: A piston ground to a slightly oval shape which, under the heat of operation, becomes round.

CAMSHAFT: Shaft containing lobes or cams which operate engine valves.

CAPACITANCE: The property which opposes any change in voltage in an electrical circuit.

CAPACITOR: A device which possesses capacitance (stores electricity). In simple state consists of two metal plates separated by an insulator.

CARBON: A common nonmetallic element which is an excellent conductor of electricity. It also forms in combustion chamber of an engine during burning of fuel and lubricating oil.

CARBON MONOXIDE: Gas formed by incomplete combustion. Colorless, odorless, very poisonous.

CARBURETOR: A device for automatically mixing fuel in proper proportion with air to produce a combustible gas.

CARBURETOR "ICING": A term used to describe formation of ice on a carburetor throttle plate during certain atmospheric conditions.

CASE HARDEN: To harden the surface of steel.

CELL: Part of a battery containing a group of positive and negative plates along with electrolyte.

CELL CONNECTOR: Lead bar or link connecting the pole of one battery cell to the pole of another.

CELSIUS: A measurement of temperature. Zero on Celsius scale is 32 deg. on Fahrenheit scale.

CENTRIFUGAL FORCE: A force which tends to move a body away from its center of rotation. Example: a whirling weight attached to a string.

CHAMFER: A bevel or taper at the edge of a hole.

CHARGE (or Recharge): Passing an electrical current through a battery to restore it to activity.

CHECK VALVE: A gate or valve which allows passage of gas or fluid in one direction only.

CHOKE: A reduced passage, such as a valve placed in a carburetor air inlet to restrict the volume of air admitted.

CIRCUIT: The path of electrical current, fluids or gases. Examples: for electricity, a wire; for fluids and gases, a pipe.

CLEARANCE: The space allowed between two parts, such as between a journal and a bearing.

CLOCKWISE ROTATION: Rotation in same direction as hands of clock.

COEFFICIENT OF FRICTION: A measurement of the amount of friction developed between two surfaces pressed together and moved one on the other.

COIL: An electrical device made up of a number of coils in spiral form to provide electrical resistance.

COMBUSTION: The process of burning.

COMBUSTION SPACE OR CHAMBER: Volume of cylinder above piston with piston on top center.

COMMUTATOR: A ring of copper bars in a generator or electric motor providing connections between armature coils and brushes.

COMPOUND: A mixture of two or more ingredients.

COMPRESSION: The reduction in volume or the "squeezing" of a gas. As applied to metal, such as a coil spring, compression is the opposite of tension.

COMPRESSION RATIO: Volume of combustion chamber at end of compression stroke as compared to volume of cylinder and chamber with piston on bottom center.

COMPONENTS: The parts that constitute a whole.

CONCENTRIC: Two or more circles having a common center.

CONDENSATION: The process of vapor becoming a liquid; the reverse of evaporation.

CONDENSER: Applied to an electric circuit; a device for temporarily collecting and storing a surge of electrical current for later discharge.

CONDUCTION: The flow of electricity through a conducting body.

CONDUCTOR: A material along or through which electricity will flow with slight resistance; silver, copper and carbon are good conductors.

CONNECTING ROD: Rod that connects the piston to the crankshaft.

CONTACT BREAKER: A device for interrupting an electrical circuit; often automatic and may be known as a "circuit breaker," "interrupter," "cut-out" or "relay."

CONTACT POINTS: Also called breaker points. Two separable points usually faced with silver, platinum or tungsten which interrupt the primary circuit in the distributor or magneto for the purpose of inducing a high tension current in the ignition system.

CONTRACTION: A reduction in mass or dimension; the opposite of expansion.

CONVECTION: A transfer of heat by circulating heated air.

CONVERTER: As used in connection with liquefied petroleum gas, it is a device which converts or changes LP-Gas from liquid to vapor state for use in the engine.

CORRODE: To eat away gradually, especially by chemical action, such as rust.

COUNTERCLOCKWISE ROTATION: Rotating the opposite direction of the hands on a clock. Anti-clockwise rotation.

COUPLING: A connecting means for transferring movement from one part to another; may be mechanical, hydraulic or electrical.

CRANKCASE: The housing for the crankshaft and other related internal parts.

CRANKCASE DILUTION: Under certain conditions of operation, unburned portions of the fuel find their way past the piston rings into the crankcase and oil reservoir where they dilute or "thin" the engine lubricating oil.

CRANKSHAFT: The main shaft of an engine which, working with the connecting rods, changes the reciprocating motion of the pistons into rotary motion.

CRANKSHAFT COUNTER-BALANCE: A series of weights attached to or forged integrally with the crankshaft so placed as to offset the reciprocating weight of each piston and rod assembly.

CRUDE OIL: Oil as it comes from the ground. Unrefined.

CU. IN.: Cubic inch.

CURRENT: As used in automobiles, the flow of electricity.

CYLINDER: A round chamber of some depth bored to receive a piston; also sometimes referred to as "bore" or "barrel."

CYLINDER BLOCK: The largest single part of an engine. The basic or main mass of metal in which the cylinders are bored or placed.

CYLINDER HEAD: A detachable portion of an engine fastened securely to the top of the cylinder block which contains all or a part of the combustion chamber.

CYLINDER SLEEVE: A liner or tube placed between the piston and the cylinder wall or cylinder block to provide a readily renewable wearing surface of the cylinder.

DEAD CENTER: The extreme upper or lower position of the crankshaft throw at which the piston is not moving in either direction.

DEGREE: Abbreviated deg. or indicated by a small "o" placed alongside of a figure; may be used to designate temperature readings or may be used to designate angularity, one degree being 1/360 part of a circle.

DEMAGNETIZE: To remove the magnetization of a pole which has previously been magnetized.

DENSITY: Compactness; relative mass of matter in a given volume.

DETERGENT: A compound used in engine oil to remove engine deposits and hold them in suspension in the oil.

DETONATION: A too-rapid burning or explosion of the mixture in the engine cylinders. It becomes audible through a vibration of the combustion chamber walls and is sometimes confused with a "ping" or spark knock.

DIAPHRAGM: A flexible partition or wall separating two cavities.

DIE CASTING: An accurate and smooth casting made by pouring molten metal or composition into a metal mold or die under pressure.

DILUENT: A fluid which thins or weakens another fluid. Example: gasoline dilutes oil.

DILUTION: See Crankcase Dilution.

DIRECT CURRENT: Electric current which flows continuously in one direction. Example: a storage battery.

DISCHARGE: With reference to a battery, the flow of electric current from the battery; the opposite of charge.

DISPLACEMENT: See Piston Displacement.

DISTORTION: A warpage or change in form from the original shape.

DOWEL PIN: A pin inserted in matching holes in two parts to maintain those parts in fixed relation one to the other.

DOWN-DRAFT: Used to describe a carburetor type in which the mixture flows downward to the engine.

DROP FORGING: A piece of steel shaped between dies while hot.

DWELL PERIOD: The number of degrees the breaker cam rotates from the time the breaker points close until they open again. Also known as cam angle.

DYNAMOMETER: A machine for measuring the actual power produced by an internal combustion engine.

ECCENTRIC: One circle within another circle, each having a different center. Example: a cam on a camshaft.

ELECTRODE: Usually refers to the insulated center rod of a spark plug. It is also sometimes used to refer to the rods attached to the shell of the spark plug.

ELECTROLYTE: A mixture of sulphuric acid and distilled water used in storage batteries of the wet type.

ELEMENT: One set of positive plates and one set of negative plates complete with separators assembled together.

EMF (Electromotive Force or Voltage): See Electrical Section.

ENERGY: The capacity for doing work.

ENGINE: The term applies to the prime source of power generation.

ENGINE DISPLACEMENT: The sum of the displacement of all the engine cylinders. See Piston Displacement.

EPITROCHOID: A geometric path followed by a specific point located in a generating circle which is rolled around the periphery of a base circle.

ETHYL GASOLINE: Gasoline to which Ethyl fluid has been added. Ethyl fluid is a compound of tetraethyl lead, ethylene dibromide and ethylene dichloride. The purpose of the material is to slow down and control the burning rate of the fuel in the cylinder to produce an expansive force rather than an explosive force. Thus, detonation or "knocking" is reduced.

EVAPORATION: The process of changing from a liquid to a vapor, such as boiling water to produce steam; evaporation is the opposite of condensation.

EXHAUST: The spent fuel after combustion takes place in an internal combustion engine.

EXHAUST PIPE: The pipe connecting the engine to the muffler to conduct the exhausted or spent gases away from the engine.

EXPANSION: An increase in size. Example: when a metal rod is heated, it increases in length and perhaps also in diameter. Expansion is the opposite of contraction.

FAHRENHEIT (F): A scale of temperature measurement ordinarily used in English-speaking countries. The boiling point of water is 212 deg. Fahrenheit, as compared to 100 deg. Celsius.

FEELER GAUGE: A metal strip or blade finished accurately with regard to thickness used for measuring the clearance between two parts; such gauges ordinarily come in a set of different blades graduated in thickness by increments of .001 in.

FERROUS METAL: Metals which contain iron or steel and are therefore subject to rust.

FIELD: In a generator or electric motor, the area in which magnetic flow occurs.

FIELD COIL: A coil of insulated wire surrounding the field pole.

FILLET: A rounded filling such as a weld between two parts joined at an angle.

FILTER (Oil, Water, Gasoline, etc.): A unit containing an element, such as a screen of varying degrees of fineness. The screen or filtering element is made of various materials depending upon the size of the foreign particles to be eliminated from the fluid being filtered.

FIT: A kind of contact between two machined surfaces.

FLANGE: A projecting rim or collar on an object for keeping it in place.

FLASHOVER: Tendency of current to travel down the outside of a spark plug instead of through the center electrode.

FLASH POINT: The temperature at which an oil when heated will flash and burn.

FLOAT: A hollow tank which is lighter than the fluid in which it rests and which is ordinarily used to operate automatically a valve controlling the entrance of the fluid.

FLOATING PISTON PIN: A piston pin which is not locked in the connecting rod or the piston, but is free to turn or oscillate in both the connecting rod and the piston.

FLOAT LEVEL: The predetermined height of the fuel in the carburetor bowl, usually regulated by means of a suitable valve.

"FLUTTER" OR "BOUNCE": As applied to engine valves, refers to a condition arising from the valve not being held tightly on its seat during the time the cam is not lifting it.

FLYWEIGHTS: Special weights which react to centrifugal force to provide automatic control of other mechanisms such as accelerators or valves.

FLYWHEEL: A heavy wheel in which energy is absorbed and stored by means of momentum.

FOOT POUND (or ft. lbs.): This is a measure of the amount of energy or work required to lift 1 lb. 1 ft.

FORCE-FIT: Also known as a press-fit, interference or drive-fit. This term is used when the shaft is slightly larger than the hole and must be forced in place.

FORGE: To shape metal while hot and plastic by hammering.

FOUR CYCLE ENGINE: Also known as Otto cycle. An explosion occurs every other revolution of the crankshaft, a cycle being considered as a half revolution of the crankshaft. These strokes are (1) intake; (2) compression; (3) power and; (4) exhaust stroke.

FRICTION: Resistance to motion created when one surface rubs against another.

FUEL KNOCK: Same as detonation.

FULCRUM: The support on which a lever turns in moving a body.

GAS: A substance which can be changed in volume and shape according to the temperature and pressure applied to it. Example: air is a gas which can be compressed into smaller volume and into any shape desired by pressure. It can also be expanded by the application of heat.

GASKET: Anything used as a packing, such as a non-metallic substance placed between two metal surfaces to act as a seal.

GASSING: The bubbling of the battery electrolyte which occurs during the process of charging a battery.

GEAR RATIO: The number of revolutions made by a driving gear as compared to the number of revolutions made by a driven gear of different size. Example: if one gear makes three revolutions while the other gear makes one revolution, the gear ratio would be 3 to 1.

GENERATOR: A device consisting of an armature, field coils and other parts which when rotated will generate electricity. It is usually driven by a belt from the engine crankshaft.

GLAZE: As used to describe the surface of the cylinder, an extremely smooth or glossy surface such as a cylinder wall highly polished over a long period of time by the friction of the piston rings.

GLAZE BREAKER: A tool for removing the glossy surface finish in an engine cylinder.

GLOW PLUG: A device with a fine wire connected in series with an electrical circuit for the purpose of creating resistance and heat enough to ignite fuel in a combustion chamber. Used in place of spark plugs on some engines.

GOVERNOR: A mechanical, hydraulic or electrical device to control and regulate speed.

GRID: The metal framework of an individual battery plate in which the active material is placed.

GRIND: To finish or polish a surface by means of an abrasive material.

GUM: In automotive fuels, this refers to oxidized petroleum products which accumulate in the fuel system, carburetor or engine parts.

HEAT TREATMENT: A combination of heating and cooling operations timed and applied to a metal in a solid state in a way that will produce desired properties.

HIGH TENSION: As used in electricity, it refers to the secondary or induced high voltage electrical current; includes the wiring from the cap of the ignition distributor to the coil and to each of the spark plugs.

HONE: An abrasive tool for correcting small irregularities or differences in diameter in cylinder, such as an engine cylinder.

HOT SPOT: Refers to a comparatively thin section or area of the wall between the inlet and exhaust manifold of an engine, the purpose being to allow the hot exhaust gases to heat the comparatively cool incoming mixture. Also used to designate local areas of the cooling system which have above average temperatures.

HP: Horsepower, the energy required to lift 550 lbs. 1 ft. in 1 second.

HYDROCARBON: Any compound composed entirely of carbon and hydrogen, such as petroleum products.

HYDROCARBON ENGINE: An engine using petroleum products, such as gas, liquefied gas, gasoline, kerosene or fuel oil as a fuel.

HYDROMETER: An instrument for determining the state of charge in a battery by finding the specific gravity of the electrolyte.

ICEI: Internal Combustion Engine Institute, Inc.

ID: Inside Diameter.

IDLE: Refers to the engine operating at its slowest practical speed.

IGNITION DISTRIBUTOR: An electrical unit usually containing the circuit breaker for the primary circuit and providing a means for conveying the secondary or high tension current to the spark plug wires as required.

IGNITION SYSTEM: The means for igniting the fuel in the cylinders; includes spark plugs, wiring, ignition distributor, ignition coil and source of electrical current supply.

IHP: Indicated horsepower developed by an engine and a measurement of the pressure of the explosion within the cylinder expressed in pounds per square inch.

IMPELLER: A rotor or wheel with blades to pump water or propel objects through water or other fluids.

INDUCTION: The influence of magnetic fields of different strength not electrically connected to one another.

INDUCTION COIL: Essentially a transformer which through the action of induction creates a high tension current by means of an increase in voltage.

INERTIA: A physical law which tends to keep a motionless body at rest or also tends to keep a moving body in motion; effort is thus required to start a mass moving or to retard or stop it once it is in motion.

INHIBITOR: A material to restrain or hinder some unwanted action, such as a rust inhibitor which is a chemical added to cooling system to retard the formation of rust.

INLET VALVE OR INTAKE VALVE: A valve which permits a fluid or gas to enter a chamber and seals against exit.

INSULATION: Any material which does not conduct electricity; used to prevent the flow or leakage of current from a conductor. Also used to describe a material which does not conduct heat readily.

INSULATOR: An electrical conductor covered or shielded with a non-conducting material, such as a copper wire within a rubber tube.

INTAKE MANIFOLD OR INLET PIPE: The tube used to conduct the gasoline and air mixture from the carburetor to the engine cylinders.

INTEGRAL: Formed as a unit with another part.

INTENSIFY: To increase or concentrate, such as to increase the voltage of an electrical current.

INTERFERENCE FIT: Difference in angle between mating surfaces of a valve and a valve seat.

INTERMITTENT: Motion or action that is not constant but occurs at intervals.

INTERNAL COMBUSTION: The burning of a fuel within an enclosed space.

JOURNAL: The part of a shaft or crank which rotates inside a bearing.

JUMP SPARK: A high tension electrical current which jumps through the air from one terminal to the other.

KEY: A small block inserted between the shaft and hub to fasten a pulley or gear to a shaft.

KEYWAY OR KEYSEAT: A groove or slot cut to permit the insertion of a key.

KNURL: To indent or roughen a finished surface.

KNOCK: A general term used to describe various noises occurring in an engine; may be used to describe noises made by loose or worn mechanical parts, preignition or detonation.

LACQUER: A solution of solids in solvents which evaporate with great rapidity.

LAMINATE: To build up or construct out of a number of thin sheets. Example: the laminated core in a magneto coil.

LAPPING: The process of fitting one surface to another by rubbing them together with an abrasive material between the two surfaces.

LEAD: A short connecting wire which makes electrical contact between two points.

LINER: Usually a thin section placed between two parts, such as a replaceable cylinder liner in an engine.

LINKAGE: Any series of rods, yokes, and levers, etc., used to transmit motion from one unit to another.

LIQUID: Any substance which assumes the shape of the vessel in which it is placed without changing volume.

LOBE: An off-center or eccentric enlargement on a shaft which converts rotary motion to reciprocating. Also called a cam.

LOST MOTION: Motion between a driving part and a driven part which does not move the driven part. Also see Backlash.

LP-GAS: Made usable as a fuel for internal combustion engines by compressing volatile petroleum gases to liquid form. When so used, must be kept under pressure or at low temperature in order to remain in liquid form.

MAGNET (Permanent): A piece of hard steel often bent into a "U" shape so as to have opposite poles and which can be charged with and retain magnetic power.

MAGNETIC FIELD: The flow of magnetic force or magnetism between the opposite poles of a magnet.

MAGNETO: An electrical device which generates current when rotated by an outside source of power; may be used for the generation of either low tension or high tension current.

MANIFOLD: A pipe with multiple openings used to connect various cylinders to one inlet or outlet.

MECHANICAL EFFICIENCY (Engine): The ratio between the indicated horsepower and the brake horsepower of an engine.

MEMA: Motor and Equipment Manufacturers Association.

MEWA: Motor and Equipment Wholesalers Association.

MICROMETER: A measuring instrument for either external or internal measurement in thousandths and sometimes tenths of thousandths of inches.

MILL: To cut or machine with rotating tooth cutters.

MISFIRING: Failure of an explosion to occur in one or more cylinders while the engine is running; may be continuous or intermittent failure.

MODULE: A packaged functional assembly of wired electronic components

for use with other such assemblies.

MOTOR: Actually this term should be used in connection with an electric motor and should not be used when referring to the engine.

MUFFLER: A chamber attached to the end of the exhaust pipe which allows the exhaust gases to expand and cool. It is usually fitted with baffles or porous plates and serves to subdue much of the noise created by the exhaust.

NEEDLE BEARING: An anti-friction bearing using a great number of small diameter rollers of greater length; also known as a quill type bearing.

NEGATIVE POLE: The point from which an electrical current flows through the circuit. It is designated by a minus sign (−).

NEON TUBE: An electric "bulb" or tube filled with a rare gas, used on ignition test instruments.

NONFERROUS METALS: This designation includes practically all metals which contain no iron or very little iron and are therefore not subject to rusting.

NORTH POLE: The pole of a magnet where the lines of force start; the opposite of south pole.

NSPA: National Standard Parts Association.

OD: Outside diameter.

OHM: A measurement of the resistance to the flow of an electrical current through a conductor.

OIL PUMPING: A term used to describe an engine which is using an excessive amount of lubricating oil.

OPEN CIRCUIT: A break or opening in an electrical circuit which interferes with the passage of the current.

OSCILLATE: To swing back and forth like a pendulum.

OTTO CYCLE: Also called four-stroke cycle. Named after the man who adopted the principle of four cycles of operation for each explosion in an engine cylinder. They are (1) intake stroke, (2) compression stroke, (3) power stroke, (4) exhaust stroke.

OXIDIZE: To combine an element with oxygen or convert into its oxide. The process is often accomplished by a combination, for example, when carbon burns, it combines with oxygen to form carbon dioxide or carbon monoxide; when iron rusts, the iron has combined with oxygen from the air to form rust (oxide of iron).

PETROLEUM: A group of liquid and gaseous compounds composed of carbon and hydrogen which are removed from the earth.

PHILLIPS SCREW: A screw head having a cross instead of a slot for a corresponding type of screwdriver.

PISTON: A cylindrical part closed at one end which is connected to the crankshaft by the connecting rod. The force of the explosion in the cylinder is exerted against the closed end of the piston causing the connecting rod to move the crankshaft.

PISTON COLLAPSE: An abnormal reduction in diameter of the piston skirt due to heat or stress.

PISTON DISPLACEMENT: The volume of air moved or displaced by moving the piston from one end of its stroke to the other.

PISTON HEAD: The part of the piston above the rings.

PISTON LANDS: Those parts of a piston between the piston rings.

PISTON PIN: The journal for the bearing in the small end of an engine connecting rod which also passes through piston walls; also known as a wrist pin.

PISTON RING: An expanding ring placed in the grooves of the piston to seal it against the passage of fluid or gas.

PISTON RING EXPANDER: A spring placed behind the piston ring in the groove to increase the pressure of the ring against the cylinder wall.

PISTON RING GAP: The clearance between the ends of the piston ring.

PISTON RING GROOVE: The channel or slots in the piston in which the piston rings are placed.

PISTON SKIRT: That part of the piston below the rings and the bosses.

PISTON SLAP: Rocking of loose fitting piston in a cylinder, making a hollow bell-like sound.

PIVOT: A pin or short shaft upon which another part rests or turns, or upon and about which another part rotates or oscillates.

PLATINUM: An expensive metal having an extremely high melting point and good electrical conductivity. Often used in magneto breaker points.

POLARITY: The positive or negative terminal of a battery or an electric circuit; also the north or south pole of a magnet.

POLARIZE: Give polarity to an electric circuit so current will flow in proper direction.

POPPET VALVE: A valve structure consisting of a circular head with an elongated stem attached in the center which is designed to open and close a circular hole or port.

PORCELAIN: General term applied to the material or element used for insulating the center electrode of a spark plug.

PORT: In engines, the openings in the cylinder block for valves, exhaust and inlet pipes, or water connections. In two cycle engines the openings for inlet and exhaust purposes.

POSITIVE POLE: The point to which an electrical current returns after passing through the circuit. This is designated by a plus sign (+).

POST: The heavy circular part to which the group of plates is attached and which extends through the cell cover to provide a means of attachment to the adjacent cell or battery cable.

POTENTIAL: An indication of the amount of energy available.

POTENTIAL DIFFERENCE: A difference of electrical pressure which sets up a flow of electric current.

PREIGNITION: Ignition occurring earlier than intended. Example: an explosive mixture could fire in a cylinder by a flake of incandescent carbon before the electric spark occurs.

PRESS-FIT: Also known as a force-fit or drive-fit. Fit accomplished by forcing a shaft into a hole slightly smaller than the shaft.

PRIMARY WINDING: In an ignition coil or magneto armature, a wire which conducts the low tension current which is to be transformed by induction into high tension current in the secondary winding.

PRIMARY WIRES: The wiring circuit used for conducting the low tension or primary current to the points where it is to be used.

PRONY BRAKE: A machine for testing the power of an engine while running against a friction brake.

PROPANE: A petroleum hydrocarbon compound which has a boiling point about −44 deg. F. and which is used as an engine fuel. It is loosely referred to as LP-Gas and often combined with Butane.

PSI: A measurement of pressure in pounds per square inch.

PUSH ROD: A connecting link in an operating mechanism, such as the rod interposed between the valve lifter and rocker arm on an overhead valve engine.

RATIO: The relation or proportion that one number bears to another.

R.C. ENGINE: Rotating combustion engine. Same as rotary engine.

REAM: To finish a hole accurately with a rotating fluted tool.

RECIPROCATING: A back and forth movement, such as the action of a piston in a cylinder.

RECTIFIER: Device used to convert alternating current to pulsating direct current.

REED VALVE: A flat springy valve covering ports between carburetor and crank chamber in two cycle engine.

REGULATOR: An automatic pressure reducing or regulating valve.

RESISTANCE: Quality of an electric circuit, or any part of it, that opposes flow of current through it.

RETARD: When used with reference to an ignition distributor, means to cause the spark to occur at a later time in the cycle of engine operation. Opposite of spark advance.

ROLLER BEARING: An inner and outer race upon which hardened steel rollers operate.

ROTARY VALVE: A valve construction in which ported holes come into and out of register with each other to allow entrance and exit of fluids or gases.

ROTARY ENGINE: An engine which transmits energy from burning fuel directly to a rotating mechanical member.

ROTOR: A rotating valve or conductor for carrying fluid or electrical current from a central source to the individual outlets as required. In Wankel rotary engines, the triangular rotating member which transmits energy from burning fuel to an eccentric shaft.

RPM: Revolutions per minute.

RUBBER: An elastic vibration-absorbing material of either natural or synthetic origin.

RUNNING-FIT: Where sufficient clearance has been allowed between the shaft and journal to allow free running without overheating.

SAE: Society of Automotive Engineers.

SCALE: A flaky deposit occurring on steel or iron or the mineral and metal build-up in a cooling system.

SCORE: A scratch, ridge or groove marring a finished surface.

SEAT: A surface, usually machined, upon which another part rests or seats. Example: the surface upon which a valve face rests.

SECONDARY WINDING: In an ignition coil or magneto armature, a wire in which a secondary or high tension current is created by induction due to the interruption of the current in the adjacent primary winding.

SEDIMENT: Active material of the battery plates which is gradually shed and accumulates in a space provided below the plates.

SEIZE: When one surface moving on another binds, then sticks. Example: a piston will seize in a cylinder because of lack of lubrication or overexpansion due to excessive heat.

SEPARATORS: Sheets of rubber or wood inserted between the positive and negative plates of a cell to keep them out of contact with each other.

SHIM: Thin sheets used as spacers between two parts, such as the two halves of a journal bearing.

SHORT CIRCUIT: To provide a shorter path; often used to indicate an accidental ground in an electrical device or conductor.

SHRINK-FIT: An exceptionally tight fit achieved by heating and/or cooling of parts. The outer part is heated above its normal operating temperature or the inner part chilled below its normal operating temperature and assembled in this condition.

SHROUD: A light cover over the flywheel which shields the flywheel and helps direct flow of air over the engine to carry away heat.

SHUNT: To bypass or turn aside. In electrical apparatus, an alternate path for the current.

SHUNT WINDING: An electric winding or coil of wire which forms a bypass or alternate path for electric current. When applied to electric generators or motors, each end of the field winding is connected to an armature brush.

SIDE SEALS: The spring loaded seals located on the sides of the triangular rotor in the Wankel engine.

SILLMENT SEAL: Compacted powder that helps insure permanent assembly of a spark plug and eliminates compression leakage under operating conditions.

SLIDING-FIT: Clearance between a shaft and journal sufficient to allow free running without overheating.

SLUDGE: As used in connection with engines, a composition of oxidized petroleum products along with an emulsion formed by the mixture of oil and water. This forms a pasty substance, clogs oil lines and passages and interferes with engine lubrication.

SOLENOID: An iron core surrounded by a coil of wire which moves due to magnetic attraction when an electrical current is fed to the coil; often used to actuate mechanisms by electrical means.

SOLVENT: A solution which dissolves some other material. Example: water is a solvent for sugar.

SOUTH POLE: The pole of a magnet to which the lines of force flow; the opposite of north pole.

SPARK: An electrical current possessing sufficient pressure to jump an air gap from one conductor to another.

SPARK ADVANCE: When used with reference to an ignition distributor: causing the spark to occur at an earlier time in the cycle of engine operation; opposite of retard.

SPARK GAP: The space between the electrodes of a spark plug across which the spark jumps. Also a safety device in a magneto to provide an alternate path for the current when it exceeds a safe value.

SPARK KNOCK: See Preignition.

SPARK PLUG: A device inserted into the combustion chamber of an engine containing an insulated central electrode for conducting the high tension current from the ignition distributor or magneto. This insulated electrode is spaced a predetermined distance from the shell or side electrode in order to control the dimensions of the gap which the spark must jump.

SPECIFIC GRAVITY: The relative weight of a substance as compared to water. Example: if a cubic inch of acid weighs twice as much as a cubic inch of water, the specific gravity of the substance would be 2.

SPIRAL BEVEL GEARS: A gear and pinion wherein the mating teeth are curved and placed at an angle with the pinion shaft.

SPLINE: A long keyway.

SPURT-HOLE: A hole drilled through a connecting rod and bearing which allows oil under pressure to be squirted out of the bearing for additional lubrication of the cylinder walls.

SQ. FT.: Square feet.

SQ. IN.: Square inch.

STAMPING: A piece of sheet metal cut and formed into the desired shape with the use of dies.

STARTER: An electric motor attached by gearing to an engine to provide power to turn it over for starting.

STATOR: Stationary coils of an alternating current generator.

STRESS: The force or strain to which a material is subjected.

STROBOSCOPE: A term applied to an ignition timing light which by being attached to the distributor points gives the effect of making a marking on a rapidly rotating wheel, such as a flywheel, appear to stand still for observation.

STROKE: In an engine, the distance moved by the piston.

STUDS: A rod with threads cut on both ends such as a cylinder stud which screws into the cylinder block on one end and has a nut placed on the other end to hold the cylinder head in place.

SUCTION: Suction exists in a vessel when the pressure is lower than the atmospheric pressure. Also see Vacuum.

SULPHATED: When a battery is improperly charged or allowed to remain in a discharged condition for some length of time, the plates will have an abnormal amount of lead sulphate. The battery is then said to be "sulphated."

SUMP: A part of the block in a small four-stroke engine which holds and collects the lubricating oil.

SUPERCHARGER: Device for increasing the volume of the air charge for an internal combustion engine.

SYNCHRONIZE: To cause two events to occur in unison or at the same time.

TACHOMETER: A device for measuring and indicating the rotational speed of an engine.

TAP: To cut threads in a hole with a tapered, fluted, threaded tool.

TAPPET: The adjusting screw for varying the clearance between the valve stem and the cam. May be built into the valve lifter or into the rocker arm on an overhead valve engine.

TDC: Top dead center.

TENSION: Effort devoted towards elongation or "stretching" of a material.

TERMINAL: In electrical work, a junction point where connections are made, such as the terminal fitting on the end of a wire.

THERMAL EFFICIENCY: A gallon of fuel contains a certain amount of potential energy in the form of heat when burned in the combustion chamber. Some of this heat is lost and some is converted into power. The thermal efficiency is the ratio or work accomplished compared to the total quantity of heat contained in the fuel.

THERMOSTAT: A heat controlled valve used in the cooling system of an engine to regulate the flow of water or used in the electrical circuit to control the current.

THROW: With reference to an engine, usually the distance from the center of the crankshaft main bearing to the center of the connecting rod journal.

TIMING CHAIN: Chain used to drive camshaft and accessory shafts of an engine.

TIMING GEARS: Any group of gears which are driven from the engine crankshaft to cause the valves, ignition and other engine driven apparatus to operate at the desired time during the engine cycle.

TOLERANCE: A permissible variation between the two extremes of a specification of dimensions.

TORQUE: An effort devoted toward twisting or turning.

TORQUE WRENCH: A special wrench with a built-in indicator to measure the applied force.

TRANSFER PORT: In two cycle engines, an opening in the cylinder wall permitting fuel to enter from the crankcase.

TROUBLESHOOTING: Refers to a process of diagnosing or determining the source of the trouble or troubles from observation and testing.

TUNE-UP: With reference to an engine, a process of accurate and careful adjustments to obtain the utmost in engine performance.

TURBULENCE: A disturbed or disordered, irregular motion of fluids or gases.

TWO CYCLE ENGINE: An engine design permitting a power stroke once for each revolution of the crankshaft.

UP-DRAFT: Used to describe a carburetor type where the mixture flows upward to the engine.

UPPER CYLINDER LUBRICATION: A method of introducing a lubricant into the fuel or intake manifold in order to permit lubrication of the upper cylinder, valve guides and other parts.

VACUUM: A perfect vacuum has not been created as this would involve an absolute lack of pressure. The term is ordinarily used to describe a partial vacuum, that is, a pressure less than atmospheric pressure; in other words a suction.

VACUUM GAUGE: An instrument designed to measure the amount of vacuum existing in a chamber.

VALVE: A device for alternately opening and sealing an aperture.

VALVE CLEARANCE: Air gap allowed between end of valve stem and valve lifter or rocker arm to compensate for expansion due to heat.

VALVE FACE: That part of a valve which mates with and rests upon a seating surface.

VALVE GRINDING: A process of mating the valve seat and valve face performed with the aid of an abrasive.

VALVE HEAD: The portion of the valve upon which the valve face is machined.

VALVE KEY OR VALVE LOCK: The key, washer or other device which holds the valve spring cup or washer in place on the valve stem.

VALVE LIFTER: A push rod or plunger placed between the cam and the valve on an engine; is often adjustable to vary the length of the unit.

VALVE MARGIN: On a poppet valve, the space or rim between the surface of the head and the surface of the valve face.

VALVE OVERLAP: An interval expressed in degrees where both valves of an engine cylinder are open at the same time.

VALVE SEAT: The matched surface upon which the valve face rests.

VALVE SPRING: A spring attached to a valve to return it to the seat after it has been released from the lifting or opening operation.

VALVE STEM: That portion of a valve which rests within a guide.

VALVE STEM GUIDE: A bushing or hole in which the valve stem is placed which allows only two-way motion.

VANES: Any plate, blade or the like attached to an axis and moved by or in air or a liquid.

VAPORIZER: A device for transforming or helping to transform a liquid into a vapor; often includes the application of heat.

VAPOR LOCK: A condition wherein fuel boils in the fuel system forming bubbles which retard or stop the flow of liquid fuel to the carburetor.

VENTURI: Two tapering streamlined tubes joined at their small ends so as to reduce the internal diameter.

VIBRATION DAMPER: A device to reduce the torsional or twisting vibration which occurs along the length of the crankshaft used in multiple cylinder engines; also known as a harmonic balancer.

VISCOSITY: The resistance to flow or adhesiveness characteristics of an oil.

VOLATILITY: The tendency for a fluid to evaporate rapidly or pass off in the form of vapor. Example: gasoline is more volatile than kerosene as it evaporates at a lower temperature.

VOLT: A unit of electrical force which will cause a current of one ampere to flow through a resistance of one ohm.

VOLTAGE REGULATOR: An electrical device for controlling or regulating voltage.

VOLTMETER: An instrument for measuring the voltage in an electrical circuit.

VOLUME: The measure of space expressed as cubic inches, cubic feet or other units of linear measure.

VOLUMETRIC EFFICIENCY: A combination between the ideal and actual efficiency of an internal combustion engine. If the engine completely filled each cylinder on each induction stroke the volumetric efficiency of the engine would be 100 percent. In actual operation, however, volumetric efficiency is lowered by the inertia of the gases, the friction between the gases and the manifolds, the temperature of the gases and the pressure of the air entering the carburetor. Volumetric efficiency is ordinarily increased by the use of large valves, ports and manifolds and can be further increased with the aid of a supercharger.

WANKEL, FELIX: German inventor of the Wankel rotary engine.

WATER COLUMN: A reference term used in connection with a Manometer.

WATT: A measuring unit of electrical power obtained by multiplying amperes by volts.

WIRING DIAGRAM: A detailed drawing of all wiring, connections and units in an electrical circuit.

WRIST PIN: The journal for the bearing in the small end of an engine connecting rod which also passes through piston walls; also known as a piston pin.

INDEX